FRONTIERS IN NEUROSCIENCE

Series Editors
Sidney A. Simon, Ph.D.
Miguel A.L. Nicolelis, M.D., Ph.D.

Published Titles

Apoptosis in Neurobiology
Yusuf A. Hannun, M.D., Professor of Biomedical Research and Chairman/Department of Biochemistry and Molecular Biology, Medical University of South Carolina
Rose-Mary Boustany, M.D., tenured Associate Professor of Pediatrics and Neurobiology, Duke University Medical Center

Methods for Neural Ensemble Recordings
Miguel A.L. Nicolelis, M.D., Ph.D., Professor of Neurobiology and Biomedical Engineering, Duke University Medical Center

Methods of Behavioral Analysis in Neuroscience
Jerry J. Buccafusco, Ph.D., Alzheimer's Research Center, Professor of Pharmacology and Toxicology, Professor of Psychiatry and Health Behavior, Medical College of Georgia

Neural Prostheses for Restoration of Sensory and Motor Function
John K. Chapin, Ph.D., Professor of Physiology and Pharmacology, State University of New York Health Science Center
Karen A. Moxon, Ph.D., Assistant Professor/School of Biomedical Engineering, Science, and Health Systems, Drexel University

Computational Neuroscience: Realistic Modeling for Experimentalists
Eric DeSchutter, M.D., Ph.D., Professor/Department of Medicine, University of Antwerp

Methods in Pain Research
Lawrence Kruger, Ph.D., Professor of Neurobiology (Emeritus), UCLA School of Medicine and Brain Research Institute

Motor Neurobiology of the Spinal Cord
Timothy C. Cope, Ph.D., Professor of Physiology, Emory University School of Medicine

Nicotinic Receptors in the Nervous System
Edward D. Levin, Ph.D., Associate Professor/Department of Psychiatry and Pharmacology and Molecular Cancer Biology and Department of Psychiatry and Behavioral Sciences, Duke University School of Medicine

Methods in Genomic Neuroscience
Helmin R. Chin, Ph.D., Genetics Research Branch, NIMH, NIH
Steven O. Moldin, Ph.D, Genetics Research Branch, NIMH, NIH

Methods in Chemosensory Research
Sidney A. Simon, Ph.D., Professor of Neurobiology, Biomedical Engineering, and Anesthesiology, Duke University
Miguel A.L. Nicolelis, M.D., Ph.D., Professor of Neurobiology and Biomedical Engineering, Duke University

The Somatosensory System: Deciphering the Brain's Own Body Image
Randall J. Nelson, Ph.D., Professor of Anatomy and Neurobiology, University of Tennessee Health Sciences Center

The Superior Colliculus: New Approaches for Studying Sensorimotor Integration
William C. Hall, Ph.D., Department of Neuroscience, Duke University
Adonis Moschovakis, Ph.D., Institute of Applied and Computational Mathematics, Crete

New Concepts in Cerebral Ischemia
Rick C. S. Lin, Ph.D., Professor of Anatomy, University of Mississippi Medical Center

DNA Arrays: Technologies and Experimental Strategies
Elena Grigorenko, Ph.D., Technology Development Group, Millennium Pharmaceuticals

Methods for Alcohol-Related Neuroscience Research
Yuan Liu, Ph.D., National Institute of Neurological Disorders and Stroke, National Institutes of Health
David M. Lovinger, Ph.D., Laboratory of Integrative Neuroscience, NIAAA

In Vivo Optical Imaging of Brain Function
Ron Frostig, Ph.D., Associate Professor/Department of Psychobiology, University of California, Irvine

Primate Audition: Behavior and Neurobiology
Asif A. Ghazanfar, Ph.D., Primate Cognitive Neuroscience Lab, Harvard University

Methods in Drug Abuse Research: Cellular and Circuit Level Analyses
Dr. Barry D. Waterhouse, Ph.D., MCP-Hahnemann University

Functional and Neural Mechanisms of Interval Timing
Warren H. Meck, Ph.D., Professor of Psychology, Duke University

Biomedical Imaging in Experimental Neuroscience
Nick Van Bruggen, Ph.D., Department of Neuroscience Genentech, Inc., South San Francisco
Timothy P.L. Roberts, Ph.D., Associate Professor, University of Toronto

The Primate Visual System
John H. Kaas, Department of Psychology, Vanderbilt University
Christine Collins, Department of Psychology, Vanderbilt University

Neurosteroid Effects in the Central Nervous System
Sheryl S. Smith, Ph.D., Department of Physiology, SUNY Health Science Center

Modern Neurosurgery: Clinical Translation of Neuroscience Advances
Dennis A. Turner, Department of Surgery, Division of Neurosurgery, Duke University Medical Center

MODERN NEUROSURGERY

CLINICAL TRANSLATION OF NEUROSCIENCE ADVANCES

EDITED BY

Dennis A. Turner M.A., M.D.

Departments of Surgery (Neurosurgery) and Neurobiology
Center for Neuroengineering
Duke University Medical Center
Durham Veterans Affairs Medical Center
Durham, North Carolina

CRC PRESS

Boca Raton London New York Washington, D.C.

Library of Congress Cataloging-in-Publication Data

Modern neurosurgery : clinical translation of neuroscience advances / edited by Dennis A. Turner.
p. cm. -- (Frontiers in neuroscience)
Includes bibliographical references and index.
ISBN 0-8493-1482-8 (alk. paper)
1. Nervous system--Surgery.
[DNLM: 1. Neurosurgery--trends. 2. Neurosurgical Procedures--trends. WL 368 M6885 2004] I. Turner, Dennis A. II. Title. III. Series: Methods & new frontiers in neuroscience series.

RD593.M627 2004
617.4′8--dc22 2004045731

Direct all inquiries to CRC Press LLC, 2000 N.W. Corporate Blvd., Boca Raton, Florida 33431.

International Standard Book Number 0-8493-1482-8
Library of Congress Card Number 2004045731
Printed in the United States of America 1 2 3 4 5 6 7 8 9 0
Printed on acid-free paper

Foreword and Scope

Advances in clinical neuroscience often arise from a better understanding of brain function and hypotheses based at the cellular, system, or organ level. Recent emphasis is on translating functions or structure-based hypotheses into clinical treatment schemes. This process of translational research depends on a number of critical steps, and in most cases, a clinical market that would make commercialization worthwhile financially. Rather than focus on current treatment schemes, this volume will critically discuss treatments in the process of development, particularly those that have arisen or will arise from advances in neuroscience knowledge. The three categories of such treatments are: (1) treatments, aids, and techniques currently in clinical trials or pending U.S. Food and Drug Administration (FDA) approval and new indications for older approved drugs and devices; (2) advances in the promising preclinical stages that may lead to a rapid progression to initial human trials over the next 5 to 10 years; and (3) approaches that failed at the clinical application level, but still offer insights into whether the initial hypothesis was invalid or significantly flawed in some respect.

Many of these advances are hypothesis-based, particularly the pharmacological approaches. However, as a surgical specialty, neurosurgery also has experienced many technical advances, both in terms of treatment and also for both diagnostic approaches and aids that enhance the technical performance of surgical procedures. Such technical advances have led the FDA to devise new methods of approval for approaches that do not directly entail treatment, for example, aids to performance of the surgery. Such aids include stereotactic frames, frameless computer-guided approaches, diagnostic ultrasound, operating microscopes, and many other devices that highlight the dominant role that technological advances continue to exert in translating neuroscience into clinical practice. However, even the application of a new technology requires the identification of a hypothesis. Clear specification of the underlying hypothesis and associated supportive data may lead logically to identifying required testing and enhancement of data both for and against a concept.

This book intends to examine the interface between neuroscience progress and clinical neuroscience advances by assessing the hypotheses that drive this evolution. With this hypothesis-based approach, this book will review the relevant neuroscience underpinnings of new neurosurgical techniques, treatments, and conceptual approaches that are likely to shape clinical neuroscience over the next decade. This dynamic approach is a radical departure from more descriptive books on the topic of 21st century neurological sciences that focus on reviews of current techniques or treatment schemes with timelines to clinical application greater than 10 years.

The specific charge to all the chapter authors was to outline and discuss advances in clinical neurosciences that may occur over the next 5 to 10 years, but are not yet clinical realities. This horizon includes treatment schemes that may be in early stages of clinical adaptation, but the goal is to depart from a review of current clinical practice. As these advances progress in their translation into clinical practice, clearly many may not pass the critical steps of possessing sufficient safety, efficacy, market potential, and usefulness to become marketable items or common practices. Many excellent concepts developed over the past 10 years failed to generate impacts as clinical solutions because of unanticipated problems arising in the translation, even though the underlying hypotheses driving the concepts were excellent. Such concepts include multiple forms of percutaneous discectomy approaches, the clinical use in surgery of laser tumor removals and intraventricular glial-derived neurotrophic factor (GDNF) for Parkinson's disease. We are hopeful that we have chosen wisely — that we will not highlight a collection of "white elephant" approaches, but rather will illustrate broader principles of hypothesis-based neuroscience advances.

Acknowledgments

I thank Dr. Dragan Dimitrov, whose insight and assistance led to the genesis of this volume. I am indebted to my patient famiy (Annika, Brita, and Kathleen), whose tolerance is highly apprciated.

Editor

Dr. Dennis A. Turner was born in 1950, and received a combined M.D./M.A degree from Indiana University in 1975. After a residency in neurological surgery from 1976 to 1981, he was a postdoctoral fellow at the University of Oslo from 1981 to 1982.

Contributors

David Corey Adamson, M.D., M.P.H., Ph.D.
Duke University Medical Center
Durham, North Carolina
Adams022@mc.duke.edu

Michael J. Alexander, M.D.
Duke University Medical Center
Durham, North Carolina
Michael.Alexander@duke.edu

Simon J. Archibald, Ph.D.
Integra Neurosciences
Plainsboro, New Jersey
sarchibald@Integra-LS.com

Christopher J. Beaver, Ph.D.
Duke University Medical Center
Durham, North Carolina
cjbeaver@duke.edu

Dragan F. Dimitrov, M.D.
787 Pacific Street
Monterey, California
dragan.dimitrov@chomp.org

Kelley A. Foster, Ph.D.
Duke University Medical Center and
Durham Veterans' Administration
Medical Center
Durham, North Carolina
FosterKA@duke.edu

Timothy M. George, M.D.
Duke University Medical Center
Durham, North Carolina
Georg017@mc.duke.edu

Larry B. Goldstein, M.D.
Duke University Medical Center
Durham, North Carolina
Golds004@mc.duke.edu

Michael M. Haglund, M.D., Ph.D.
Duke University Medical Center
Durham, North Carolina
Haglu001@mc.duke.edu

Jeffrey S. Henn, M.D.
University of Florida
Gainesville, Florida
jhenn@neurosurgery.uf.edu

Kenneth M. Little, M.D.
Duke University Medical Center
Durham, North Carolina
Littl002@mc.duke.edu

Roger D. Madison, Ph.D.
Duke University Medical Center
Durham, North Carolina
Madis001@mc.duke.edu

Cheryl A. Miller, Ph.D.
Duke University Medical Center
Durham, North Carolina
Nikla001@mc.duke.edu

Kent C. New, M.D., Ph.D.
Duke University Medical Center
Durham, North Carolina
kentnew@duke.edu

Miguel A.L. Nicolelis, M.D., Ph.D.
Duke University Medical Center
Durham, North Carolina
nicolei@neuro.duke.edu

Laura Niklason, M.D., Ph.D.
Duke University Medical Center
Durham, North Carolina
Nikla001@mc.duke.edu

Parag G. Patil, M.D., Ph.D.
Duke University Medical Center
Durham, North Carolina
Patil003@duke.edu

Ricardo Pietrobon, M.D., Ph.D.
Duke University Medical Center
Durham, North Carolina
rpietro@duke.edu

Ashutosh A. Pradhan, M.D.
Duke University Medical Center
Durham, North Carolina
Pradh002@mc.duke.edu

William J. Richardson, M.D.
Duke University Medical Center
Durham, North Carolina
William.Richardson@duke.edu

John Sampson, M.D., Ph.D.
Duke University Medical Center
Durham, North Carolina
John.Sampson@duke.edu

Lee Selznick, M.D.
Duke University Medical Center
Durham, North Carolina
Selzn001@mc.duke.edu

Ashok K. Shetty, Ph.D.
Duke University Medical Center
Durham, North Carolina
Ashok.Shetty@duke.edu

Cheryl Smith, Ph.D.
Duke University Medical Center
Durham, North Carolina
Smith467@mc.duke.edu

Dennis A. Turner, M.A., M.D.
Duke University Medical Center
Durham, North Carolina
dennis.turner@duke.edu

Kevan Van Landingham, M.D.
Duke University Medical Center
Durham, North Carolina
Vanla001@mc.duke.edu

Osama O. Zaidat, M.D.
Duke University Medical Center
Durham, North Carolina
Zaida001@mc.duke.edu

Ali Zomorodi, M.D.
Duke University Medical Center
Durham, North Carolina
Zomor001@mc.duke.edu

Table of Contents

1 Neuroscience Hypotheses and Translation into Neurosurgery Practice

Dennis A. Turner and Simon J. Archibald

CONTENTS

1.1 INTRODUCTION

1.1.1 Concept of Translational Neuroscience

The concept of "translational" research is based on the effective rendering of research ideas into actual clinical practice — in other words "translating" the research finding into clinical usefulness.[1,2] This concept has many different definitions, depending upon the location along a continuous axis that extends from preclinical experimental

0-8493-1482-8/05/$0.00+$1.50

work to what would be considered purely clinical research. In a recent review, translational research is considered to occur when an endpoint is measured in a patient rather than via a preclinical experiment.[2] A team is usually involved, including the preclinical scientists developing the idea for the treatment strategy, the clinicians involved with providing care to patients and using the treatment, and the formal clinical trial necessities such as trial statisticians and research nurses. However, a broader definition is provided by The American Physiological Society, which defines translational research as "the transfer of knowledge gained from basic research to new and improved methods of preventing, diagnosing or treating disease, as well as the transfer of clinical insights into hypotheses that can be tested and validated in the basic research laboratory."[3]

The team concept has evolved through necessity because basic neuroscience tends to be very focused on cellular and genetic mechanisms, whereas clinical trials and subsequent applications to humans are now unique specialties in their own rights, with clinicians and neurosurgeons providing patient care in the middle course between the new specialties. In a way, almost all biological research is translational, but the timelines for clinical applications may differ considerably, i.e., 1 year, 10 years, 100 years, etc.

Neurosurgeons tend to prefer more rapid relevance (within 1 or 2 years) because of inherent impatience. However, many of the great clinical advances stem from older basic science research findings applied outside their original fields. For example, microbiologists were cataloguing bacteria for years before the process had any clinical relevance. The advent of antibiotics created an immediate need to classify and understand organisms due to their differing sensitivities to antibiotics.

The three aspects of translational research include first a great preclinical idea developed often from laboratory findings that appear to have significant potential (Figure 1.1). This seminal idea may have been tested in rodents or in nonhuman primates and hopefully a side effect profile and likely therapeutic range are in hand. For many neurosurgery devices and products, no dose-response curve is available or applicable, and often Phase I trials for side effects in normal patient populations are obviated and the device or product instead moves rapidly into Phase I/II trials in the target patient population. Thus, information obtained about side effects and dosage (if applicable) from preclinical studies can be very helpful.

The preclinical idea must then have an enthusiast or sponsor willing to do the work to begin human testing. The sponsor must obtain a relevant investigational device exemption (IDE) for a device or an investigational new drug (IND) application for a drug from the U.S. Food and Drug Administration (FDA), and usually requires an industry cosponsor to handle manufacturing compliance, assist in defraying costs, and set up the initial testing format. Individuals and companies may have varying motivations to proceed to initial clinical testing, but usually the motivation is a mix of altruism (to improve some medical condition) and a profit incentive based upon the possibility of a marketable product at some point in the future.[4]

Many pitfalls are present in the transition from idea or preclinical data into a clinical concept. In many cases a "translator" person may serve as an intermediary. Such a person is familiar both with the basic science and clinical concepts inherent to the product and the business aspects related to marketability. From the view of

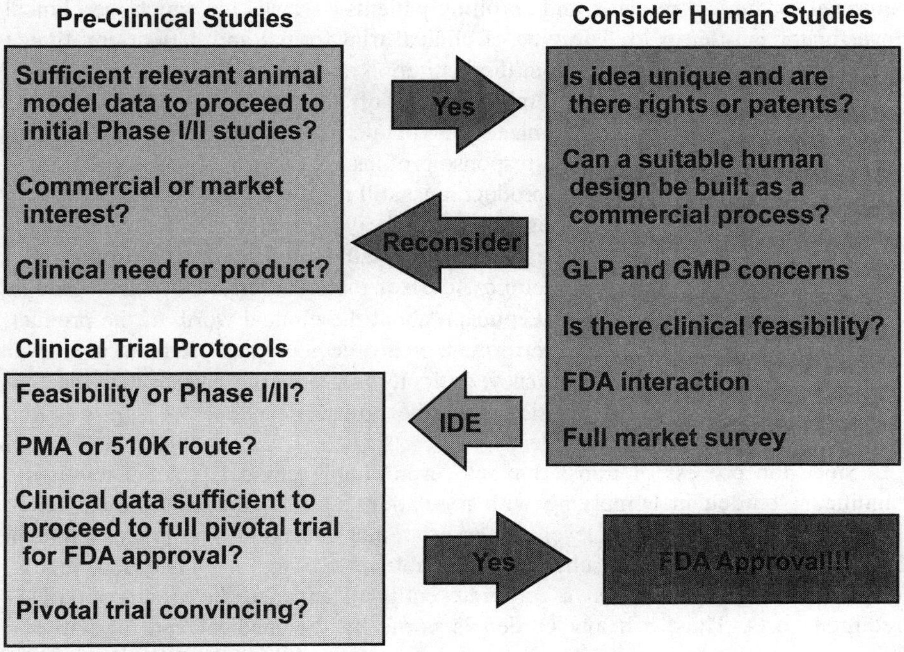

FIGURE 1.1 The transition path of a great basic science idea from preclinical studies to product. First, clinical need, commercial interest, and animal data must be sufficient before one can consider proceeding to initial human protocols. Initial consideration of human studies requires further evaluation of other factors including intellectual property rights, ability to devise a suitable human manufacturing process, market surveys to indicate clinical marketing likelihood, and initial FDA interactions. If initial feasibilities of Phase I/II clinical trials are approved, an investigation device exemption (IDE) must be obtained and the manufacturer must choose a premarket approval (PMA) or 510 K approval pathway. A PMA requires considerable additional data and clinical studies for an entirely new treatment approach; a 510 K links the product to existing FDA-approved products to show equivalence. If a pivotal trial is convincing, the FDA may finally approve marketing; then the product can be sold freely for FDA-approved indications.

the translator, preclinical data may not necessarily be needed to evaluate a hypothesis but may be needed to decrease the risk of failure of the product through more extensive testing. The involvement of a basic science person in this transition may sometimes be difficult because such scientists often do not have a full picture of clinical relevance and marketing aspects and, in many cases, may not understand why a product is not developed or is suppressed by a company for business reasons or because of side effects.

The second aspect then is to look for (usually) academic neurosurgery collaborators willing to apply the device or product to patients in the proscribed clinical trial format.[5] This initiation of clinical testing in humans also requires significant paperwork and oversight, including obtaining institutional review board (IRB) approval for human research and enrolling patients — tasks for which the clinical investigator is often paid. The type of clinical trial format and patient enrollment are then closely monitored by both the institution and the FDA.

The third aspect is to then assemble the data, often from multiple sources, decide on a Phase III format for a randomized trial, obtain FDA approval to proceed based on the initial side effect and dose-response profiles, and then perform the definitive trial. Even after FDA approval, a product must still meet the FDA burden of registry of late side effects and long-term issues, particularly for implanted surgical products that may involve unknown consequences years after implantation. At some point after FDA approval, more open clinical trials are often initiated by other groups, usually those with considerable skepticism about the clinical worth of the product. After these external trials are performed, many devices and products are never commonly used due to lack of efficacy, difficulty in use by those other than the core enthusiasts, or because of unanticipated side effects despite FDA approval and availability.

Since the process of translation into even simple clinical practice requires a significant burden in complying with regulations at both the institutional (IRB, ethical board, etc.) and FDA levels, the translational process also requires a clinician familiar with the treatment scheme and capable of delving adequately into regulations for approval and often a corporate entity to advance the significant FDA-required costs. Thus, a heavy burden is borne by the medical care system and clinicians who participate in the process of translational medicine.[6-8]

Translational research also implies a mechanistic understanding at the molecular, cellular, and systemic levels of the function or action of the therapy in relation to the disease mechanism or target. Many clinical advances are not translational or hypothesis-based, but rather are evolutionary or simply empiric. Thus, the concept of translational research is usually applied primarily to a situation where a hypothesis is generated, tested at the preclinical level, and then applied in sequence to initial and then final stages of clinical testing for human use. In many instances, drug development has followed this approach whenever possible, although many notable failures occurred as well.[4]

Translational research depends critically upon an animal model of the disease for preclinical testing of the proposed therapy, and the translational process can fail, for example, by applying results from an inappropriate animal model to a human disease. Thus, translational research continues to evolve and, in many cases, further understanding as to why a treatment does not work in an intact individual may lead to the opening of additional preclinical research avenues.

A recent review on translational neuroscience[4,6] focused on the concept of designer drugs generated by hypotheses and new information about the nervous system and the mechanisms by which such preclinical hypotheses can be applied to human medical care. The field of neuroscience has developed many promising new treatment strategies now in the process of testing, and arising from basic neuroscience advances. In many ways such a hypothesis- and data-driven approach contrasts with traditional drug screening in which many compounds are subject to blind screening via a validated technique. However, in both empiric and hypothesis-driven treatment development, many pitfalls and development problems can arise, particularly unforeseen side effects, when new therapeutics are applied to clinical treatment schemes.

All applications into the clinical arena depend upon ample (and willing) supplies of patients for testing and clinicians willing and sufficiently enthusiastic to spend their time (beyond ordinary clinical care) for such testing. Some clinicians may also have sufficient understanding of the therapy at both the basic science and clinical levels to serve as a bridge to facilitate the transfer of the treatment to clinical care.[1]

1.1.2 Translational Neurosurgery versus Translational Neuroscience

In many ways, neurosurgery is very different from neuroscience in general. First, it involves far less emphasis on systemic drug treatment, although, of course, considerable crossover and use occur, as in the cases of anticonvulsants, antibiotics, chemotherapeutic agents, and drugs in general medical care. Second, in addition to therapeutics, a whole field of devices, most of which require FDA approval, serve as aids to surgeons performing procedures and are not directed at patient therapy. Third, many therapeutics in development and use are devices and permanent implants that may require an invasive form of delivery. The safety and efficacy requirements that must be met for FDA approval may be quite different for such therapeutics from requirements for oral or systemic drug delivery.

Because the FDA treats devices very differently from drugs, the requirements for specific types of clinical trials also differ and the entire process of translation from a preclinical state to clinical use requires different forms of expertise and knowledge of clinical trials. The focus of most of this volume is on these various categories of therapeutics, devices, and approaches to translation of preclinical advances into clinical usefulness — in other words issues more relevant to ordinary neurosurgical practice and research. These issues are rarely covered in print because the number of devices and their applicability are far fewer than medical applications for new drugs.

Another category of clinical development and advances includes the rationalization of existing therapy. For example, most neurosurgery procedures such as craniotomy and laminectomy involve a few standard approaches that have been in development for more than a century. Since the FDA regulates surgeons in contrast to drugs and devices, little data exists on many neurosurgery procedures, their relative efficacy and safety, and their indications.

This lack of data perplexes both rational care providers and medical consumers because many different approaches to the same clinical problem may be suggested by various surgical specialists. Because the neurosurgical literature mostly involves anecdotal case series and little data generated by randomized controlled trials, few guidelines based on such data apply to management of typical problems, particularly complex issues such as brain tumors and spine therapy.

Where therapy has been rationally studied, as in the case of carotid endarterectomy and cerebral aneurysm, less contention exists, but many technical and timing issues remain. While many neurosurgical procedures will never be thoughtfully studied because of insufficient patient populations or lack of contention about treatment choice, many treatment options could be studied rationally and various formats of clinical trials continue to percolate and develop, particularly those that go beyond traditional randomized clinical trials.

New clinical advances depend on the ability to rationally and efficiently translate new understanding of brain function into clinical neuroscience practice. Currently, most clinical neuroscience advances are purely empiric, and often are subject to clinical testing without full identification of the cellular mechanisms involved. Thus much time, energy, and money have been allocated to new treatments with minimal examination of their scientific bases and applicability. However, for many reasons, it is critical to define the hypotheses underlying the application of neuroscience to clinical use.

This definition may lead to reexamination of the data underlying advances in terms of the adequacy of support of the hypotheses and may lead to a fresh approach. However, in spite of a rational approach, the transition from preclinical studies to clinical medicine may still be difficult because of unanticipated potential side effects, clinical trial flaws or inadequacies, inappropriate disease translation, and lack of sufficient market potential.

Most current neurosurgery procedures developed from both clinical hypotheses and practice-related outcome measures to assess the worth of the hypotheses. Many stable and confirmed clinical hypotheses are common in the practice of neurosurgery, particularly the concept that "mass effect" or pressure, if relieved, may improve brain, spinal cord, or peripheral nerve functioning. However, such simple hypotheses do not work for more complex abnormalities, such as intrinsic brain tumors that involve both infiltration and mass effect. As a result, more complex hypotheses often encompassing cellular, systemic, and organ level concepts have been developed.

In many situations, neurosurgery is moving away from the simplistic mass effect hypothesis that has dominated clinical thinking for many years and into specific mechanistic approaches requiring further insight into anatomical, physiological, and pathological factors unique to the brain. This book focuses on such fresh approaches in a variety of neurosurgical fields.

Compared to pharmaceutical mechanisms of translational research, neurosurgery presents many challenges. The first is a small market throughout the world for most conditions under the neurosurgery umbrella, particularly compared to neuroscience diagnoses not involving surgical treatment, for example, Alzheimer's disease and cardiovascular disease. This is particularly true for clinical products intended for neurosurgery centers rather than for patients (surgical instruments, diagnostics, and other intraoperative aids). This small market may preclude effective development and commercialization because its potential is often insufficient.

A second factor is that experimental surgical procedures are far more expensive to study than experimental pharmaceuticals. A typical price for an experimental surgery, for example, a cell transplant procedure, may reach $150,000 in direct costs in addition to the great amount of liability coverage required and the need for sham or placebo surgical implant procedures.[9] The cost per patient is much higher than the cost of testing experimental drugs and a high level of preliminary efficacy must be demonstrated prior to engaging in clinical trials. Other requirements are substantial financial backing and significant market potential.

Because of these burdens, rarely has an experimental surgical procedure been developed commercially, except as a direct derivative of an existing procedure for which clinical payment coverage may be obtained. Examples are pain or deep brain stimulators. Rarely has a sponsoring company paid clinical study expenses except for the costs of the devices because in most cases the patients may have obtained some benefits. Obviously, this clinical coverage scheme would not be workable for a randomized, placebo-controlled clinical trial.

1.1.3 Examples of Translational Products

Collagen nerve guide tubes — For years, neuroscientists have tried to improve the recovery capability and ease of repair of peripheral nerves. A number of different nerve guide tubes were developed. The first used autologous materials (such as arteries and veins) because many injuries allowed insufficient autologous peripheral nerve for cable grafting of a long lesion. A simple collagen nerve guide tube was developed by Archibald et al.[10] to aid in regrowth of peripheral nerves, with the advantage of absorption over time (see Chapter 3). This absorption obviated some of the problems of permanent materials such as silicone that eventually became restrictive to the nerves. After extensive testing in nonhuman primate median nerves across large gaps, the nerve guide tubes were also compared with conventional cable grafts. However, the initiation of human feasibility trials was difficult in the U.S. and European trials that were conducted first. After several years, a corporate sponsor became interested, pursued additional clinical trials, and eventually the product became FDA-approved for nerve injury repair. This time span from bench to bedside application exceeded 15 years, and the device clearly was a hypothesis-based translational product.

Frameless stereotactic devices — While stereotactic frames have been in common human use since the early 1950s, the difficulties in using a frame and the discomfort to the patient led to consideration of other techniques for surgical navigation. As digital scans such as magnetic resonance imaging (MRI) and computerized tomography (CT) and algorithms to reconstruct the scans and provide three-dimensional representations became more readily available, it became possible to align a patient's brain in the three-dimensional space of an operating room with the patient's own computed images. The critical pieces needed to accomplish this alignment are rapid three-dimensional representations of computerized brain images and an accurate and robust three-dimensional digitizer.[11,12] A variety of three-dimensional digitizing systems were developed and continue to evolve to accomplish intraoperative navigation. FDA approval of frameless stereotactic devices has been expanded to require evidence of clinical usefulness because the devices are used solely by surgeons to aid intraoperative navigation. The devices have evolved into clinical products in wide use and include the Stealth (Medtronics) and BrainLAB systems. Both systems were built upon rapid advances in three-dimensional localizer technology, computer systems, and graphics. The entire laboratory-to-operating-room translational process took less than 10 years.

GDNF for Parkinson's disease — Glial-derived neurotrophic factor (GDNF) was the primary dopamine growth factor discovered in the 1980s and purified as a recombinant human protein. After several years of experimentation, nonhuman primate experiments showed considerable promise for GDNF in initiating regrowth of dopaminergic collaterals within the striatum. Initial human clinical trials were begun in 1996,[13] but ended prematurely due to unexpected severe side effects. After further work in nonhuman primates with both direct GDNF infusion into the putamen and gene therapy for GDNF transfection, initial human clinical trials with both methods of administration are in progress.[14] GDNF continues to show significant promise and further pivotal trials will likely be conducted for at least one of these two novel methods of administration (see Chapter 8). Although FDA approval has not yet been granted, GDNF is another example of a hypothesis-driven bench-to-bedside product.

1.2 CATEGORIES OF NEUROSURGERY ADVANCES

Neurosurgery advances can be divided into three basic categories. The first category involves drugs and devices that are therapeutic and typically involve obtaining an IND or IDE for initial human use and some form of clinical trial sequence prior to full FDA approval. Examples of drugs include the Gliadel wafer (BCNU; 1,3-bis(2-chloroethyl)-1-nitroso urea), which is directly deposited into a brain tumor cavity at the time of craniotomy, the intracerebral infusion of GDNF for the treatment of Parkinson's disease, and adenovirus vector delivery of gene therapy for GDNF enhancement in the striatum.[15]

Many of these new approaches are based on neural regeneration, biological plasticity, tissue grafts, and new engineering approaches.[16] Examples of devices include implants such as cerebrospinal fluid (CSF) shunts, hardware for spine fixation, and deep brain stimulating (DBS) electrodes for movement disorders. Because this category generally involves permanent implants for therapy and the devices are highly invasive, extra consideration is usually involved to ensure long-term safety.

The second category involves aids to treatment and the performance of procedures. These devices are not directly involved in therapy; they are diagnostic and surgical tools are meant to facilitate the surgeon's application of the patient's primary therapy. For example, diagnostic tools include MRI scanning of the nervous system in radiology and potentially in the operating room suite, ultrasound for intraoperative diagnosis, and newer computer-based tools such as the Stealth computer-aided intraoperative navigation system and other devices. Other diagnostic tools recently approved by the FDA include microelectrodes and associated physiological recording apparatus for movement disorder surgery, evoked potential devices, and Licox (Integra Neurosciences) oxygen recording catheters for intracerebral use. Most of the general instrumentation (retractors, scissors, clamps, etc.) used during procedures is not individually FDA approved; the manufacturer may have a general FDA approval for manufacturing techniques in a global sense. Many other common devices are not FDA approved. One example is the operating microscope that is so important to most surgeons. Because neurosurgery involves procedures, it also involves many devices used by surgeons as therapeutic approaches. Thus, the FDA recently added a new category of approval for surgeon's aids; the primary criterion is usefulness.

The third category is rationalization of established products the FDA has already approved and older procedures already in common clinical practice for which efficacy was not established.

Proceeding with a clinical trial for established surgery requires significant contention about the worth of the procedure, as when carotid endarterectomy began to be carefully scrutinized. For example, the reasons for the carotid endarterectomy trials in the 1980s included the high cost to society for the multiple procedures performed, the lack of any valid data regarding what would happen if the procedure was not performed (contemporary natural history studies), and the prophylactic nature of the procedure, i.e., to prevent a bad event despite the risks of the procedure. Many common products have been applied to additional disease mechanisms without adequate studies of the appropriateness or risk-to-benefit ratio, and thus the great need to establish the proper patient population and determine whether existing procedures are indeed efficacious, safe, and appropriate continues to exist.

1.3 CRITICAL QUESTIONS IN TRANSLATIONAL NEUROSURGERY

1.3.1 When Is Preclinical Data Sufficient to Proceed to Human Experimentation?

Because devices and products involving surgical application or implantation usually involve more risk than drugs, manufacturers have somewhat higher burdens to demonstrate product worth in preclinical research before they move on to clinical trials. The pace at which a preclinical treatment or device is applied to initial patient experimentation is often dictated more by the entity developing the treatment rather than by any rational approach to quality of the preclinical data. For drug development, the FDA has established a rigorous process of IND application and development. For devices, it uses a parallel structure of IDE approval prior to clinical trial. However, many aspects of neurosurgery and translational research are not covered by these regulatory pathways if they do not involve drugs or devices. The definition of experimental surgery has always been a complex issue for surgeons. Standard clinical practice varies considerably among surgeons. Thus, is a slight difference in surgical technique an "experiment"?

An example of a new surgical development was the rapid application of autologous adrenal medullary tissue grafts to patients with Parkinson's disease, beginning in 1987, after a report from a foreign medical center that they were beneficial. Although linked indirectly to preclinical research suggesting benefits from this procedure, no preclinical studies supported the transition of the procedure to initial clinical experimentation. However, because the tissue was autologous (came from the patient) and the procedure involved only slight variation from an ordinary craniotomy, no regulatory agencies or issues were directly involved. The primary critical issue was whether sufficient informed consent could be obtained for such a blatant experimental procedure.

If a procedure is clearly labeled experimental, insurance carriers usually do not pay for it. This places a great burden on the patient to provide payment for an experimental procedure and assume significant risk without a known benefit. There are many examples of such deviations from surgical practice that to many observers clearly represent experimental surgery and to others constitute only small departures from current practice. The rapidity of clinical application of a new advance continues to be a highly contentious issue, clearly requiring institutional backup from the IRB and ethical support whenever a neurosurgeon engages in some form of human research.

If a corporate entity is involved with a therapeutic, then the rapidity of entry into clinical trials may be more dictated by the need for a marketable product than necessarily the quality of the preclinical evidence. Examples include many of the proprietary neural tissue graft trials.[9,16] Companies have developed both porcine embryonic cell transplants (as replacements for human embryonic cells) and tumor-derived cell lines for human neural grafting protocols. These cases usually proceed from initial rodent preclinical data directly to clinical trial because of the cost and time required to adequately assess the therapy in nonhuman primates.

One example is a trial of cultured human neuronal neurons for deep hemorrhagic stroke; they were applied to humans in early clinical trials after only a few rodent studies showed cell survival and presence of grafted cells.[9,17] Because the therapy may be headed for FDA approval, the promoters usually have a burden to demonstrate safety and efficacy before the therapy reaches final clinical trials. However, if a corporate entity is less involved or is not as enthusiastic as the investigators, as was the case with GDNF gene therapy trials for Parkinson's disease, much more careful nonhuman primate work may be performed before initial human experimentation is considered.[13,14] Thus, the source of the therapeutic, the need for clinical product development, and marketing for later commercialization may all dictate both the manner and pace of the translational process.

Do neurosurgeons jump too quickly to human experimentation without proper consideration for appropriate human risk and benefit? Clearly, the example discussed earlier of autologous adrenal medulla transplants for Parkinson's disease involved premature human application, and fairly rapid abandonment of the procedure because of significant risk and lack of efficacy. Thus, convincing preclinical data in a validated animal model of the disease (whenever possible) is required to support the transition to initial clinical trials.

Occasionally the enthusiasm of a corporate entity to bring a therapeutic to market needs to be curbed by clinicians investigating the background and rationale for the transition. Because the data required for initial FDA approval to proceed to clinical trials are quite different in quality and type from those usually required for peer-reviewed publication (the information is often proprietary and difficult to access), external peer-reviewed scrutiny is rarely possible. Thus, neurosurgeons are commonly perceived as being on the edge of ethics, poorly defining experimental surgery as such, and stretching or breaking the (unwritten) rules as to when therapies should proceed to initial human testing.[18]

1.3.2 Who Is Involved with Translational Neurosurgery?

Ideally, the application of devices, drugs, and surgeons' aids should be performed by clinicians with sufficient background to understand and critically assess the value of a new approach and suggest optimal application.[5,19] Because a preclinical team may have minimal clinical experience or knowledge, obvious clinical problems in the translational process may easily be avoided by an astute clinician who has the necessary experience. Also, clinicians involved in the translational process should understand FDA procedures and regulations, IRBs, and ethics and should have the expertise to fully validate initial clinical trials. Particularly for new approaches fresh from a preclinical scheme, knowledge of the goal of the underlying application may considerably aid the translational process because appropriate translation usually requires changes and scaling from preclinical applications.

Thus translational effort of taking a therapy or device from discovery to clinical application is generally provided by a team, and rarely by a lone neurosurgeon. The FDA requires rigorous consideration of the manufacture and construction of devices by a company with knowledge of human applications and materials to preclude "garage" level implementations. Thus, a team may include preclinical scientists who foster an idea and clinicians who are knowledgeable about the basic science side and the initial clinical application and who have access to appropriate patients, statisticians, trial designers, research nurses, and database and analysis personnel. This team usually requires outside funding to fully implement the translational advance, either through a grant or from a corporate entity with visions of a marketable and profitable product within a specific timeline.

Considering that training may be needed in a variety of disciplines beyond neurosurgery, the typical academic neurosurgeon often may be overwhelmed by the needs of even a simple clinical trial. The degree of paperwork, oversight, and IRB approval is astounding without significant administrative help, and often the design of a clinical trial from an industry-funded approach is insufficient to answer a scientific question even though it may be sufficient for FDA approval.

Academic neurosurgeons interested in translational approaches must be aware of and understand many of the basic neuroscience implications of the research, have captive patient groups who can be recruited, have the necessary clinical skills to adequately institute the methodology, and be aware of the relevant clinical trial needs for the study. This is a wide range of skills and the training to obtain most of them (beyond ordinary clinical skills) is not readily available through medical school or neurosurgery training — it requires additional time.[19,20] As is the case for most academic medicine efforts, developing a team is critical, and a shortage of neurosurgeons interested in and sufficiently enthusiastic about such research to become translational "bridges" continues to exist.[1]

1.3.3 Mechanisms of Translational Neurosurgery: University and Corporate Involvement

For a medical material such as an antibiotic or antiviral, the typical pathway consists of a scientific discovery that works well in a laboratory setting followed by translation through clinical trial into a treatment. Usually this approach is sponsored by a drug company, and therapies are developed because of market forces. An initial market survey is necessary to determine how much a drug would cost to develop, patent, and manufacture, how much profit is required to recapture development costs, and the size of the potential market. However, a large number of drug and device companies often take ideas from academics and then perform the translational work to prepare for clinical trials without actually proceeding on to clinical trials due to the cost. Instead, the marketable preclinical products may then be sold to more traditional drug firms or the translational companies themselves converted or sold into a different entity for further product development.

The drug field provides many examples of successes and failures. Many start-up drug companies went bankrupt in the search for new drugs, often at an initial level because the research was too far from a direct path to clinical development. In some cases, such as drug trials for stroke, a drug appeared promising in animal trials, but failed at the initial clinical trial level. This may have been caused by incorrect application of the animal model to the human situation (exploiting drugs that work in focal stroke to treat global ischemia), failure to understand how a drug may work in an intact system, side effects (a psychotropic profile for N-methyl-D-aspartate [NMDA] antagonist), or failure of clinical trial design. However, for devices and particularly for experimental surgeries, translational research often means something completely different. Obviously, a device may be patented, but it may be difficult ethically and legally to patent a surgical procedure.

1.3.4 Device Development Process

Specific device or therapeutic development begins with a great idea, but a complex process usually involving commercialization must be followed before considering human application of the concept (Figure 1.1). Commercialization of an idea for clinical use involves consideration of many critical issues before development proceeds further. The critical issues include exclusiveness and the availability of patent rights, market size and access to markets, and the feasibility of commercial production. Once a process is deemed feasible for production using accepted standards for devices and drugs (good laboratory and manufacturing practices), a number of parties may decide whether to proceed with initial human feasibility trials. For most devices, no equivalent of Phase I volunteer testing in healthy subjects exists, so most initial trials for feasibility follow Phase I/II in patient populations relevant to the product.

Considerable interaction with the FDA is required during design and performance of feasibility and then pivotal clinical trials. Finally, the path to FDA approval can include a full premarket application (PMA) or a comparison of equivalence to an existing device or therapeutic (510K application). Defining the selected indications for use and patient populations for potential use are critical in order to obtain the widest FDA approval possible.

After FDA approval is received, postapproval market selection proceeds. Percutaneous discectomy devices (numerous after chymopapain approval in the 1980s) suffered rapid fall-offs in clinical use after FDA approval because of difficulty of use or perceived (or real) lack of efficacy in many surgeons' hands. Thus, many products are essentially dormant despite FDA approval. It may take several years for a product to find a market niche, even though the FDA approved it. Further rigorous clinical studies for postmarket approval may be required to fully define indications, risks, and efficacy.

1.3.5 Guidelines for Efficacious Treatment Schemes

In many instances, a product becomes a standard aspect of clinical treatment schemes, for example, immunization vaccines for many childhood diseases. Multiple studies confirmed high degrees of efficacy and the clinical evidence was considerable. Clinical guidelines developed for many medical care situations and diseases usually incorporate treatments that are generally agreed to be efficacious as parts of the standard clinical treatment scheme. However, a second type of translation[8] involves convincing practitioners to routinely provide care based on guidelines. This requires considerable education of practitioners. However, family practitioners now face so many guidelines, particularly for long-term health care maintenance, that the time required merely to follow the guidelines is considerably more than that required to provide ordinary health care. In a way, postmarket guideline development is critical for the transference of information from tight clinical trials with rigorous patient populations to the general patient population at large.

1.4 OUTLINE OF TOPICS OF NEUROSURGERY ADVANCES

This volume is not meant to be all-inclusive. It is intended to provide an overview of several exciting areas of neurosurgery. We selected fields that lend themselves to translational work and discuss examples of translational products. For example, little translational science is now ongoing in the field of skull base surgery, but functional neurosurgery to treat epilepsy and movement disorders, insert neuroprosthetics, and perform neural grafting is well represented because of many hypotheses at the preclinical level. Neurosurgery involves a broad range of subdisciplines, but is generally divided into a few basic categories or mechanisms of disease:

Brain tumors and meoplasias — Examples of translational research include new brain tumor therapeutics such as antibodies, radiation, infusions, drugs, Gliadel wafers, and many promising new approaches (see Chapter 9).

Pediatrics and congenital — Examples include folic acid for prevention of meningomyelocele, clinical trials underway on fetal surgery, and development of new CSF shunt techniques (see Chapter 10).

Head injury, peripheral nerve regeneration, and trauma — A significant amount of work is focused on central and peripheral regeneration, improved intensive care unit (ICU) therapy, and enhanced recovery after injury (see Chapter 2, Chapter 3 and Chapter 13).

Stereotactic and functional — A number of new devices and treatments for epilepsy and movement disorders have reached preclinical and clinical development along with continued improvements in cell implants, such as stem cells. A host of new neuroprosthetics devices including new stimulators and pumps for medicine delivery are also in development (see Chapter 2 and Chapter 5 through 8).

Cerebrovascular, stroke, and endovascular — Challenging topics include treatment of stroke, delayed cerebral vasospasm, new approaches for endovascular treatments such as catheters, balloons, and coils, and improved postoperative care in ICU settings (see Chapter 4 and Chapter 11 through 13).

Spine and peripheral nerves — A large number of spine implants have been approved and are now in common use; their usefulness is poorly characterized. Spine surgery needs considerable rationalization as to when it is appropriate, and what exactly should be done under various circumstances (see Chapter 3 and Chapter 14).

Although the subject is not disease based, the history of neurosurgery has also become a very popular topic at neurosurgery meetings over the past 20 years. While the history of neurosurgery is clearly beyond the scope of this volume, interesting failures of the past may provide excellent guidance as to how not to proceed in the future. Other topics are aspects of clinical trials as applied to developing translational approaches (see Chapter 15) and new approaches to the teaching of neurosurgery skills (see Chapter 16).

The focus is the discussion of advances beyond the current clinical domain for two main reasons. The first is that current and conventional treatments are well treated in many aspects of neurosurgery literature — both books and journal articles. The second is that much can be learned about flawed hypotheses, pitfalls of the translational approach in general, and untested hypotheses and clinical advances from past (and now unused) treatments. This volume focuses on promising new concepts still in preclinical development, those already applied to some clinical trial phase, and those that failed during application to patients. It should provide a worthwhile opportunity to review whether the basic mechanisms, animal models, and translational processes were flawed. Clearly, many of these topics fall outside the scope of a traditional, detailed, and comprehensive neurosurgery textbook.

1.5 LEVELS OF NERVOUS SYSTEM FUNCTIONING: CELLULAR TO SYSTEMS

Diseases are caused by mechanisms and affect levels ranging from subcellular (genetic, mitochondrial, etc.) to organ. In general, most surgical disciplines primarily consider therapeutics aimed at the organ level (e.g., resecting brain tumors), whereas medical disciplines often test cellular or subcellular approaches involving medicines and drugs. Because knowledge about brain function is accumulating at every level, the approaches at different levels such as subcellular (genetic, molecular, organelle), cellular (electrical integration, channels, and regeneration), local circuits, systems, and organs should be considered. Eventually all translation from preclinical findings to clinical testing depends in one form or another on clinical hypotheses and appropriate clinical trial design for adequate testing of hypotheses and devices where a hypothesis rests at one particular level.

Although pharmaceutical development is usually at the cellular level of functioning, most neurosurgery treatments stem from system or organ level structures or functions. In many cases translational research in neurosurgery can take advantage of all three levels of brain functioning loosely defined as follows:

1. Brain function at the cellular level — The primary cell type and basic functional building block of the nervous system is the neuron. Neurons are assembled together in local circuits. Important additional types include glial cells and Schwann cells. Several disorders are based on disturbed aspects of cellular and local circuit function (epilepsy, movement disorders, demyelinating diseases, and aberrant regeneration). In many situations, treatment schemes for these disorders may include pharmacotherapy directed at single neurons or circuits, gene therapy to alter individual cell functioning, cellular replacement and transplantation, and other forms of restorative treatment.

2. Brain function at the system level — Systems within the nervous system include various local circuits and regions working together for a common modality. Examples of modalities include a variety of motor planning, modulation, and execution systems, sensory systems dedicated to particular types of inputs, cognitive and memory systems, and basic systems that control alertness, respiration, and cardiac status. Each system forms a unit of functioning based on a certain modality and assembled for cooperative nervous system functioning. When a system dysfunctions, for example with movement disorders, treatment such as deep brain stimulation may be directed at the systems level to alter the function of the system.

3. Brain structure and function at the organ level — Regardless of the function of the nervous system, the brain remains an organ that requires adequate nutrition, blood flow, oxygenation, removal of waste products, and mechanical support from the skull and spinal column. It can also develop mass lesions such as tumors. The treatment of such disorders, although specific to the brain, is similar to other clinical treatments at the organ level, i.e., resecting of mass lesions, enhancement of blood flow, CSF diversion, and mechanical restoration of the spinal column. Many of these treatments are empirical and may require assessment of their clinical efficacy separate from any cellular basis for the treatment scheme.

1.6 CONCLUSIONS

Neurosurgeons have long been eager to develop "designer" surgical procedures to solve specific problems and perform "experimental" surgical procedures. Much medical knowledge has been gained from these (usually) rational approaches. For example, most of our current dermatome maps were derived from single root and multiple root dorsal rhizotomy procedures performed in the 1920s and 1930s to eliminate cancer pain; sensory losses were carefully mapped postoperatively. Likewise, many of the "fad" surgical procedures performed to help Parkinson's disease patients and most of our clinical knowledge of the function of the basal ganglia derive from a long line of experimental procedures (see Chapter 8). However, many newer advances resulted from the basic neuroscience that blossomed over the past 20 years and are being tested more rationally. This volume covers many of these exciting new advances, with the caveat that trying to peer into the future is not necessarily an exact science, and many of the products and devices mentioned may fail in application or development and some will succeed.

REFERENCES

1. Bast, R.C., Mills, G.B., and Young, R.C., Translational research: traffic on the bridge, *Biomed. Pharmacother.*, 55, 565–571, 2001.
2. Birmingham, K., What is translational research? *Nature Med.,* 8, 647, 2002.
3. Hall, J.E., The promise of translational physiology, *Am. J. Physiol. Heart Circ. Physiol.*, 51, 803–804, 2002.
4. Choi, D.W., Exploratory clinical testing of neuroscience drugs, *Nature Neurosci.,* 5S, 1023–1025, 2002.
5. Carrel, T., The relationship between surgeon and basic scientist, *Transplant Immunol.*, 9, 331–337, 2002.
6. Finkelstein, R., Miller, T., and Baughman, R., The challenge of translational research: a perspective from the NINDS, *Nature Neurosci.*, 5S, 1029–1030, 2002.
7. Pober, J.S., Neuhauser, C.S., and Pober, J.M., Obstacles facing translational research in academic medical centers, *FASEB J.*, 15, 2303–2313, 2001.
8. Sung, N.S. et al., Central challenges facing the national clinical research enterprise, *JAMA*, 289, 1278–1287, 2003.
9. Kondziolka, D. et al., Transplantation of cultured human neuronal cells for patients with stroke, *Neurology,* 55, 565–569, 2000.
10. Archibald, S.J. et al., Monkey median nerve repaired by nerve graft or collagen nerve guide tube, *J. Neurosci.,* 15, 4109–4123, 1995.
11. Barnett, G.H. et al., Intraoperative localization using an armless, frameless stereotactic wand: technical note, *J. Neurosurg.*, 78, 510-514, 1993.
12. De la Porte, C., Technical possibilities and limitations of stereotaxy, *Acta Neurochir.*, 124, 3-6, 1993.
13. Nutt, J.G. et al., Randomized, double-blind trial of glial cell line-derived neurotrophic factor (GDNF) in PD, *Neurology,* 60, 69–73, 2003.
14. Gill, S.S. et al., Direct brain infusion of glial cell line-derived neurotrophic factor in Parkinson's disease, *Nature Med.*, 9, 589–595, 2003.
15. Janson, C.G. et al., Viral-based gene transfer to the mammalian CNS for functional genomic studies, *Trends Neurosci.,* 24, 706–712, 2001.
16. Hodges, C.J. and Boakye, M., Biological plasticity: the future of science in neurosurgery, *Neurosurgery,* 48, 2–16, 2001.
17. Zivin, J.A., Cell transplant therapy for stroke: hope or hype? *Neurology,* 55, 467, 2000.
18. Sugarman, J. and McKenna, W.G., Ethical hurdles for translational research, *Radiation Res.*, 160, 1–4, 2003.
19. Nathan, D.G., Careers in translational clinical research: historical perspectives, future challenges, *JAMA*, 287, 2424–2427, 2002.
20. Campbell, E.G. et al., Status of clinical research in academic health centers, *JAMA*, 286, 800–806, 2001.

2 Clinical Prospects for Neural Grafting Therapy for Cortical Lesions

Ashutosh A. Pradhan, Ashok K. Shetty, and Dennis A. Turner

CONTENTS

2.1 INTRODUCTION

The goal of this chapter is to assess how close the hypothesis of neural grafting to enhance nervous system function may be to clinical reality, and the problems yet to be resolved before it is applicable clinically.[1–5] This chapter focuses on neural

0-8493-1482-8/05/$0.00+$1.50

grafting for lesions of the cerebral cortex arising from epilepsy, stroke, and head injury. Additional chapters will include further information on neural grafting for spinal cord injury (Chapter 3) and Parkinson's disease (Chapter 8). Although clinical neural grafting has been performed primarily in the context of Parkinson's disease,[6] many issues relevant to this disease do not necessarily transfer to lesions of the cerebral cortex and hippocampus.

For example, Parkinson's disease involves grafting of a dopaminergic phenotypic cell that possesses a large, diffuse axonal elaboration with a modulatory function, rather than specific synaptic relay systems as noted in glutaminergic synapses within the cortex and hippocampus. Thus, the goal of grafting in cortical lesions is usually to functionally replace part of a highly organized relay system, whereas in Parkinson's disease the goal is to replace dopaminergic innervation nonspecifically to the striatum.

Clinical treatments for acute lesions of the cerebral cortex and hippocampus such as head injury and stroke have focused almost exclusively on cytoprotection and prevention of secondary damage within the early period after the lesion.[4] While a moderate reduction in the number of neurons damaged may facilitate recovery, in many instances this format of treatment was clearly insufficient clinically because recovery was less than optimal. Moreover, the spontaneous neuronal replacement that transpires via proliferation of endogenous stem/progenitor cells after injury appears to be very restricted, ephemeral, and nonfunctional.[7]

Few available treatments are aimed at enhancing recovery of function of surviving neuronal elements.[2] Additionally, recovery can be accompanied by aberrant axonal plasticity of surviving neurons, characterized by inappropriate innervation of denervated synaptic regions.[8,9] One consequence of such inappropriate recovery is the late occurrence of epilepsy due in many instances to isolation of hyperexcitable regions, but which may still exert an untoward effect on the intact brain.[8,10] Restoration of normal afferent brain control over autonomous, hyperexcitable regions may be critical to both restore function and alleviate epilepsy. Exogenous transplants of multipotent progenitor or stem cells may play a role not only in epilepsy, but also in head injury, stroke, and degenerative disease.[11–17]

At both early and late time points after hippocampal or cortical lesions, one method to enhance recovery and restore function may be grafting of committed embryonic neural cells even though the goals may differ.[1,4] Neural grafting acutely after a lesion may provide additional unformed neuronal elements that may then insinuate and become integrated into the host circuitry, potentially enhancing overall recovery.[1,18,19] Early grafting may also change the acute milieu, decreasing death among host cells. Late after a lesion, when the damage is stable, neural grafts may be competent to enhance actual appropriate circuitry reconstruction.[3] This is likely accomplished by:

1. Providing correct target neurons for host axons
2. Furnishing proper afferent axons to host neurons
3. Inducing withdrawal of aberrantly formed synaptic contacts

These events together may suppress hyperexcitability and restore afferent control in autonomous regions. Embryonic neural grafts have the dual advantage of surviving the transplantation trauma and anoxia and possessing competence for considerable axonal growth into the adult host CNS.[1,3,20] In Parkinson's disease, for example, neural grafts have been used to treat a stable, long-term disorder by adding ectopic but critical dopaminergic re-innervation.[6]

The goal of circuitry reconstruction with neural grafts requires appropriate neuronal elements for the host region that are capable of becoming functionally integrated within the host. Many other possible goals and mechanisms can be achieved by neural transplantation including release of neurotrophic factors or neurotransmitters and replacement of glial cells.[16,21–23] However, the specific requirement for circuitry reconstruction leads to a hypothesis as to what an ideal graft may be.[3] An ideal graft would have certain characteristics:

1. Adequate survival of the transplanted neurons within the host environment (at least 20% of grafted neurons)
2. Appropriate dispersion or migration of the transplanted cells to restore host neuronal cell layers (leaving few cells at the transplant site)
3. Normal cellular development including acquisition of region-specific dendritic complexity, synapses and intrinsic characteristics
4. Appropriate elaboration of both local circuit and long-distance axons for synaptic connectivity into the host
5. Attraction of a significant number of specific afferent axons from the host

While these requirements are rigorous, the quantitative measurement of these characteristics may lend credence to exertion by the graft of a specific, defined role in the host, as opposed to a nonspecific or non-neuronal effect.[1,3]

Grafting into cerebral cortex or hippocampus to facilitate circuitry reconstruction may be radically different from the grafting treatment of Parkinson's disease. For example, grafts into the striatum of dopamine-enriched tissue are intentionally ectopic, and do not appear to develop long-distance axonal growth despite the fact that embryonic dopaminergic axons are inherently capable of such growth.[6] The other major difference is the type of neuron that is grafted and its neurotransmitter type because dopamine neurons possess much more diffuse and larger axonal terminal synaptic fields than the more typical glutamatergic neurons and GABAergic neurons considered in hippocampal or cortical grafting. Thus, only some parallels may be noted between the two different regions, but issues of graft tissue survival and integration remain paramount for both.[3,24,25]

The hippocampus represents a critical model region for cerebral cortex in general for the analysis and testing of grafting treatments because all the elements present throughout the neocortex are noted in some form in the hippocampus, including the various types of principal cells and interneurons and the intervening neuropil. The purpose of this chapter is to first describe the preclinical data for neural grafting. Second, the clinical situations to which hippocampal or cortical neural grafting may be applicable will be analyzed, in addition to potential graft sources and their limitations. Finally, the bridge between preclinical research and clinical usefulness and applicability will be discussed.

2.2 GRAFT CELL INTEGRATION: PRECLINICAL STUDIES

We defined graft integration into the host on a quantitative, cellular basis specifically to assess circuitry reconstruction.[26–28] Neural grafting has many other goals, for example, provision of an enzyme or neurotransmitter, furnishing cells to form myelin sheaths for host axons, and production of growth factors or metabolic products. Our hypothesis of cellular integration applies primarily to the goal of making a graft an integral part of synaptic circuitry within the brain. The specific measurable aspects of integration include:

1. Cell survival, directly comparing the number of cells transplanted and those recovered *in vivo* at different postgrafting time points
2. Cell dispersion and migration away from the graft site
3. Graft cell differentiation into region-specific neuronal phenotypes
4. Graft cell local and long-distance efferent synaptic interactions with the host neurons
5. Graft cell afferent connectivity with appropriate host axons

Graft integration may be differentially analyzed for various cell types, including embryonic neurons and immature stem cells.[28,29] Figure 2.1 is a schematic of the results of these preclinical studies.

2.2.1 Graft Cell Labeling

Assessment of graft integration requires a unique label for the grafted neurons so that their survival, migration, and differentiation fate after transplantation may be followed.[30] Genetically engineered cells may be labeled with a permanent, gene-based label (such as green fluorescent protein or beta galactosidase).[31] Prior to harvesting embryonic postmitotic cells, embryonic neurons may be efficiently labeled with a DNA label such as the thymidine analog 5-bromodeoxyuridine (BrdU)[26] by injecting the maternal host during times of neurogenesis for those cells. Because the cells are postmitotic and committed after embryonic harvesting, the neurons retain the BrdU label permanently.

After harvesting embryonic cells, fluorescent labels such as rhodamine dextran (RDA) may be used.[30] Serial sections through the host can define the location and developmental fate of the grafted cells and the percent of survival and degree of migration and/or dispersion can be calculated.[26] The label also allows confirmation of the graft cells when double-labeled with a second marker specific for the graft cell phenotype, long-distance connectivity, or physiology. For analysis purposes, the placement of micrografts (10,000 to 30,000 cells) is much more definitive than the use of larger but more therapeutic macrografts of $>1 \times 10^6$ cells. The smaller number of cells within micrografts can be explicitly counted and tracked using

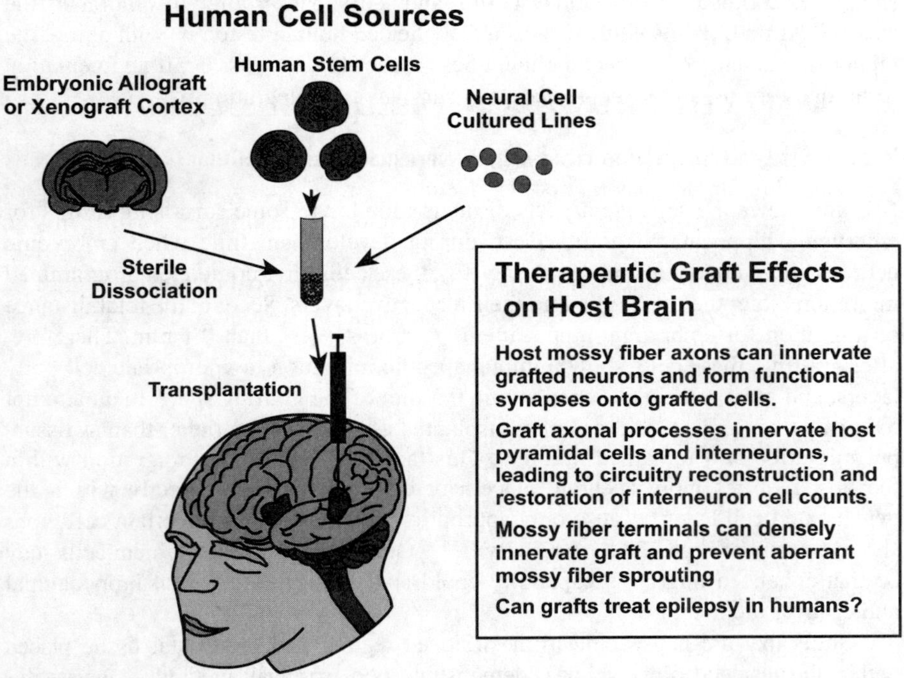

FIGURE 2.1 (See color insert following page 146.) Cortical grafting studies. First, cell sources include human or porcine embryonic cortex or hippocampus, various types of progenitor or stem cells, or cultured cell lines, most derived from neuronal tumors. After dissociation and transplantation, the fate of the transplanted cells can be assessed for synaptic integration within the host. In rodent models, therapeutic graft effects on the host include the formation of mossy fiber synapses onto grafted neurons, amelioration of postlesion interneuron loss, and prevention of aberrant mossy fiber sprouting. Whether grafts can ameliorate epilepsy remains to be analyzed in rodents and humans although the framework has been established.

unbiased cell counting methods. Unique labels form the critical basis for evaluation of graft integration within the host to unambiguously identify the grafted cells within the host.[26]

2.2.2 Graft Cell Survival

Cell survival can be assessed in terms of the number of cells grafted compared to those recovered later *in vivo*. In normal or intact hosts and grafts performed late after lesions, only 18 to 30% of grafted hippocampal cells survived.[26] In contrast, dopaminergic grafts in models of Parkinson's disease showed far poorer survival ranging from 3 to 20% in various studies. However, at early time points following lesions, the degree of survival was much greater (60 to 80%), particularly in young adults.[26] This enhanced survival is due primarily to the enhanced neurotrophic factor environment present up to 10–14 days following a lesion and the potential effects of denervation.[32] Grafts soon after lesions were well tolerated and considerably enhanced, compared to the intact cortex and the situation after resolution of the lesion.[26] At time points equivalent to a fully healed human lesion or with aging, the hippocampus and cortex become much less receptive to grafts. Graft augmentation techniques are required to enhance graft survival and integration.[33,34]

2.2.3 Graft Cell Migration and Dispersion

Breaking down the separate aspects of integration led to some surprising results for embryonic hippocampal grafts. First, during development time when embryonic neurons are removed (at embryonic day 19), these cells have completed programmed migration along the radial glia into their respective layers. Second, the total distance of migration for embryonic hippocampus is short — less than 0.1 mm. Therefore, after grafting, these cells show minimal specific migration to appropriate cell body layers, and most remain clumped within 0.5 mm of the grafting site.[26] In an attempt to enhance dispersion, grafts were transplanted as a suspension rather than as tissue, but migration still remained minimal. This lack of capability for migration within the host requires more accurate placement of multiple grafts directly within the degenerated cell layer because only appropriately placed grafts of certain cell types show capacity for specific connectivity.[27,35,36] One of the promises of stem cells may be that enhanced migration capability could lead to recapitulation of hippocampal and cortical architecture.[15]

Grafts that are nonspecific to the lesioned region, such as striatal tissue placed within the lesioned hippocampus, demonstrate poor survival and little capacity for integration.[9] Thus, it is critical that appropriate cellular elements are placed into locations most suitable for their development. They differ from dopaminergic grafts for Parkinson's disease that appear to function better when placed ectopically within the striatum, and grafted as strands rather than as a dissociated cell suspension.[6] In the hippocampus, the various types of functions that improve with appropriate placement include afferent circuitry (mossy fiber terminals on grafted CA3 neurons) and efferent circuitry to both the deafferented CA1 region and commissural projection areas.[9,27,35]

2.2.4 Synaptic Graft–Host Interactions

Connectivity between graft and host requires that host fiber tracts have access to the graft and that graft axons are able to recognize and follow host axon guidance pathways.[3,27,36] A graft placed as a chunk of tissue, rather than as dissociated cells, encourages local connections to form within the graft, discouraging connections between the surrounding brain and the graft.[37] Axon guidance pathways differ, depending on the inherent wiring and pattern of connections.[20,27] For example, hippocampal CA3 cell grafts can demonstrate contralateral commissural efferent projections if located near the degenerated CA3 cell layer, which apparently provides the appropriate molecular signals for such long-distance connectivity. We have termed this capability the *axon guidance pathway*, which for commissural connections appears to be specific.

Fibers from many regions of the hippocampus terminate in the septum. Therefore grafts placed in most regions of the hippocampus can robustly send efferent fibers into the septum. These embryonic graft neuron axons demonstrate competence to follow innate host axon guidance pathways and are not susceptible to inhibitory molecules such as myelin-associated glycoproteins along the host pathways, unlike adult host axons. The locations of these axon guidance pathways in the host may be highly specific, requiring accurate placement of the grafts to achieve access. Afferents into the graft develop readily, particularly mossy fiber ingrowths, if the grafted cells are their natural target cells (i.e., CA3 cells), demonstrated both physiologically by direct slice neuronal recordings and anatomically by Timm's histochemical staining.[9,20,30] Likewise, short-distance outgrowth from embryonic grafts was demonstrated to be both dense and appropriate (from CA3 grafts to denervated regions of the CA1 subfield and the dentate gyrus), indicating that embryonic grafts can develop both appropriate afferent and efferent connections in the hippocampus.[28]

2.2.5 Preclinical Graft Cell Sources

Aspects of integration have been well defined for embryonic hippocampal cells, as discussed above.[26] However, these cells are not optimal in the sense that they are not "ideal" grafts, particularly because of limited supply, ethical issues, and lack of ability to migrate within the host after transplantation. Therefore, hippocampal stem cells with pyramidal neuronal phenotypes have also been analyzed as alternatives to embryonic cells.[12,28,29] These cells likely arise from the posterior subventricular zone and form neurospheres *in vitro* in the presence of mitogenic factors such as epidermal growth factor (EGF) or fibroblast growth factor (FGF).[29,38–40]

Neurospheres are large collections of undifferentiated cells that develop in specific culture conditions from clonal stem cells removed from *in vivo* subventricular zones and contain both stem cells and their progeny. However, these cells show limited differentiation into neurons *in vitro* and *in vivo*, and may require conditioning with appropriate neurotrophic factors to enhance neuronal differentiation both prior to and after transplantation.[29] For example, physiological development and fiber outgrowth may be limited, even for cells resembling pyramidal neurons, due to their limited differentiation and axon growth. Further, the milieu of the injured brain could adversely affect differentiation of stem cells into neurons as a result of inadequate positional cues. Thus, hippocampal stem cells (and neural stem cells in general) are very promising, but will clearly require priming into partially differentiated region-specific neurons prior to grafting to fully achieve their differentiation and connectivity specific to the site of grafting. Whether this differentiated phenotype will be maintained after grafting, particularly for prolonged periods, will require further research.[11,13,15,38–44]

Immortalized cell lines have also been developed to obviate logistical problems from the use of fetal and embryonic stem cells. However, cell lines are limited by their potential to form tumors and degree of differentiation into true neurons capable of integration into the host. In addition, another goal of therapy using cell lines (NT2N cells) is to produce exogenous proteins needed in the CNS for particular disorders instead of completely integrating into existing circuitry. The human embryonal carcinoma cell line NT2N exhibits many properties of neuroepithelial precursor cells.[45,46]

2.3 CLINICAL CONTEXTS

This chapter focuses on clinical entities — primarily lesions of the cerebral cortex including the hippocampus. Grafting proposed for spinal cord treatment will be discussed in Chapter 3; subcortical grafting for Parkinson's disease is discussed in Chapter 8.

2.3.1 Cortical Lesions: Early Postlesion Grafting

Common lesions of the cerebral cortex (including the hippocampus) include head injuries, particularly cerebral contusions, and cerebral infarcts. Both head injury and stroke may be accompanied by extensive tissue damage and early neuronal replacement via grafts may facilitate structural and functional recovery by adding unformed neural elements to assist in circuitry reconstitution.[18,19,47] Such facilitation could consist of enhancing regional recovery and also preventing aberrant regeneration that may accompany cortical recovery in the form of compensatory sprouting of neighboring axons and inappropriate innervation of denervated synaptic sites.[9] The relatively unformed neuronal characteristics of neural grafts and their enhanced short- and long-distance axonal collateral growth in the adult, host CNS, may facilitate host recovery far beyond what would be obtained with either innate axonal regrowth alone or endogenous stem/progenitor cell proliferation and differentiation.[7]

The environment, within a few days after the lesion, appears particularly conducive to graft survival and integration.[26] This favorable host environment may be due to an enhancement in the level of neurotrophic factors in the vicinity of the lesion[32] as a result of astrocytic hypertrophy, microglial activation, and enhanced neurotrophic gene expression in surviving neurons.

For clinical grafting purposes, hippocampal grafts could be placed directly within the appropriate cell layers by stereotactic injection. However, neocortical suspension grafts may require multiple small injections into the neocortex on the border of the damaged region because direct injection into a severely damaged (or ischemic) area may provide minimal tissue nutrition and support for initial growth of axons. Preclinical studies of grafts into ischemic regions suggest excellent integration of embryonic cells into the appropriate tissue.[18,47] Histological demonstration of surviving cells and behavioral changes are considered to define "graft" effect in many studies. Depending on the goal of the graft, if circuitry reconstitution is desired, clear evidence of actual restoration of the damaged circuitry at the cellular level of analysis should be present. In other words, graft cell presence in the host does not necessarily imply circuitry reconstruction or appropriate synaptic interactions with host neurons. Physiological study of the grafted neurons and their synaptic interactions with the host is critical to fully define mechanisms of graft action on circuitry.[30]

Grafting may have other goals beyond circuitry reconstitution and these goals should be fully specified for each type of transplant. Embryonic grafts placed within a few days after an acute lesion have been shown to provide several clearly beneficial effects for the host. First, there is a clear anatomical and physiological demonstration of afferent connectivity from the host onto grafted cells when cells specific to the lesioned site are grafted.[26,27,30,35] Second, early appropriate grafts can permanently prevent the development of aberrant supragranular mossy fiber sprouting following development of mossy fiber sprouts into the graft.[9] Third, the apparent down-regulation of glutamate decarboxylase and calcium binding proteins in GABAergic interneurons following a CA3 lesion can be reversed by placement of appropriate (as opposed to inappropriate or control grafts) embryonic hippocampal grafts at 45 days postlesion.[35] These graft influences on the host strongly indicate that cellular graft integration can exert a positive influence on host lesion recovery.

2.3.2 Cortical Lesions: Late Postlesion Grafting

Most preclinical studies of graft integration focused on early transplantation after a lesion, particularly 10 to 14 days postlesion due to the propitious effects of the host environment on graft survival.[1,26,32] However, in many clinical situations such as chronic epilepsy and Parkinson's disease, the host environment many years after the lesion has occurred is resistant to graft integration (similar to normal cortex).[26] This host resistance may worsen with age.[3,33,34] In contrast to immediate results after a lesion, when the extracellular environment, postsynaptic cells, and presynaptic axons are all ready to attempt circuitry restoration, late after a lesion all three critical elements have returned to a more quiescent and less facilitating state.

Thus, grafts placed late after a lesion may require significant enhancement of the number of cells transplanted, their readiness to re-innervate the host, and critical placement.[34] It may be possible to also prepare the host prior to the graft with a small lesion sufficient to induce a glial reaction (for example, placement of a probe 7 to 10 days ahead of time and subsequent placement of the cells) or with pregraft infusion of neurotrophic factors. Because of the difficulty with graft integration at such late postlesion times, fewer preclinical investigations focused on overcoming this resistance have been performed although these barriers to graft cell survival are important to analyze clinically.

2.3.3 Neural Grafts for Treatment of Epilepsy

The lifespan incidence of seizures shows a dramatic increase at the extremes of young and old ages, particularly seizures due to lesions of the brain including those arising from head injuries, strokes, tumors, and Alzheimer's disease. In younger patients one of the most common seizure types is partial complex, resulting from mesial temporal sclerosis (MTS).[8] Most types of lesions that lead to later epilepsy involve neuronal and tissue loss and this is exemplified by MTS.

One concept of lesions resulting in hyperexcitability and eventually epilepsy is that an autonomous region becomes disconnected from the normal afferent host control. This autonomous region persists in demonstrating intrinsic hyperactivity, possibly manifest as an interictal focus, and can under some conditions lead to seizure propagation within the remainder of the brain.[8] One hypothesized role of grafts is to reconnect the autonomous area directly to the host. Another hypothesis involves modulation of other systems that can suppress seizures, for example, noradrenergic, serotonergic, midbrain, or cholinergic inputs.[11,21,48–50]

Animal models of convulsions and epilepsy reflect large numbers of types and categories of the human disease. For example, numerous mutants show early onset of generalized seizures,[48,51,52] kindling models of seizures,[16,21,49,53] and many models of irritants that can lead to convulsions. Several hippocampal models may reflect some features of partial complex seizures and anatomically may resemble aspects of MTS.[3,8,9,32,53] Late manifestations in these animal lesions often resemble the human situation and include aberrant mossy fiber sprouting, permanent down-regulation of calcium binding proteins in the CA1 subfield and dentate gyrus, and loss of glutamate decarboxylase within major fractions of interneurons.[35]

Embryonic grafts into animals following kainic acid lesions demonstrate a number of positive effects on the host that are indicative of a high degree of graft integration. Hippocampal grafts receive afferents from the host dentate granule cells (mossy fibers).[20] If the grafted CA3 cells are sufficiently numerous, the result can be amelioration of aberrant mossy fiber sprouting, indicating that these axons prefer an appropriate rather than inappropriate target.[9] Based on *in vivo* studies, the CA3 grafts develop long-distance connections, including to the contralateral CA3 region and to the septum. Long-term *in vitro* tissue studies using organotypic hippocampal cultures indicated a dense local connectivity established to the CA1 region. These graft efferents can also lead to a powerful re-innervation of the CA1 region, restoring glutamate decarboxylase in the GABAergic interneuron population as compared to lesion-only hosts.[35] All these beneficial effects confirm that graft integration may be sufficient to reconstruct the hippocampal circuitry after a kainic acid lesion. However, *in vivo* EEG and behavioral studies are needed to confirm beneficial effects on lesion recovery and host electrographic or clinical seizures. A stable rodent model of seizures following a lesion is clearly needed to assess how well these grafts may ameliorate seizures.[7,12]

Other types of grafts have been proposed for amelioration of kindling-induced seizures, particularly locus coeruleus grafts that contain norepinephrine-producing cells[21] and cholinergic grafts.[49,54] Further, the antiepileptogenic outcome of specific neural grafting in the latter studies was linked to the degree of graft-derived noradrenergic or cholinergic innervation of the stimulated brain region. Nevertheless, these findings are not clinically relevant for application of neural grafting to epilepsy, particularly MTS, because lesions of noradrenergic or cholinergic neurons are not present in the human condition. GABAergic grafts (those containing inhibitory interneurons predominantly) have also been suggested[16,25,55] due to the considerable inhibitory effect of GABA (the main inhibitory neurotransmitter) on seizures.[50]

Some hippocampal grafts have been shown to function as a heterotopia, particularly when chunks (rather than suspensions) of hippocampal tissue are placed.[37,54] These heterotopic grafts are inherently epileptogenic and can actually induce seizures in the host — a highly undesirable situation. When hippocampal tissue is placed as a chunk graft, the internal recurrent circuitry tends to form within the graft instead of an external connection between the host and graft through the graft–host interface. These internal connections within a graft may reinforce the innate hippocampal tendency toward hyperexcitability and seizures. Thus, integration of the graft into the host circuitry and appropriate afferent control over the graft are critical for both lesion recovery enhancement and possible treatment of the epilepsy. If grafts are placed early after a lesion, then prevention of development of an epileptic focus, a true anti-epileptogenic therapy, could possibly result.[9,35]

2.3.4 Neural Grafts for Treatment of Stroke

Stroke is one of the leading causes of death and causes severe disability. Treatments are limited to prevention and the acute setting. After a cerebrovascular accident has occurred, only supportive therapies are available. Grafting would seem ideal to replace lost cells.[56] Currently, stroke is one of the most active areas for cell transplantation, with small Phase I and Phase II trials completed for select basal ganglia (deep) strokes in humans.[5,19] The basis of these initial human studies lies in the success of animal studies although synaptic integration has not been fully analyzed in such a deep model of hemorrhage. Many stroke models exist, but the rodent model of the middle cerebral artery (MCA) occlusion has many clinical similarities to conditions seen in humans during cerebral ischemia. Much success has been achieved in this area with experimental functional recovery.[18]

Transplantation in stroke models has shown improvement in behavioral dysfunction as early as 1 month after grafting using an immortalized cell line as the cell source although animals required immunosuppressants to maintain the robust effects of the grafts. However, even non-immunosuppressed animals showed improvement in comparison to control animals. These studies have justified the use of human-derived NT2N neurons in stroke by showing functional improvements in the animals. Finally, the grafts produced no obvious deleterious effects.

2.3.5 Neural Grafts for Treatment of Severe Head Injury

Traumatic brain injuries and head injuries have very limited treatment. Most therapies are aimed at controlling intracranial pressure in the acute phase, but such supportive measures aim to decrease secondary cell loss rather than enhance recovery. Cell transplantation has a two-fold strategy: (1) to decrease the initial inflammatory reaction that leads to cell death and (2) to replace lost cells from the primary and secondary injuries. Initial studies showed poor survival of grafts in injured areas, but animal models of lesions showed considerably enhanced survival within a few days of the initial injury.[3,26] In addition, multiple sources of cells have also been used in this paradigm. Strategies have evolved to improve survival of grafts and integration, such as cografting of neural stem cells with supportive cells or substrates. Partial functional recovery has been shown with cografts including marrow-derived stromal cells and fibronectin.[57,58] In addition, developed cell lines have been successfully grafted.[17]

The primary model of neural grafting for treatment of head injury has focused on early addition of unformed elements to cortical areas (and potentially to areas of white matter shear injury), then allowing these elements to participate in the overall recovery and rehabilitation of the patient. However, measurement of improved outcome with the wide range of severity of head injury may be very difficult compared to measuring stroke outcomes.

2.4 CLINICALLY APPROPRIATE GRAFT CELL SOURCES

Potential sources of cells include various forms of embryonic neural cells, non-neural cells, tissue cultures cell lines, and pluripotent stem cells. All these groups can be further subdivided depending on the times they are acquired and the sources from which they are initially derived including autografts (from individual patients), allografts (from another individual of the same species), and xenografts (from a different species, for example, another mammal). All the initial human Parkinson's grafting studies, for example, were performed with autografts (adrenal medulla harvested at the same surgical session and reimplanted into the brain) or allografts from mixed embryonic cadaver donors.[6]

Embryonic porcine xenograft cells have also undergone clinical trials because mammalian embryonic cells appear to substitute well across species.[1,59,60] All these cells (with perhaps the exception of directly derived autografts) now require extensive FDA approval for reimplantation strategies. The approval requires that sterility and safety be ensured during processing of any *in vitro* maintenance or tissue culture and during direct reimplantation, particularly since these cells require direct brain implantation.

2.4.1 Embryonic Neural Cells

No ideal graft donor cells currently exist.[3] All tissue sources show significant limitations from both scientific and ethical viewpoints.[61] While embryonic neural cells currently demonstrate the best integration, specific migration of postmitotic neurons is highly limited, thus impeding appropriate cell distribution in the host. Additionally, embryonic allografts and xenografts impose ethical burdens because abortion is the tissue source, because they alter the innate human characteristics of brain and mind, and because of the risk of rejection.[6,59,60,61] However, xenograft embryonic cells have the advantage of availability in large numbers, particularly from porcine sources. They appear capable of substituting for human cells of similar origin based on equivalent neuronal sizing and lengths of projections.[60]

Other advantages of embryonic neural tissue are the ready, appropriate growth of embryonic axons into the host CNS, the known, postmitotic fates of the cells, and their excellent survival in a relatively anoxic host environment directly after grafting. However, because human trials of allografts for Parkinson's disease have shown marginal improvement and unexpected side effects, enthusiasm for any form of embryonic cell transplant is now considerably diminished.[6]

2.4.2 Cultured Cell Lines

Various types of precursor cell populations have been immortalized using oncogenes or telomerase. These cells offer the potential benefit of generating clonally identical cells, with innate genetic rules (such as temperature elevation) for inhibition of further mitoses.[1,11,13,16,22,23,45] In addition, spontaneously generated tumor (hNT) cells have been subcloned and characterized, and these tumor lines have indefinite capability for mitotic activity. Some show differentiation with retinoic acid, but these are usually not inherent CNS cells and quantitative assessment of their actual (rather than projected) integration into the CNS remains limited. Also, the tumor genotype remains, and long-term questions about tumor reversion *in vivo* remain in spite of short-term differentiation *in vitro*.

Thus, these tumor cell lines have the advantage of ready availability and unlimited numbers but the worries about appropriateness for various grafting purposes and the residual risk of tumor escape remain, although such escape does not appear to have occurred during short-term preclinical testing. One initial clinical trial using such cells involved deep intracerebral hemorrhage.[19] For most of these cell lines, their abilities to differentiate into neurons capable of CNS integration have not been fully tested in ways similar to the ways embryonic cells have been tested. Thus, histological evidence of cell survival has often led to the premature conclusion that the cells are integrated within the CNS circuitry; for most cell lines, this analysis remains to be done. Although they are convenient and available in large numbers, such cell lines may not readily behave as CNS neurons due to their origins as immature cells.

2.4.3 Non-Neural Cells

The idea that differentiated cells have limited choices of progeny has recently been challenged. The concepts of transdifferentiated and dedifferentiated CNS cells have potentially supplied new populations of cells for transplantation.[62–65] Transdifferentiation involves taking cells such as bone marrow stromal cells and forming neural progenitor cells. Dedifferentiation is exemplified by glial progenitor cells that can form neurons. Many questions still surround this cell source.

Other cell types include transfected fibroblasts, glial cells, and multiple types of non-neural systemic cells such as lymphocytes that may transfer specific functions to the nervous system even though they are not able to function as neurons. Interest in most available non-neural cells has waned as newer forms of cells have become available, particularly various forms of stem cells.

2.4.4 Pluripotent Progenitor and Stem Cells

True embryonic stem (ES) cells can differentiate into all embryonic derivatives.[40] However, these cells require the highest number of cues for subsequent differentiation. They are isolated very early from the inner cell masses of embryonic blastocysts. Relatively few experiments have been done using this very primitive cell population because of the great difficulty in forcing differentiation along various lines. These cells offer several advantages, particularly their general CNS fate capability and rapid cell division to create large numbers of cells, although access to blastocysts is highly limited in the U.S. today. These cells give rise to slightly more differentiated and regional stem cells derived from ventricular and subventricular zones, and then further along the path of commitment are progenitor cells.[40] All these cell types are actively being pursued for transplantation paradigms for recovery of function.[12,14,15,28,29,41–44,66]

Various types of multipotent, self-renewing neural progenitor or stem cells show considerable promise but differentiation into a specific lineage remains difficult to control before and after grafting particularly when grafted into a lesioned CNS.[13,28,31,42,44] It was initially thought possible that a lesioned brain might direct specific differentiation of otherwise unformed cell transplants, but it was realized subsequently that differentiation is difficult to maintain after grafting. Furthermore, the exact differentiation potential of neural stem cells obtained from distinct brain regions after grafting into different areas of the lesioned adult brain is mostly unknown — particularly whether neural stem cells from different brain areas produce neurons specific to their region of origin or specific to the site of their grafting. Addressing these concerns directly will help determine whether we must use different kinds of neural stem cells to treat different types of neurodegenerative disorders based on the area of the brain afflicted. For example, hippocampal stem cells may be specific for repair of a lesioned hippocampus in epilepsy, mesencephalic stem cells for Parkinson's disease, and striatal stem cells for Huntington's disease, etc.

The overall differentiation into neurons improves with progenitor cells that are more rather than less differentiated (i.e., subventricular zone cells versus embryonic stem cells from blastocysts), but still remains a critical issue. Thus, characterization of molecular mechanisms that control the fate of neural stem/progenitor cells after grafting into different regions of the lesioned adult CNS in experimental models is necessary prior to their routine clinical use as treatments for neurodegenerative disorders.[11,12,15,38,43] While the promise remains that stem cells may eventually be directed to function as neurons *in vivo*, this promise has yet to be clearly realized in preclinical studies.[15]

2.5 SIDE EFFECTS OF NEURAL GRAFTS

Unlike a medical therapy that can be suspended, graft treatments currently are irreversible because no clinically applicable method to destroy or eliminate a neural graft *in vivo* has been developed.[6] Potential deleterious side effects of grafts include increased seizures,[54] transmission of a virus or tumor to the host, induction of rejection, and difficult-to-treat problems related to the disease, for example, dyskinesias noted with neural grafts in Parkinson's disease.[6]

In all such instances, it may be helpful to have a method to noninvasively remove the graft or alter it selectively without damage to the host brain. This problem is peculiar to neural grafts because they normally require diffuse placement as cell suspensions or chunks of cells and thus cannot be removed surgically without causing extensive damage. This is particularly true if the grafts are capable of migration to specific regions, in which case their diffusion and insinuation into the brain preclude direct forms of removal.

In animals, graft cells can be labeled before transplantation with a triggerable stealth toxin that releases singlet oxygen only when specifically triggered (chlorin E6).[67] Without the appropriate trigger, the cells develop normally and are indistinguishable from control grafts. However, upon illumination with even a low level of infrared light (at 720 nm), the chlorin E6 releases massive singlet oxygen that can destroy the grafted cells selectively *in situ* and show minimal host damage. Other methods of selectively destroying grafts include allografts and immunotoxins that may attack xenografts selectively. Some forms of triggerable genes may also be transfected into graft cells to allow initiation of selective cell death *in situ* without host damage. These methods may be helpful to extinguish any side effects from grafted cells by virtually destroying the cells selectively within the milieu of the brain.

2.6 CLINICAL APPLICABILITY AND CHALLENGES IN TRANSLATION

How far an experimental surgical treatment must be developed prior to initial human application remains a very difficult, almost unregulated, and contentious question. One set of guidelines generally outlines preclinical studies needed along with human experimentation requirements.[68] As an example, grafting of cultured tumor cells into deep basal ganglia lesions after intracerebral hemorrhages was performed in patients[19] following extensive preclinical testing.[18] Another example is the application of porcine embryonic cells to humans for Parkinson's disease and potentially for hippocampal or cortical use.[60] Clearly, a cell source should be FDA-approved for initial human trials of cell lines in terms of safety and freedom from transmissible diseases and neoplasias. Preclinical evidence should support a specific intended use. These requirements have clearly been met for treatment of Parkinson's disease, as confirmed by the large number of proposed and performed clinical trials for embryonic cell grafts.[1,4,6]

Methods are likewise needed to enhance graft functioning at late, stable postlesion phases likely used to treat neurological disorders. Such methods could include enhancing the extent of survival of grafted cells using pretreatment of donor or host cells with distinct neurotrophic factors and other factors such as caspase inhibitors[24,34,69] that suppress the apoptotic deaths of grafted cells during the early postgrafting period. At this juncture, the most appropriate donor cells for hippocampal grafting may be porcine embryonic cells from the age of gestation directly after hippocampal neurogenesis (10 to 12 weeks of gestation in the human, slightly earlier in the porcine model).[60] Since the FDA has now imposed extensive requirements for processing implanted cells, a method should be established to determine appropriate sterility, cell numbers, and presence of contaminants. These requirements may facilitate the standardization of grafting studies.

While much of this chapter discussed the mechanisms underlying graft integration into a host, by the time when human grafting experiments are pursued for neurological disorders, these principles will not be known in human subjects although they presumably will have been developed in animal models. Most medications helpful in treating seizures, head injuries, and strokes now have known bases from laboratory studies but this was not true at their market introduction. Thus, there is no need to have actual mechanistic understanding for a treatment to go forward and become FDA approved. On a scientific basis, however, such mechanistic underpinnings are critical to understand and improve treatments. In summary, the neurobiology of graft integration, survival, and differentiation is not yet fully mapped or understood. Assuming an appropriate graft cell source becomes available for further human testing, a critical approach to host integration of the graft and mechanistic treatments of neurological disorders will be needed if this form of restorative neurosurgery is to become a long-term, viable treatment option.[1,4,41]

REFERENCES

1. Bjorklund, A. and Lindvall, O., Cell replacement therapies for central nervous system disorders, *Nature Neurosci.*, 3, 537–544, 2000.
2. Hodges, C.J. and Boakye, M., Biological plasticity: the future of science in neurosurgery, *Neurosurgery*, 48, 2–16, 2001.
3. Turner, D.A. and Shetty, A.K., Clinical prospects for neural grafting therapy for hippocampal lesions and epilepsy, *Neurosurgery*, 52, 632–644, 2003.
4. Thompson, T., Lunsford, L.D., and Kondziolka, D., Restorative neurosurgery: opportunities for restoration of function in acquired, degenerative and idiopathic neurological diseases, *Neurosurgery*, 45, 741–752, 1999.
5. Kondziolka, D. et al., Neuronal transplantation for motor stroke: from the laboratory to the clinic, *Phys. Med. Rehab. Clin. N. Am.*, 14, S153–S160, 2003.
6. Freed, C.R. et al., Transplantation of embryonic dopamine neurons for severe Parkinson's disease, *NEJM*, 344, 710–719, 2001.
7. Arvidsson, A. et al., Neuronal replacement from endogenous precursors in the adult brain after stroke, *Nature Med.*, 8, 963–970, 2002.
8. Dudek, F.E. and Spitz, M., Hypothetical mechanisms for the cellular and neurophysiological basis of secondary epileptogenesis, *J. Clin. Neurophysiol.*, 14, 90–101, 1997.
9. Shetty, A.K. and Turner, D.A., Fetal hippocampal cells grafted to kainate-lesioned adult hippocampus suppress aberrant supragranular sprouting of host mossy fibers, *Exp. Neurol.*, 143, 231–245, 1997.
10. Aiken, S.P. and Brown, W.M., Treatment of epilepsy: existing therapies and future developments, *Frontiers Biosci.*, 5, E124–E152, 2000.
11. Cao, Q., Benton, R.L., and Whittemore, S.R., Stem cell repair of central nervous system injury, *J. Neurosci. Res.*, 68, 501–510, 2002.
12. Gage, F.H., Mammalian neural stem cells, *Science*, 287, 1433–1438, 2000.
13. Gray, J.A. et al., Conditionally immortalized, multipotential and multifunctional neural stem cell lines as an approach to clinical transplantation, *Cell. Transplant.*, 9, 143–168, 2000.
14. Liu, C.Y., Apuzzo, M.L.J., and Tirrell, D.A., Engineering of the extracellular matrix: working toward neural stem cell programming and neurorestoration, *Neurosurgery*, 52, 1154–1167, 2003.
15. Temple, S., Stem cell plasticity: building the brain of our dreams, *Nature Rev. Neurosci.*, 2, 513–520, 2001.
16. Thompson, K. et al., Conditionally immortalized cell lines engineered to produce and release GABA, modulate the development of behavioral seizures, *Exp. Neurol.*, 161, 481–489, 2000.
17. Riess, P. et al., Transplanted neural stem cells survive, differentiate, and improve neurological motor function after experimental traumatic brain injury, *Neurosurgery*, 51, 1043–1053, 2002.

18. Borlongan, C.V. et al., Transplantation of cryopreserved human embryonal carcinoma-derived neurons (NT2N Cells) promotes functional recovery in ischemic rats, *Exp. Neurol.,* 149, 310–321, 1998.
19. Kondziolka, D. et al., Transplantation of cultured human neuronal cells for patients with stroke, *Exp. Neurol.,* 55, 565–569, 2000.
20. Field, P.M. et al., Selective innervation of embryonic hippocampal transplants by adult host dentate granule cell axons, *Neuroscience,* 41, 713–727, 1991.
21. Bengzon, J. et al., Host regulation of noradrenaline release from grafts of seizure-suppressant locus coeruleus neurons, *Exp. Neurol.,* 111, 49–54, 1991.
22. Sinden, J.D. et al., Recovery of spatial learning by grafts of a conditionally immortalized hippocampal neuroepithelial cell line into the ischemia-lesioned hippocampus, *Neuroscience,* 81, 599–608, 1997.
23. Virley, D. et al., Primary CA1 and conditionally immortal MHP36 cell grafts restore conditional discrimination learning and recall in marmosets after excitotoxic lesions of the hippocampal CA1 field, *Brain,* 122, 2321–2335, 1999.
24. Boonman, Z. and Isacson, O., Apoptosis in neuronal development and transplantation: role of caspases and trophic factors, *Exp. Neurol.,* 156, 1–15, 1999.
25. Jacoby, D.B. et al., Long-term survival of fetal porcine lateral ganglionic eminence cells in the hippocampus of rats, *J. Neuroscience Res.,* 56, 581–594, 1999.
26. Shetty, A.K. and Turner, D.A., Enhanced cell survival in fetal hippocampal suspension transplants grafted to adult rat hippocampus following kainate lesions: a three-dimensional graft reconstruction study, *Neuroscience,* 67, 561–582, 1995.
27. Shetty, A.K. and Turner, D.A., Development of long-distance efferent projections from fetal hippocampal grafts depends upon pathway specificity and graft location in kainate-lesioned adult hippocampus, *Neuroscience,* 76, 1205–1219, 1997.
28. Shetty, A.K. and Turner, D.A., Neurite outgrowth from progeny of epidermal growth factor-responsive hippocampal stem cells is significantly less robust than from fetal hippocampal cells following grafting onto organotypic hippocampal slice cultures: effect of brain-derived neurotrophic factor, *J. Neurobiol.,* 38, 391–413, 1999.
29. Shetty, A.K. and Turner, D.A., In vitro survival and differentiation of neurons derived from epidermal growth factor-responsive postnatal hippocampal stem cells: enhancing and inducing effects of brain-derived neurotrophic factor, *J. Neurobiol.,* 35, 395–425, 1998.
30. Pyapali, G.K., Turner, D.A., and Madison, R.D., Anatomical and physiological localization of prelabeled grafts in rat hippocampus, *Exp. Neurol.,* 116, 133–144, 1992.
31. Shihabuddin, L.S., Holets, V.R., and Whittemore, S.R., Selective hippocampal lesions differentially affect the phenotypic fate of transplanted neuronal precursor cells, *Exp. Neurol.,* 139, 61–72, 1996.
32. Lowenstein, D.H., Seren, M.S., and Longo, F.M., Prolonged increases in neurotrophic activity associated with kainate-induced hippocampal synaptic reorganization, *Neuroscience,* 56, 597–604, 1993.
33. Wagner, A.P. et al., Brain plasticity: to what extent do aged animals retain the capacity to coordinate gene activity in response to acute challenges, *Exp. Gerontol.,* 35, 1211–1227, 2000.
34. Zaman, V. and Shetty, A.K., Combined neurotrophic supplementation and caspase inhibition enhances survival of fetal hippocampal CA3 cell grafts in lesioned CA3 region of the aging hippocampus, *Neuroscience,* 109, 537–553, 2002.
35. Shetty, A.K. and Turner, D.A., Fetal hippocampal transplants containing CA3 cells restore host hippocampal glutamate decarboxylase-positive interneurons numbers in a rat model of temporal lobe epilepsy, *J. Neurosci.,* 20, 8788–8801, 2000.

36. Shetty, A.K., Zaman, V., and Turner, D.A., Pattern of long distance projections from fetal hippocampal field CA3 and CA1 cell grafts in lesioned CA3 of adult hippocampus follows intrinsic character of respective donor cells, *Neuroscience,* 99, 243–255, 2000.
37. Stafekhina, V.S., Bragin, A.G., and Vinogradova, O.S., Integration of hippocampal suspension grafts within host neocortex, *Neuroscience,* 64, 643–651, 1995.
38. Keirstead, H.S., Stem cell transplantation into the central nervous system and the control of differentiation, *J. Neurosci, Res.,* 63, 233–236, 2001.
39. Shih, C.C. et al., Identification of a candidate human neurohematopoietic stem-cell population, *Blood,* 98, 2412–2422, 2001.
40. Pevny, L. and Rao, M.S., The stem-cell menagerie, *Trends Neurosci.,* 26, 351–359, 2003.
41. Bruce, J.N. and Parsa, A.T., Why neurosurgeons should care about stem cells, *Neurosurgery,* 48, 243–244.
42. Flax, J.D. et al., Engraftable human neural stem cells respond to developmental cues, replace neurons, and express foreign genes, *Nature Biotechnol.,* 16, 1033–1039, 1998.
43. Grisolia, J.S., CNS stem cell transplantation: clinical and ethical perspectives, *Brain Res. Bull.,* 57, 823–826, 2002.
44. Mehler, M.F. and Gokhan, S., Postnatal cerebral cortical multipotent progenitors: regulatory mechanisms and potential role in the development of novel regenerative strategies, *Brain Pathol.,* 9, 515–526, 1999.
45. Pleasure, S.J. and Lee, V.M., Ntera 2 cells: a human cell line which displays characteristics expected of a human committed neuronal progenitor cell, *J. Neurosci. Res.,* 35, 585–602, 1993.
46. Pleasure, S.J., Page, C., and Lee, V.M., Pure, postmitotic, polarized human neurons derived from Ntera 2 cells provide a system for expressing exogenous proteins in terminally differentiated neurons, *J. Neurosci.,* 12, 1802–1815, 1992.
47. Mudrick, L.A., Baimbridge, K.G., and Peet, M.H., Hippocampal neurons transplanted into ischemically lesioned hippocampus: electroresponsiveness and re-establishment of circuitries, *Exp. Brain Res.,* 86, 233–247, 1989.
48. Coleman, J.R. et al., Tectal graft modulation of audiogenic seizures in Long–Evans rats, *Exp. Neurol.,* 164, 139–144, 2000.
49. Ferencz, I. et al., Suppression of kindling epileptogenesis in rats by intrahippocampal cholinergic grafts, *Eur. J. Neurosci.,* 10, 213–220, 1998.
50. Gale, K., Mechanisms of seizure control mediated by GABA: role of the substantia nigra, *Fed. Proc.,* 44, 2414–2424, 1985.
51. Clough, R. et al., Fetal raphe transplants reduce seizure severity in serotonin-depleted GEPRS, *NeuroReport,* 8, 241–246, 1996.
52. Holmes, G.L. et al., Effects of neural transplantation on seizures in the immature genetically epilepsy-prone rat, *Exp. Neurol.,* 116, 52–63, 1992.
53. Holmes, G.L. et al., Effect of neural transplants on seizure frequency and kindling in immature rats following kainic acid, *Dev. Brain Res.,* 64, 47–56, 1991.
54. Buzsaki, G. et al., Suppression and induction of epileptic activity by neuronal grafts, *PNAS,* 85, 9327–9330, 1988.
55. Loscher, W. et al., Seizure suppression in kindling epilepsy by grafts of fetal GABAergic neurons in rat substantia nigra, *J. Neuroscience Res.,* 51, 196–209, 1998.
56. Onteniente, B. et al., Molecular pathways in cerebral ischemia: cues to novel therapeutic strategies, *Mol. Neurobiol.,* 27, 33–72, 2003.
57. Lu, D. et al., Neural and marrow-derived stromal cell sphere transplantation in a rat model of traumatic brain injury, *J. Neurosurg.,* 97, 935–940, 2002.

58. Tate, M.C. et al., Fibronectin promotes survival and migration of primary neural stem cells transplanted into the traumatically injured mouse brain, *Cell. Transplant.,* 11, 283–295, 2002.
59. Brevig, T., Holgersson, J., and Widner, H., Xenotransplantation for CNS repair: immunological barriers and strategies to overcome them, *TINS,* 23, 337–344, 2000.
60. Jacoby, D.B. et al., Fetal pig neural cells as a restorative therapy for neurodegenerative disease, *Artificial Organs,* 21, 1192–1198, 1997.
61. Turner, D.A. and Kearney, W., Scientific and ethical concerns in neural fetal tissue transplantation, *Neurosurgery,* 33, 1031–1037, 1993.
62. Mahmood, A. et al., Treatment of traumatic brain injury in female rats with intravenous administration of bone marrow stromal cells, *Neurosurgery,* 49, 1196–1203, 2001.
63. Mahmood, A. et al., Intracerebral transplantation of marrow stromal cells cultured with neurotrophic factors promotes functional recovery in adult rats subjected to traumatic brain injury, *J. Neurotrauma,* 19, 1609–1617, 2002.
64. Liu, Y. and Rao, M.S., Transdifferentiation: fact or artifact? *J. Cell. Biochem.,* 88, 29–40, 2003.
65. Chen, J.L. et al., Therapeutic benefit of intravenous administration of bone marrow stromal cells after cerebral ischemia in rats, *Stroke,* 32, 1005–1011, 2001.
66. Liu, S. et al., Embryonic stem cells differentiate into oligodendrocytes and myelinate in culture and after spinal cord transplantation, *PNAS,* 97, 6126–6131, 2000.
67. Shetty, A.K. et al., Selective laser-activated lesioning of prelabeled fetal hippocampal grafts by intracellular photolytic chromophore, *Neuroscience,* 69, 407–416, 1995.
68. Redmond, D.E. and Freeman, T., The American Society for Neural Transplantation and Repair considerations and guidelines for studies of human subjects, *Cell. Transplant.,* 10, 661–664, 2001.
69. Schultz, J.B., Weller, M., and Moskowitz, M.A., Caspases as treatment targets in stroke and neurodegenerative diseases, *Ann. Neurol.,* 34, 421–429, 1999.

3 Advances in Treatment of Spinal Cord and Peripheral Nerve Injury

Ali Zomorodi and Roger D. Madison

CONTENTS

3.1 INTRODUCTION

An estimated 400,000 people in the U.S. have permanent paralysis as a consequence of spinal cord injury and an additional 10,000 are injured each year. Patients with spinal cord injuries (SCI) can live 25 to 30 years after their initial injuries. Each patient must cope with a lifetime of neurological dysfunction including paralysis, bowel and bladder dysfunction, sexual dysfunction, spasticity, deafferentation pain, loss of skin integrity, and autonomic dysfunction.[1] Thus, SCI can be a devastating neurological disorder in terms of the years of disability caused and the associated physical and psychological complications. However, patients can remain highly functional with the use of modern aids, such as wheelchairs; they can participate fully in work, sports, and activities of daily living despite the obvious disability associated with the loss of function.

0-8493-1482-8/05/$0.00+$1.50

For thousands of years, physicians have been pessimistic in their approach to treating SCI because of the lack of innate recovery and secondary complications such as infections that usually ensue. Writing around 1700 BCE in the Edwin Smith Papyrus, an ancient Egyptian physician described SCI as a "disease not to be treated."[2] Now, almost 4000 years later, the treatment of SCI remains largely palliative: preventing injury progression; implementing bowel and bladder regimens; managing complications of sensory loss and skin breakdown; treating spasticity, dysautonomia, and deafferentation pain syndromes; and teaching patients how to cope with their disabilities. While such palliative care is highly successful and now results in nearly normal lifespans and functional capabilities, most affected patients still would like to enhance their mobility and regain more normal function. Fortunately, ongoing advances in neurobiology coupled with initiatives to facilitate the translation of this research into medical therapy promise to change this paradigm from palliation to cure.

Broadly speaking, current approaches to the treatment of SCI fall into one of four categories: (1) the prevention of secondary injury and delayed demyelination or axon loss (neuroprotection), (2) the repair or replacement of interrupted neural circuitry (spinal cord repair), (3) the use of aggressive rehabilitation techniques to optimize recovery through residual spinal cord plasticity (rehabilitation), and (4) the augmentation of function through prostheses (prosthetics). In this chapter, we will review advances in each discipline, the current hypotheses, and their future applications. Figure 3.1 outlines these potential treatment options.

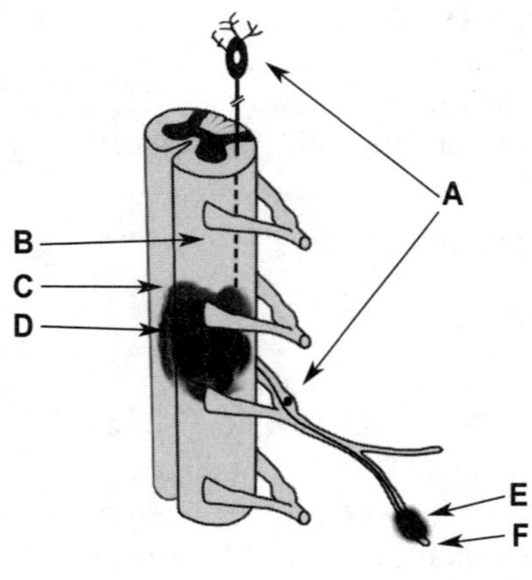

FIGURE 3.1 (See color insert following page 146.) Targets for the treatment of spinal cord and peripheral nerve injuries. A: Genetic or small molecule treatments to induce a proregeneration response in the cell body. B: Inhibition of myelin associated growth inhibitors and chondroitin sulfate proteoglycans at the site of injury. C: Neuroprotectants to prevent the spread of injury through secondary mechanisms. D: Transplantation of stem cells, olfactory ensheathing glia, or Schwann cell bridges to span the area of injury and replace lost cell populations; transplantation of macrophages or neurotrophin-secreting cells to prevent cell loss and promote regrowth. E: Use of synthetic grafts infused with Schwann cells, extracellular matrix, or neurotrophins to span the area of injury. F: Infusion of neurotrophins or use of electrical stimulation to improve the pace and accuracy of axonal regeneration.

3.2 SPINAL CORD NEUROPROTECTION

3.2.1 MATURATION OF SPINAL CORD INJURY

SCI can occur when the spinal cord is lacerated or macerated by a sharp penetrating force, contused or compressed by a blunt force, or infarcted by a vascular insult. Blunt force injuries are the most common causes of SCI, accounting for up to 50% of cases.[1] This form of SCI has been well modeled in animals, using weights dropped onto the exposed spinal cord. Initial histopathological studies suggested that secondary events that unfold after the mechanical injury enlarge the contusion and are responsible for a substantial portion of the ultimate functional deficit that results — the so-called secondary injury. From this came the hypothesis that identifying the components of secondary injury could provide rational targets for pharmaceutical interventions that could significantly limit the morbidity of SCI. This approach has dominated SCI research for much of this century and spawned many promising therapies. However, in reality, patients rarely lose additional function after they present with initial levels of injury, suggesting that, in practical terms, very little secondary injury occurs and most of the damage results from the initial impact.

In experimental models of blunt SCI, the initial mechanical force delivered to the cord results in a necrotic core that involves the spinal grey matter and spares a rim of white matter around the contusion site. Electrophysiological experiments have shown that neurons that survive the initial trauma become hyperexcitable and fire repeated salvos of action potentials. Intracellular and extracellular electrolyte concentrations are altered measurably as a consequence. Intracellular concentrations of calcium and sodium and extracellular concentrations of potassium increase significantly, making normal neuronal activity impossible. Clinically this is manifested as a flaccid motor paralysis below the level of injury and can last several months ("spinal shock"); it is eventually replaced by spasticity as the spinal cord slowly recovers innate tone. As a result of the flaccidity, a systemic hypotension ("neurogenic shock") may ensue. Meanwhile, petechial hemorrhages and progressive edema develop around the injury site along with a collapse of the microcirculation with a measurable reduction in spinal cord perfusion. As cells lyse, excitatory neurotransmitters reach toxic levels in the extracellular fluid and free oxygen radicals are elaborated. The consequent lipid peroxidation hastens cell death and promotes formation of cytokines — major components of the inflammatory cascade. Neutrophils enter the contusion within 24 hours, followed closely by lymphocytes. These cells start cleaning the debris while elaborating more cytokines and chemokines that reach measurable levels within 48 hours and continue the inflammatory cascade. Whether this form of inflammation is restorative and necessary (to clean up debris) or destructive in some manner remains highly contentious.

Meanwhile, apoptosis occurs in cells surrounding the initial core of necrosis, causing the lesion to grow further. Neutrophils are eventually replaced by macrophages and fibroblasts. With time, areas of extensive necrosis are replaced by the classic glial scar. Areas of milder injury develop scars rich in astrocytes; areas of large hemorrhage are replaced by glial-lined areas of myelomalacia that can sometimes lead to late (years later) post-traumatic syrinx.

Secondary processes continue to play a role in the clinical features of SCI even chronically. Robust local sprouting of injured and uninjured axons within the spinal cord segments produces circuits implicated in spasticity. Changes in the distribution and excitability of ion channels along with changes in excitatory and inhibitory inputs cause permanent hyperexcitability in some cell populations, possibly leading to chronic pain syndromes and hyperactivity causing motor spasticity; chronic demyelination can block signaling in other pathways.

The response to SCI can be divided into acute and chronic phases. Acutely, cell loss occurs due to the mechanical injury associated with excitotoxicity, lipid peroxidation, and inflammation, as the lesion is cleaned up. Chronically, cells that survive the initial events may go on to regenerate in a limited and imprecise way or they may succumb to apoptosis or demyelination.[3,4] These insights suggest several promising targets for pharmacologic intervention, assuming that the primary injury can be overcome.

3.2.2 Secondary Neuroprotection Treatment Schemes

To date, the only clinical treatment to emerge from neuroprotection research is high dose methylprednisolone (MP) therapy. Since the 1960s, corticosteroids have been used in the treatment of SCI. Initially, these agents were used for their ability to limit inflammation and spinal cord edema. Optimal therapeutic schemes with steroids involved pretreatment prior to injury, which provided better benefits than treatment after injury (which obviously is beneficial for spinal cord surgery). However, the initial National Acute Spinal Cord Injury Study (NASCIS) a nonplacebo controlled comparison of high-dose versus low-dose MP failed to show any benefit in the treatment of SCI.[5]

In the early 1980s, it was shown that key components of secondary injury included post-traumatic alteration of spinal cord blood flow, elaboration of free radical oxygen, and peroxidation of membrane lipids.[6,7] Trials in rodents, cats, and dogs demonstrated that MP can improve functional recovery from SCI by categorizing these processes, but it must be administered in intravenous doses of 30 mg/kg — much higher doses than those used in the NASCIS trial.[8–10] Incorporating some of these findings, the second NASCIS trials found that sustained high doses of MP administered within 8 hours of injury caused a statistically significant improvement in neurologic function although new SCI scales were required to measure this improvement.[11–14] Again, no placebo control was used and the initially determined outcome measures were abandoned and a new system for evaluating neurologic function in SCI was devised to show the benefits of treatment.

It is not apparent whether the statistically significant improvements translated into clinical benefit. Although the design and statistical analysis of the trials were widely challenged,[15] the high dose "Solu-Medrol Protocol" is almost universally applied in the emergency room management of SCI.[16] This is perhaps more reflective of physicians' desperation to offer some treatment to SCI patients than of the scientific validity of the studies. However, this high dose, short-term protocol is now considered the standard of care.

These studies lent credence to the hypothesis that secondary injury mechanisms may be important in the clinical evolution of SCI and spurred the development of multiple agents, each of which has been shown in some animal models to be somewhat efficacious. Among these is the 21-aminosteroid, tirilazad mesylate (TM), that scavenges free radicals, inhibits lipid peroxidation, and maintains spinal cord blood flow in animal models. Because it lacks the glucocorticoid activity of MP, TM is a safer drug and considerable interest in its clinical efficacy was generated.[17–19] Unfortunately, the NASCIS 3 trial concluded that TM does not appear beneficial in the treatment of SCI.[12]

Based on the premise that acute inflammation is deleterious to nervous tissue, specific inhibitors of the inflammatory response have been evaluated for benefit in SCI.[20] Among these, IL-10 has been shown to limit axonal loss, contusion size, and functional deficit following SCI in rats.[21,22] So have the broad spectrum chemokine receptor antagonist, vMIPII,[23] and the selective cyclooxygenase-2 inhibitors.[24,25] These latter drugs are already approved for human use and are well tolerated; it would be relatively simple to verify their ability to provide benefit to human victims of SCI. However, in many instances, the initial inflammation after a CNS lesion may actually be considered favorable for axonal recovery and regrowth and for enhancement of cell survival, as demonstrated by placement of neural grafts into lesions at short postlesion time points (see Chapter 2). Other pathways considered for treatment options include blockade of excitotoxicity,[26–33] treatment of apoptosis,[34–40] and application of hypothermia.[41]

Although most neuroprotection research produced promising results in animal models (similar to results shown for stroke research; see Chapter 4), the natural history of human SCI suggests a limited role for neuroprotectants. First, most patients with incomplete SCI and some patients with complete SCI at the time of presentation spontaneously regain some degree of neurological function over time.[42] This spontaneous recovery creates difficulty for treatment study design because it is difficult to attribute an improvement to treatment without a randomized control group. It is also rare for a patient's neurological injury to progress significantly after presentation, i.e., an injury is at its worst at the time of presentation. This suggests that secondary injury mechanisms do not play a major role in determining the clinical extent of injury. Furthermore, only a narrow window of opportunity exists for the administration of neuroprotectants. The best results with the different agents mentioned above were obtained when animals were pretreated. As in the case of stroke treatment, the degree of clinical improvement obtained by preventing secondary injury may be minor, suggesting that neuroprotection as a clinical field may represent a failure of application of animal models to the human setting.

This is not to suggest that secondary injury is not important. It simply may be that the mechanisms at work unfold so rapidly that a patient's deficit is relatively fixed at the time of presentation. It is therefore important to develop treatments that can be administered by first responders or alternatively to develop treatments that deal with the sequelae of secondary injury mechanisms. One such treatment is 4-aminopyridine, a potassium channel blocker. It has been shown that many of the axons that escape the initial injury become demyelinated, possibly due to inflammation, excitotoxicity, and apoptotic death of oligodendrocytes. Demyelination causes redistribution of sodium channels and unmasks potassium channels, both of which interfere with the conduction of action potentials.[43–45] In laboratory studies, 4-aminopyridine re-enabled signal conduction in demyelinated and partially myelinated axons.[46–47] Following preliminary clinical evidence that 4-aminopyridine can improve motor and sensory functions in SCI patients, Accorda Therapeutics initiated Phase 3 clinical trials.[48,49] Hopefully we will soon know whether this promising drug can be added to our meager armamentarium for treating SCI.

3.3 SPINAL CORD REGENERATION

Ever since the seminal observations of Ramon y Cajal early in the 20th century, it has been known that CNS neurons have very limited abilities to regenerate following injury and primarily generate local collaterals rather than long-distance axonal regrowth. This is the reason for such impetus for the development of neuroprotective agents. Allowing even a small number of neurons to escape the initial injury could produce profound functional benefit. Conversely, inducing even a small population of neurons to regenerate effectively could restore a significant amount of neurological function.

Both neuronal and non-neuronal factors limit CNS regeneration. The neurons confined to the CNS do not upregulate the expression of growth-associated genes unless they are injured close to their cell bodies.[50–54] CNS neurons that also extend axons into the peripheral nervous system (e.g., dorsal root ganglia) can undergo proregeneration cell body responses if their peripheral processes are also injured.[54–56] These findings suggest that CNS neurons possess the genetic machinery to regenerate, but they only express the necessary genes under very limited conditions.

One approach to repairing an injured spinal cord would be to find ways to turn on the regenerative machinery and effectively enhance axonal regrowth. A second approach would be to bridge the injury gap or replace cells with neural grafts, stimulating axon regrowth across the bridge or providing new cellular elements that could promote regeneration.

3.3.1 Enhancement of Axonal Regrowth

One of the first genes shown to be involved in axonal regeneration was GAP-43.[57–61] This gene, along with CAP-23, belongs to the MARCKS family of phosphoinositide-responsive protein kinases and is important in the organization and stabilization of growth cone components. Expression of the GAP-43 and CAP-23 genes in transgenic mice is sufficient to induce a regenerative response following isolated CNS injury.[62] Our laboratory is working on gene therapy methods to deliver these proregeneration genes to adult neurons. Other researchers have found that inosine, perhaps through the activation of these same kinases, can induce the regeneration of layer 5 pyramidal axons and promote reinnervation following SCI in rats.[63]

Other efforts aimed at inducing a proregeneration state in CNS neurons revolve around the use of neurotrophins, small molecules that promote neuronal outgrowth. The most promising of these appears to be NT-3, which not only promotes the regeneration of neurons following axotomy, but also minimizes atrophy and cell loss following SCI.[64–70] Clearly, the cell body response to injury in the CNS that usually does not promote regeneration can be manipulated to increase the chances for neurological recovery following SCI.

Many researchers have shown that constituents of CNS myelin inhibit the growth of neurons. Removing myelin from the CNS or using grafts lacking central myelin, for example, are two ways to promote regeneration.[51,53,71–73] Another way to promote regeneration following CNS injury is to antagonize these inhibitory signals. Three inhibitory molecules identified thus far are all components of CNS myelin: nogo, myelin-associated glycoprotein, and oligodendrocyte myelin glycoprotein.[72,74–77] Strikingly, all these molecules bind the neuronal receptor nogo-66.[77–79] This receptor has been shown to complex with p57 and activate the rho kinase pathway.[78,80] Inhibiting the rho pathway in itself allows neurons to grow in otherwise nonpermissive environments.[81,82]

Several experiments with antibodies to nogo (IN-1) and peptide inhibitors of the nogo receptor (NEP1-40) induced axonal regeneration and provided some degree of functional recovery following CNS lesions.[83–90] Interestingly, in one of these experiments, IN-1 treatment caused the unlesioned corticospinal tracts of rats to sprout and reinnervate targets on the contralateral side. Despite this clearly erroneous regeneration, the animals regained use of their affected limbs, suggesting a role for enhanced plasticity.[84] In an important study, researchers showed administration of NEP1-40 up to 1 week after spinal cord hemisection induced growth of corticospinal tracts, upregulation of growth proteins, functional reformation of synaptic connections, and locomotor recovery.[89] This has significant implications for the treatment of human SCI because the therapy can be administered at delayed (and much more convenient) times after the injury. Hopefully, the interests of pharmaceutical companies in inhibitors of nogo, will soon bring this mode of therapy to human trials.

Another class of inhibitory signals is the chondroitin sulfate proteoglycans (CSPG). These are expressed on astrocytes, oligodendrocyte precursors, and meningeal cells, which are all avidly recruited to the site of a CNS injury.[91–94] The CSPGs commonly found in glial scars are versican, neurocan, and phosphacan. Each contains a glycosaminoglycan (GAG) domain that is essential to their function. Fortunately, several bacterial enzymes specifically target and digest sulfated sugar GAG chains. Recently, the intrathecal infusion of one such enzyme, chondroitinase ABC, following SCI in rats was shown to degrade CS-GAGs at the injury site, upregulate GAP-43 in injured neurons, and promote regeneration of both ascending sensory projections and descending corticospinal tract axons. Postsynaptic activity below the lesion was restored and significant recovery of locomotor and proprioceptive behaviors ensued.[94] These findings have been validated in other injury models,[95,96] suggesting that the antagonism of CSPGs has an important role in the treatment of SCI.

3.3.2 Cellular or Inert Bridges and Neural Grafting

An additional pathway around the problem of CNS inhibition is to use bridging materials or cellular implants to guide regenerating axons around areas of significant tissue loss and glial scarring. The peripheral nerve has been known to have growth-promoting properties since the early 20th century.[97] The Schwann cell (SC), a key component of peripheral nerves, was the first graft material candidate. In an interesting set of experiments, researchers used a PVC polymer tube filled with SCs to reattach the two stumps of a completely transected spinal cord. After 1 month, they noticed significant growth of axons into the graft from both stumps. However, they also noticed that very few corticospinal axons had grown into the graft; that virtually no axons had grown out of the graft; and that noticeable tissue loss occurred at the graft–cord interfaces on both sides. Treating with MP prevented scaring and tissue loss at the interfaces and caused limited growth of axons back into the CNS environment. Very limited corticospinal fiber growth appeared in the graft.

Treatment with BDNF and NT-3 caused brainstem nuclei to extend axons into the graft and increased the total number of axons in the graft. However, again, very limited extension of graft axons was found in the cord.[98] This may have been due in part to the lack of synaptic targets in the vicinity of the graft. For example, optic nerve axons have been shown to grow through an SC graft and extend into the superior colliculus where they can form synapses.[99] The inhibitory mechanisms reviewed earlier may be the other causes of this seemingly unidirectional growth from host to graft. Axons are simply not easily persuaded to leave whatever growth-promoting environment that might be presented to them to enter the relatively inhospitable environment of the CNS. It will be important to study the combination of these grafts with nogo and CSPG inhibitors.

Other types of cellular grafts have shown exciting results in the treatment of SCI. One study with embryonal spinal cord implants showed that when neurotrophins were delivered with the implants, some host axons grew all the way through the implants. Furthermore, host axons formed synapses with the implanted cells and axons from the grafts extended for some distance into the host spinal cords. Thus it is possible that these grafts serve as relays for regenerating axons. This produced very impressive functional recovery from complete spinal cord transaction.[68] Washington University is conducting an ongoing clinical trial with the transplantation of porcine fetal spinal cells to assess the hypothesis that grafted neural cells can either enhance regeneration (function as a bridge) or provide key cellular elements.

Recent advances in the understanding of the olfactory system have led to what might be the most promising approach to overcoming CNS growth inhibition. Neurons in the olfactory mucosa are constantly dying and are replaced by new neurons that must extend their axons into the CNS. A special group of cells known as olfactory ensheathing glia (OEG) form sheaths around these axons, express growth-promoting phenotypes, and accompany these growing axons into the CNS.[100] In one study, OEG transplantation into the site of SCI was associated with the extension of corticospinal, raphe-spinal, and coeruleospinal axons through the injury and into the caudal spinal cord for at least 1 cm. This was associated with recovery of both locomotor function and sensorimotor reflexes. Other researchers have seen similar results. One study showed significant recovery of function even when transplantation was delayed for 7 days following injury.[101–103] OEGs can also be combined with SC grafts allowing further axonal growth into the host spinal cord.[104]

Not all research with OEGs has been successful. Takami et al. transplanted SC, SC and OEG, and OEG alone into rat SCI sites. They found a higher number of myelinated axons and better functional outcomes in the SC-only grafts. However, more axons extended beyond the grafts in the OEG-containing transplants.[105] These results may represent differences in the techniques for purifying and transplanting OEGs. The body of positive results with OEG transplants cannot be overlooked. Based on positive research findings, OEGs hold great promise for the future surgical treatment of human SCI.

One other cell implantation strategy for treating SCI bears mention. Schwartz et al. felt that the immune-privileged status of the CNS played a part in its poor regenerative properties.[106] Noting the prominent role of macrophages in peripheral nervous system regeneration, they implanted homologous macrophages activated by exposure to segments of peripheral nerves into the transected spinal cords of rats. They found that when sufficient macrophages were transplanted, partial recovery of both functional and electrophysiological activities occurred.[106] Based on these findings, Proneuron is in Phase I/II feasibility clinical trials with homologous activated macrophage transplantation in Belgium and at the Weiszmann Institute in Israel. The results of these trials are eagerly awaited.

Much more work must be done before we achieve the goal of regenerating the injured spinal cord. For example, the problem remains of ensuring that correct synaptic patterns are reestablished after regeneration takes place. None of the existing studies have shown that axons are sufficiently elongated to reach targets. Achieving synaptic specificity upon reaching distal spinal cord targets may in and of itself be a very difficult challenge.

The work reviewed above highlights some of the important leads that are currently being pursued. From the preliminary evidence, it seems that the first practical applications will be with agents that remove CNS growth inhibition. After that, molecular approaches to replacing damaged cells and reestablishing severed connections will hopefully be perfected and will probably lead to new challenges in reestablishing appropriate functional connections.

3.4 REHABILITATION

It is important to consider possible noninvasive approaches to help functional recovery from SCI. Chief among these are aggressive neurorehabilitation and assisted ambulation. At least five clinical trials are currently assessing the utility of assisted ambulation with body weight-supported treadmill training in promoting locomotion after SCI.

The key concept in these trials is that the spinal cord contains local pattern generators that can function independently of descending input. This was demonstrated in cats whose spinal cords were transected at the thoracic level. Edgerton showed that appropriate limb loading and manual stepping on a treadmill for 4 weeks enabled the cats to regain the ability to support their own body weights and walk on the treadmill over a range of speeds.[107,108] It was later shown that the lumbosacral grey matter responded to locomotion-associated patterns of stimulation and started generating rhythmic patterns of activity that could initiate stepping and perhaps support ambulation.[109]

The human lumbosacral spinal cord also has the ability to respond to the sensory stimulation of locomotion and generate locomotion-like electromyographic (EMG) patterns after training.[110–112] The basis of this training is Edgerton's proposal that providing the specific sensory activity associated with a task and repetitively practicing the task can lead to motor learning and plasticity in the human spinal cord.[109] Based on this research, the University of California at Los Angeles, the University of Florida, The Miami Project to Cure Paralysis, the Ohio State University, and others are enrolling patients in assisted ambulation studies.

This approach, if successful, could be combined with invasive interventions to treat SCI. For example, transplanted cells and nogo inhibitors both may increase neuronal plasticity. These treatments could be combined with aggressive physical therapy and may stimulate the reorganization of intrinsic spinal circuits and allow coordination among multiple segments to dramatically improve locomotor function. Unfortunately, budgetary constraints may limit application of aggressive physical therapy techniques, although all patients with SCI receive intensive rehabilitation currently, and as new techniques arise, this training could be redirected to different patterns.

3.5 NEUROPROSTHETICS

Another therapeutic avenue that will play a prominent role in the treatment of SCI patients is functional electrical stimulation (FES) and the field of neuroprosthetics in general (see Chapter 7). By stimulating muscles, lower motor neurons, and peripheral nerves, FES aims to return some functional modalities to patients who have complete SCI. Surgeons are implanting phrenic nerve stimulators to preserve respiration in high cervical injuries, sacral nerve stimulators to aid bowel and bladder function, and ulnar and median nerve stimulators to allow grasping movements of the hands.

Clinical studies sponsored by the Veterans' Administration and the FDA are also evaluating systems to restore arm function, enable patients to stand, and assist them in walking. These devices have the potential to significantly improve the lives of patients with SCIs and the results achieved with such devices will improve as progress is made in the field of electronics and new ways are developed to interface nervous systems and computers. The theoretical approaches to and current research and progress with CNS–machine interfaces are reviewed Chapter 7, but in general, these approaches use external actuators instead of a patient's own musculature to provide enhanced motor function.

3.6 COMBINATION THERAPIES

Future treatment of SCI will probably involve a synthesis of the approaches described earlier. Specific interventions can activate regeneration-associated genes and antagonize inhibitory signals within the CNS milieu, allowing surviving neurons to start to reestablish severed connections (Figure 3.1). This can be augmented by the transplantation of embryonic cells, olfactory ensheathing glia, and neurotrophin secreting cells to support regenerating cells, act as relays, and replace lost cell populations. The residual functional deficit after optimal treatment could then be ameliorated further by advances in FES, aggressive rehabilitation, and improved neuroprosthesis. Thus, any improvement in axonal regrowth will likely require significant patient training and rehabilitation to achieve clinical improvement.

3.7 PERIPHERAL NERVE REPAIR

Attempts to surgically treat peripheral nerve injury have been more fruitful than attempts to repair the injured spinal cord. Additionally, the two conditions are closely related as insights into the behavior of regenerating neurons obtained from the former are being applied to the latter, and vice versa. The first reported surgical repair of injured peripheral nerve was in 1608. The wars of the last two centuries, starting with the studies of Mitchell during the American Civil War, provided much material for the study of peripheral nerve injuries and repair techniques.[2] Suture repairs of severed nerves, directly or by autografting, became standard practice by World War II. Unfortunately, the results were disappointing. The key problems were inadequate realignment of fascicles, the formation of neuromas, and the difficulty of filling large gaps with autologous peripheral nerve cable grafts.

The development of surgical microscopes helped improve these results. With good microsurgical technique, it became apparent that direct repair with microsurgical alignment of fascicles provided the best results. However, if damage to the nerve was severe enough to leave a gap greater than 2 cm, an autologous nerve graft had to be used for a tension-free repair. Unfortunately, the use of normal donor nerves from another location can be limited by tissue availability, the risk of causing secondary deformities, the failure of graft survival, and the differences in graft diameter that could complicate the repair.[113] Current research on peripheral nerve regeneration focuses on developing engineered graft materials and improving specificity of reinnervation and thus functional recovery following peripheral nerve repair.

3.7.1 NERVE GUIDE TUBES

The development of nerve guide tubes stands as a critical example of translational research in neurosurgery. The use of tubular conduits in peripheral nerve repair was proposed as early as 1964.[114] By the late 1980s, researchers had tested polytetrafluoroethylene (PTFE), silicone, polyvinylidenefuoride (PVDF), arteries, preformed mesothelial tubes, collagen, polylactate, polyesters, and polylactate/polyglycolate copolymers.[115] From these studies emerged the following criteria for useful nerve conduits; collagen was one material that met all the criteria:[116]

1. The nature of the material is important in determining whether axons can grow on it.
2. The rate of resorption of the material must be on the appropriate time scale for axon regeneration to take place.
3. The mechanical properties must be stable *in vitro*.
4. The material must have appropriate permeability properties.
5. The material must not induce a deleterious inflammatory reaction.
6. The material properties must allow for easy manufacturing of different sized conduits.

Initial studies on rodents were carried out to identify the specific permeability properties that the collagen tubules would need in order to promote nerve regeneration. Collagen derived from bovine Achilles tendon was purified, gelled, homogenized, and deposited by compression onto a rotating mandrel to form tubules. Varying the amount of compression allowed control of the amount of permeability. Researchers implanted different tubules into rodents and found that making the tubules permeable to molecules the size of bovine serum albumin allowed four times greater axonal regeneration than the less permeable tubules.[117] These results were attributed to the fact that the tubule could concentrate molecules such as growth factors and adhesion molecules within its lumen, creating a "reaction chamber" that promoted axon growth.

After these initial promising results, based on funding from the National Institutes of Health and the Department of Veterans Affairs, the researchers planned to move ahead with trials in nonhuman primates. A New Jersey biomaterials company became interested in the product, assumed responsibility for manufacturing it, and also contributed funding for the trials. Fifteen median nerves and one ulnar nerve were transected above the wrists of eight *Macaca fasicularis* monkeys; a 5-mm section was removed from each nerve. One nerve in each monkey was repaired with the collagen tubule, and another with an autologous nerve graft. Four other nerves were repaired by direct suturing in standard clinical fashion. The nerves were studied for motor and sensory conduction, response to tactile stimulation, and morphology over a period of 42 months. Researchers found similar amplitudes and latencies of tactile-evoked potentials, similar recovery rates of compound muscle action potentials, and an increase in the number of myelinated axons in the distal stumps following both nerve graft and synthetic nerve conduit repairs. Thus, a synthetic material produced results similar to autologous grafting.[118] Based on these findings, the company obtained approval for use in humans and has been marketing the conduit under the brand name of NeuraGen® (Integra LifeSciences Holding Company, Plainsboro, NJ).

This example illustrates true translational neuroscience research, beginning from a technical concept in a small laboratory to large animal research with the support of a biotechnology company, to human trials, and clinical application. However, as is the case with many FDA-approved products, additional postapproval clinical trials (now ongoing) will be critical to determine whether the product remains a useful clinical entity over time.

Current research in nerve conduits centers on many of the same interventions attempted for spinal cord regeneration. As noted earlier, SCs are critical components in peripheral nerve grafts for axonal regeneration. They express specific cell adhesion molecules and bind specific extracellular matrix molecules that allow axon extension; they produce and secrete neurotrophic factors for neuronal support and axonal growth; and they possess receptors for neurotrophic factors and may act as neurotrophin-presenting cells for axon pathfinding. Some researchers are thus attempting to incorporate SCs into nerve conduits to improve the current results.[119–123]

Other researchers are experimenting with the incorporation of extracellular matrix components into nerve tubules to promote axonal outgrowth.[124,125] Some research aimed at improving the growth of axons into nerve guide tubes and distal stumps focus on the delivery of neurotrophins within grafts[126] or by genetic manipulation of SCs to express neurotrophins distal to grafts.[127,128] Other translational research in peripheral nerve repair focuses on the technical aspects of nerve repair. Researchers are studying different types of fibrin glues, fasteners, and laser repairs for treating peripheral nerve lesions in animals.[129–133]

3.7.2 Enhancement of Specificity of Regeneration

Merely increasing the number of axons that grow into the distal stump is not sufficient. Care must be taken to promote appropriate axonal pathfinding. If axons fail to reach the correct sensory or motor end organ, patients will not achieve clinical improvement, and even worse, may be left with painful consequences. The rat femoral nerve that divides into a motor branch to the quadriceps and a sensory branch to the skin serves as a useful model for studying axon pathfinding. Researchers have found that motor axons are better at finding appropriate motor fascicles in the distal stump than are sensory axons — a process called preferential motor reinnervation.[134,135] Pruning may be the reason for this.

Following injury, regenerating axons form many (redundant) collateral sprouts, and these enter SC tubules in the distal stump in a random fashion. However, with motor axons the branches that enter distal sensory fascicles are pruned back. Sensory axon neurons, on the other hand, do not necessarily trim back branches that have inappropriately entered motor fascicles in a distal stump. The result is that over time more motor axons find their targets. This suggests that local signals within SC tubules influence axonal pathfinding and under specific conditions can significantly increase specificity of regeneration.[136]

In support of this, it has been shown that SCs in contact with motor axons express different membrane glycolipids than do SCs in contact with sensory axons.[137] Also, blocking certain myelin proteins in the distal stumps can increase preferential motor reinnervation.[138] If the local determinants of axonal pathfinding were identified, it would then be possible to manipulate the expression of these signals to improve the specificity of regeneration across synthetic grafts.

Other promising interventions include noninvasive measures to enhance peripheral nerve regeneration. Electrical stimulation is felt to be beneficial in nerve repair.[139]

Recently it was reported that stimulation of the rat femoral nerve proximal to its repair site increased the degree and specificity of motor axon regeneration.[140] These effects were shown to occur by influencing the cell body to increase expression of BDNF and its trkB receptor.[141] Since electrical stimulation is already used clinically in the treatment of orthopedic injuries (for bone regrowth), it would be easy to extend its application to the treatment of peripheral nerve injuries.

The success rate with current peripheral nerve repair techniques is still disappointing. A recent report of the largest clinical series using the latest microsurgical techniques to treat peripheral nerve injuries reported at best a 70% return of function in direct repair of the ulnar nerve.[142,143] We have been able to produce synthetic graft material that can support regeneration. Future refinements of these materials will likely incorporate cells and signaling molecules to improve the pace and accuracy of axon regeneration (Figure 3.1). We still face significant challenges in the treatment of peripheral nerve injuries. One issue still to be addressed is the prevention of end organ atrophy prior to reinnervation. Aggressive physical therapy may also be useful in this context. If we can take control of the processes of axon regeneration and pathfinding, we can get closer to the goal of full functional recovery.

3.8 CONCLUSIONS

SCI and peripheral nerve injury share the problem of long-distance axon regrowth. These problems are in many ways distinct from upper CNS regeneration schemes, in which actual neuronal cell loss may be the critical event, leading to neural grafting schemes for cortical lesions in stroke or epilepsy (see Chapter 2) and Parkinson's disease (see Chapter 8). A considerable number of research schemes are under consideration for translational approaches based on promising preclinical data. However, the major problem remaining, even after axonal regrowth is achieved clinically, will be the issues of specificity when axons reach their targets and appropriate synaptic connectivity. Perhaps rehabilitation or neuroprosthetic approaches may partially bridge this subsequent, very difficult problem.

REFERENCES

1. DeVivo, M.J., B.K. Go, and A.B. Jackson, Overview of the National Spinal Cord Injury Statistical Center database, *Journal of Spinal Cord Medicine,* 25, 335–338, 2002.
2. Porter, R., (Ed.), *The Cambridge Illustrated History of Medicine*, 1996, Cambridge University Press, New York, p. 400.
3. Dumont, A.S., R.J. Dumont, and R.J. Oskouian, Will improved understanding of the pathophysiological mechanisms involved in acute spinal cord injury improve the potential for therapeutic intervention*? Current Opinions in Neurology,* 15, 713–720, 2002.
4. Dumont, R.J. et al., Acute spinal cord injury. I. Pathophysiologic mechanisms, *Clinical Neuropharmacology*, 24, 254–264, 2001.

5. Bracken, M.B. et al., Methylprednisolone and neurological function one year after spinal cord injury: results of the National Acute Spinal Cord Injury Study, *Journal of Neurosurgery*, 63, 704–713, 1985.
6. Hall, E.D. and D.L. Wolf, Post-traumatic spinal cord ischemia: relationship to injury severity and physiological parameters, *Central Nervous System Trauma*, 4, 15–25, 1987.
7. Hall, E.D. and J.M. Braughler, Role of lipid peroxidation in post–traumatic spinal cord degeneration: a review, *Central Nervous System Trauma*, 3, 281–294, 1986.
8. Hall, E.D., J.M. Braughler, and J.M. McCall, New pharmacological treatment of acute spinal cord trauma, *Journal of Neurotrauma*, 5, 81–89, 1988.
9. Hall, E.D. and J.M. Braughler, Glucocorticoid mechanisms in acute spinal cord injury: a review and therapeutic rationale, *Surgical Neurology*, 18, 320–327, 1982.
10. Hall, E.D., The neuroprotective pharmacology of methylprednisolone, *Journal of Neurosurgery*, 76, 13–22, 1992.
11. Bracken, M.B. et al., A randomized, controlled trial of methylprednisolone or naloxone in the treatment of acute spinal-cord injury: results of the Second National Acute Spinal Cord Injury Study, *New England Journal of Medicine*, 322, 1405–1411, 1990.
12. Bracken, M.B. et al., Administration of methylprednisolone for 24 or 48 hours or tirilazad mesylate for 48 hours in the treatment of acute spinal cord injury: results of the Third National Acute Spinal Cord Injury Randomized Controlled Trial, National Acute Spinal Cord Injury Study, *JAMA*, 277, 1597–1604, 1997.
13. Young, W., NASCIS: National Acute Spinal Cord Injury Study, *Journal of Neurotrauma*, 7, 113–114, 1990.
14. Young, W., Secondary injury mechanisms in acute spinal cord injury, *Journal of Emergency Medicine*, 11, 13–22, 1993.
15. Coleman, W.P. et al., A critical appraisal of the reporting of the National Acute Spinal Cord Injury Studies (II and III) of methylprednisolone in acute spinal cord injury, *Journal of Spinal Disorders*, 13, 185–199, 2000.
16. Bracken, M.B. and T.R. Holford, Neurological and functional status one year after acute spinal cord injury: estimates of functional recovery in National Acute Spinal Cord Injury Study II from results modeled in National Acute Spinal Cord Injury Study III, *Journal of Neurosurgery*, 96, 259–266, 2002.
17. Anderson, D.K. et al., Effect of delayed administration of U74006F (tirilazad mesylate) on recovery of locomotor function after experimental spinal cord injury, *Journal of Neurotrauma*, 8, 187–192, 1991.
18. Clark, W.M., J.S. Hazel, and B.M. Coull, Lazaroids: CNS pharmacology and current research, *Drugs*, 50, 971–983, 1995.
19. Francel, P.C. et al., Limiting ischemic spinal cord injury using a free radical scavenger 21-aminosteroid and/or cerebrospinal fluid drainage, *Journal of Neurosurgery*, 79, 742–751, 1993.
20. Bethea, J.R. and W.D. Dietrich, Targeting the host inflammatory response in traumatic spinal cord injury, *Current Opinions in Neurology*, 15, 355–360, 2002.
21. Bethea, J.R. et al., Systemically administered interleukin-10 reduces tumor necrosis factor-alpha production and significantly improves functional recovery following traumatic spinal cord injury in rats, *Journal of Neurotrauma*, 16, 851–863, 1999.
22. Takami, T. et al., Methylprednisolone and interleukin-10 reduce gray matter damage in the contused Fischer rat thoracic spinal cord but do not improve functional outcome, *Journal of Neurotrauma*, 19, 653–666, 2002.

23. Ghirnikar, R.S., Y.L. Lee, and L.F. Eng, Chemokine antagonist infusion promotes axonal sparing after spinal cord contusion injury in rat, *Journal of Neuroscience Research*, 64, 582–589, 2001.
24. Hains, B.C., J.A. Yucra, and C.E. Hulsebosch, Reduction of pathological and behavioral deficits following spinal cord contusion injury with the selective cyclooxygenase-2 inhibitor NS-398, *Journal of Neurotrauma*, 18, 409–423, 2001.
25. Resnick, D.K. et al., Role of cyclooxygenase 2 in acute spinal cord injury, *Journal of Neurotrauma*, 15, 1005–1013, 1998.
26. Wada, S. et al., Apoptosis following spinal cord injury in rats and preventative effect of N-methyl-D-aspartate receptor antagonist, *Journal of Neurosurgery*, 91, 98–104, 1999.
27. Panter, S.S., S.W. Yum, and A.I. Faden, Alteration in extracellular amino acids after traumatic spinal cord injury, *Annals of Neurology*, 27, 96–99, 1990.
28. Olby, N.J. et al., Chronic and acute compressive spinal cord lesions in dogs due to intervertebral disc herniation are associated with elevation in lumbar cerebrospinal fluid glutamate concentration, *Journal of Neurotrauma*, 16, 1215–1224, 1999.
29. Liu, D. et al., Neurotoxicity of glutamate at the concentration released upon spinal cord injury, *Neuroscience*, 93, 1383–1389, 1999.
30. Liu, D., An experimental model combining microdialysis with electrophysiology, histology, and neurochemistry for studying excitotoxicity in spinal cord injury: effect of NMDA and kainate, *Molecular and Chemical Neuropathology*, 23, 77–92, 1994.
31. Li, S. and P.K. Stys, Mechanisms of ionotropic glutamate receptor-mediated excitotoxicity in isolated spinal cord white matter, *Journal of Neuroscience,* 20, 1190–1198, 2000.
32. Li, S. et al., Novel injury mechanism in anoxia and trauma of spinal cord white matter: glutamate release via reverse Na^+-dependent glutamate transport, *Journal of Neuroscience,* 19, 10–16, 1999.
33. Faden, A.I. et al., N-methyl-D-aspartate antagonist MK801 improves outcome following traumatic spinal cord injury in rats: behavioral, anatomic, and neurochemical studies, *Journal of Neurotrauma*, 5, 33–45, 1988.
34. Yakovlev, A.G. and A.I. Faden, Caspase-dependent apoptotic pathways in CNS injury, *Molecular Neurobiology,* 24, 131–144, 2001.
35. Takagi, T. et al., Caspase activation in neuronal and glial apoptosis following spinal cord injury in mice, *Neurologia Medico-Chirurgica*, 43, 20–29, 2003.
36. Shibata, M. et al., Single injections of a DNA plasmid that contains the human Bcl-2 gene prevent loss and atrophy of distinct neuronal populations after spinal cord injury in adult rats, *Neurorehabilitation and Neural Repair*, 14, 319–330, 2000.
37. Liu, X.Z. et al., Neuronal and glial apoptosis after traumatic spinal cord injury, *Journal of Neuroscience*, 17, 5395–5406, 1997.
38. Hostettler, M.E., P.E. Knapp, and S.L. Carlson, Platelet-activating factor induces cell death in cultured astrocytes and oligodendrocytes: involvement of caspase-3, *Glia,* 38, 228–239, 2002.
39. Keane, R.W. et al., Apoptotic and anti-apoptotic mechanisms following spinal cord injury, *Journal of Neuropathology and Experimental Neurology*, 60, 422–429, 2001.
40. Eldadah, B.A. and A.I. Faden, Caspase pathways, neuronal apoptosis, and CNS injury, *Journal of Neurotrauma*, 17, 811–829, 2000.
41. Inamasu, J., Y. Nakamura, and K. Ichikizaki, Induced hypothermia in experimental traumatic spinal cord injury: an update, *Journal of the Neurological Sciences*, 209, 55–60, 2003.

42. Stauffer, E.S., Neurologic recovery following injuries to the cervical spinal cord and nerve roots, *Spine*, 9, 532–534, 1984.
43. Waxman, S.G., Demyelination in spinal cord injury and multiple sclerosis: what can we do to enhance functional recovery? *Journal of Neurotrauma,* 9, S105–S117, 1992.
44. Nashmi, R., O.T. Jones, and M.G. Fehlings, Abnormal axonal physiology is associated with altered expression and distribution of Kv1.1 and Kv1.2 K^+ channels after chronic spinal cord injury, *European Journal of Neuroscience*, 12, 491–506, 2000.
45. Nashmi, R. and M.G. Fehlings, Mechanisms of axonal dysfunction after spinal cord injury with an emphasis on the role of voltage-gated potassium channels, *Brain Research Reviews,* 38, 165–191, 2001.
46. Gruner, J.A. and A.K. Yee, 4-Aminopyridine enhances motor-evoked potentials following graded spinal cord compression injury in rats, *Brain Research,* 816, 446–456, 1999.
47. Hayes, K.C. et al., 4-Aminopyridine-sensitive neurologic deficits in patients with spinal cord injury, *Journal of Neurotrauma,* 11, 433–446, 1994.
48. Potter, P.J. et al., Sustained improvements in neurological function in spinal cord injured patients treated with oral 4-aminopyridine: three cases, *Spinal Cord,* 36, 147–155, 1998.
49. Darlington, C., Fampridine: Acorda Therapeutics, *Current Opinions in Investigational Drugs*, 1, 375–379, 2000.
50. Tetzlaff, W. et al., Response of rubrospinal and corticospinal neurons to injury and neurotrophins, *Progress in Brain Research*, 103, 271–286, 1994.
51. Benfey, M. and A.J. Aguayo, Extensive elongation of axons from rat brain into peripheral nerve grafts, *Nature,* 296, 150–152, 1982.
52. Doster, S.K. et al., Expression of the growth-associated protein GAP-43 in adult rat retinal ganglion cells following axon injury, *Neuron,* 6, 635–647, 1991.
53. So, K.F. and A.J. Aguayo, Lengthy regrowth of cut axons from ganglion cells after peripheral nerve transplantation into the retina of adult rats, *Brain Research*, 328, 349–534, 1985.
54. Richardson, P.M. and V.M. Issa, Peripheral injury enhances central regeneration of primary sensory neurones, *Nature*, 309, 791–793, 1984.
55. Plunet, W., B.K. Kwon, and W. Tetzlaff, Promoting axonal regeneration in the central nervous system by enhancing the cell body response to axotomy, *Journal of Neuroscience Research*, 68, 1–6, 2002.
56. Chong, M.S. et al., Intrinsic versus extrinsic factors in determining the regeneration of the central processes of rat dorsal root ganglion neurons: the influence of a peripheral nerve graft, *Journal of Comparative Neurology*, 370, 97–104, 1996.
57. Kalil, K. and J.H. Skene, Elevated synthesis of an axonally transported protein correlates with axon outgrowth in normal and injured pyramidal tract, *Journal of Neuroscience,* 6, 2563–2570, 1986.
58. Skene, J.H., Axonal growth-associated proteins, *Annual Review of Neuroscience,* 12, 127–156, 1989.
59. Neve, R.L. et al., The neuronal growth-associated protein GAP-43 (B-50, F1): neuronal specificity, developmental regulation and regional distribution of the human and rat mRNAs, *Brain Research*, 388, 177–183, 1987.
60. Benowitz, L.I. and E.R. Lewis, Increased transport of 44,000- to 49,000-dalton acidic proteins during regeneration of the goldfish optic nerve: a two-dimensional gel analysis, *Journal of Neuroscience*, 3, 2153–2163, 1983.

61. Skene, J.H. and M. Willard, Axonally transported proteins associated with axon growth in rabbit central and peripheral nervous systems, *Journal of Cell Biology*, 89, 96–103, 1981.
62. Bomze, H.M. et al., Spinal axon regeneration evoked by replacing two growth cone proteins in adult neurons, *Nature Neuroscience*, 4, 38–43, 2001.
63. Benowitz, L.I., D.E. Goldberg, and N. Irwin, Inosine stimulates axon growth *in vitro* and in the adult CNS, *Progress in Brain Research*, 137, 389–399, 2002.
64. Bradbury, E.J. et al., NT-3, but not BDNF, prevents atrophy and death of axotomized spinal cord projection neurons, *European Journal of Neuroscience*, 10, 3058–3068, 1998.
65. Bamber, N.I. et al., Neurotrophins BDNF and NT–3 promote axonal re-entry into the distal host spinal cord through Schwann cell-seeded mini-channels, *European Journal of Neuroscience*, 13, 257–268, 2001.
66. Tuszynski, M.H. et al., NT-3 gene delivery elicits growth of chronically injured corticospinal axons and modestly improves functional deficits after chronic scar resection, *Experimental Neurology*, 181, 47–56, 2003.
67. Bradbury, E.J. et al., NT-3 promotes growth of lesioned adult rat sensory axons ascending in the dorsal columns of the spinal cord, *European Journal of Neuroscience*, 11, 3873–3883, 1999.
68. Coumans, J.V. et al., Axonal regeneration and functional recovery after complete spinal cord transection in rats by delayed treatment with transplants and neurotrophins, *Journal of Neuroscience*, 21, 9334–9344, 2001.
69. Schnell, L. et al., Neurotrophin-3 enhances sprouting of corticospinal tract during development and after adult spinal cord lesion, *Nature*, 367, 170–173, 1994.
70. Grill, R. et al., Cellular delivery of neurotrophin-3 promotes corticospinal axonal growth and partial functional recovery after spinal cord injury, *Journal of Neuroscience*, 17, 5560–5572, 1997.
71. David, S. and A.J. Aguayo, Axonal elongation into peripheral nervous system "bridges" after central nervous system injury in adult rats, *Science*, 214, 931–933, 1981.
72. Savio, T. and M.E. Schwab, Lesioned corticospinal tract axons regenerate in myelin-free rat spinal cord, *PNAS,* 87, 4130–4133, 1990.
73. Vanek, P. et al., Increased lesion-induced sprouting of corticospinal fibres in the myelin-free rat spinal cord, *European Journal of Neuroscience,* 10, 45–56, 1998.
74. Caroni, P. and M.E. Schwab, Two membrane protein fractions from rat central myelin with inhibitory properties for neurite growth and fibroblast spreading, *Journal of Cell Biology,* 106, 1281–1288, 1988.
75. Caroni, P., T. Savio, and M.E. Schwab, Central nervous system regeneration: oligodendrocytes and myelin as non-permissive substrates for neurite growth, *Progress in Brain Research,* 78, 363–370, 1988.
76. GrandPre, T. et al., Identification of the Nogo inhibitor of axon regeneration as a Reticulon protein, *Nature*, 403, 439–444, 2000.
77. Wang, K.C. et al., Oligodendrocyte-myelin glycoprotein is a Nogo receptor ligand that inhibits neurite outgrowth, *Nature*, 417, 941–944.
78. Fournier, A.E., T. GrandPre, and S.M. Strittmatter, Identification of a receptor mediating Nogo-66 inhibition of axonal regeneration, *Nature*, 409, 341–346.
79. Liu, B.P. et al., Myelin-associated glycoprotein as a functional ligand for the Nogo-66 receptor, *Science*, 297, 1190–1193, 2002.
80. Yamashita, T. and M. Tohyama, The p75 receptor acts as a displacement factor that releases Rho from Rho-GDI, *Nature Neuroscience,* 6, 461–467, 2003.

81. Borisoff, J.F. et al., Suppression of Rho-kinase activity promotes axonal growth on inhibitory CNS substrates, *Molecular and Cellular Neurosciences,* 22, 405-416, 2003.
82. Lehmann, M. et al., Inactivation of Rho signaling pathway promotes CNS axon regeneration, *Journal of Neuroscience*, 19, 7537–7547, 1999.
83. Merkler, D. et al., Locomotor recovery in spinal cord-injured rats treated with an antibody neutralizing the myelin-associated neurite growth inhibitor Nogo-A, *Journal of Neuroscience,* 21, 3665–3673, 2001.
84. Raineteau, O. et al., Functional switch between motor tracts in the presence of the mAb IN-1 in the adult rat, *PNAS*, 98, 6929–6934, 2001.
85. Schnell, L. and M.E. Schwab, Axonal regeneration in the rat spinal cord produced by an antibody against myelin-associated neurite growth inhibitors, *Nature*, 343, 269–272, 1990.
86. Blochlinger, S. et al., Neuronal plasticity and formation of new synaptic contacts follow pyramidal lesions and neutralization of Nogo-A: a light and electron microscopic study in the pontine nuclei of adult rats, *Journal of Comparative Neurology,* 433, 426–436, 2001.
87. Z'Graggen, W.J. et al., Functional recovery and enhanced corticofugal plasticity after unilateral pyramidal tract lesion and blockade of myelin-associated neurite growth inhibitors in adult rats, *Journal of Neuroscience,* 18, 4744–4757.
88. Thallmair, M. et al., Neurite growth inhibitors restrict plasticity and functional recovery following corticospinal tract lesions, *Nature Neuroscience,* 1, 124–131, 1998.
89. Li, S. and S.M. Strittmatter, Delayed systemic Nogo-66 receptor antagonist promotes recovery from spinal cord injury, *Journal of Neuroscience,* 23, 4219–4227, 2003.
90. GrandPre, T., S. Li, and S.M. Strittmatter, Nogo-66 receptor antagonist peptide promotes axonal regeneration, *Nature,* 417, 547–551, 2002.
91. Asher, R.A. et al., Neurocan is upregulated in injured brain and in cytokine-treated astrocytes, *Journal of Neuroscience*, 20, 2427–2438, 2000.
92. Asher, R.A. et al., Chondroitin sulphate proteoglycans: inhibitory components of the glial scar, *Progress in Brain Research*, 132, 611–619, 2001.
93. Morgenstern, D.A., R.A. Asher, and J.W. Fawcett, Chondroitin sulphate proteoglycans in the CNS injury response, *Progress in Brain Research,* 137, 313–332, 2002.
94. Bradbury, E.J. et al., Chondroitinase ABC promotes functional recovery after spinal cord injury, *Nature*, 416, 636–640, 2002.
95. Moon, L.D. et al., Regeneration of CNS axons back to their target following treatment of adult rat brain with chondroitinase ABC, *Nature Neuroscience,* 4, 465–466, 2001.
96. Moon, L.D., R.A. Asher, and J.W. Fawcett, Limited growth of severed CNS axons after treatment of adult rat brain with hyaluronidase, *Journal of Neuroscience Research*, 71, 23–37, 2003.
97. Ramon y Cajal, S., *Cajal's Degeneration and Regeneration of the Nervous System,* DeFelipe, J. and Jones, E.G., Eds., Oxford University Press, New York, 1991.
98. Jones, L.L. et al., Neurotrophic factors, cellular bridges and gene therapy for spinal cord injury, *Journal of Physiology*, 533, 83–89, 2001.
99. Carter, D.A., G.M. Bray, and A.J. Aguayo, Regenerated retinal ganglion cell axons can form well-differentiated synapses in the superior colliculus of adult hamsters, *Journal of Neuroscience,* 9, 4042–4050, 1989.
100. Ramon-Cueto, A. and J. Avila, Olfactory ensheathing glia: properties and function, *Brain Research Bulletin*, 46, 175–187, 1998.
101. Plant, G.W. et al., Delayed transplantation of olfactory ensheathing glia promotes sparing/regeneration of supraspinal axons in the contused adult rat spinal cord, *Journal of Neurotrauma*, 20, 1–16, 2003.

102. Li, Y., P.M. Field, and G. Raisman, Regeneration of adult rat corticospinal axons induced by transplanted olfactory ensheathing cells, *Journal of Neuroscience,* 18, 10514–10524, 1998.
103. Li, Y., P.M. Field, and G. Raisman, Repair of adult rat corticospinal tract by transplants of olfactory ensheathing cells, *Science*, 277, 2000–2002, 1997.
104. Ramon-Cueto, A. et al., Long-distance axonal regeneration in the transected adult rat spinal cord is promoted by olfactory ensheathing glia transplants, *Journal of Neuroscience*, 18, 3803–3815, 1998.
105. Takami, T. et al., Schwann cells but not olfactory ensheathing glia transplants improve hind limb locomotor performance in the moderately contused adult rat thoracic spinal cord, *Journal of Neuroscience,* 22, 6670–6681, 2002.
106. Schwartz, M. et al., Potential repair of rat spinal cord injuries using stimulated homologous macrophages, *Neurosurgery*, 44, 1041–1045, 1999.
107. Lovely, R.G. et al., Weight-bearing hind limb stepping in treadmill-exercised adult spinal cats, *Brain Research*, 514, 206–218, 1990.
108. Lovely, R.G. et al., Effects of training on the recovery of full-weight-bearing stepping in the adult spinal cat, *Experimental Neurology*, 92, 421–435, 1986.
109. Edgerton, V.R. et al., Potential of adult mammalian lumbosacral spinal cord to execute and acquire improved locomotion in the absence of supraspinal input, *Journal of Neurotrauma*, 9 (Suppl. 1), S119–S128, 1992.
110. Dietz, V., Central pattern generator [comment], *Paraplegia*, 33, 739, 1995.
111. Dietz, V. et al., Locomotor capacity of spinal cord in paraplegic patients [comment], *Annals of Neurology*, 37, 574–582, 1995.
112. Dietz, V., G. Colombo, and L. Jensen, Locomotor activity in spinal man, *Lancet*, 344, 1260–1263, 1994.
113. Millesi, H., G. Meissl, and A. Berger, Further experience with interfascicular grafting of the median, ulnar, and radial nerves, *Journal of Bone and Joint Surgery*, 58, 209–218, 1976.
114. Kline, D.G. and G.J. Hayes, The use of a resorbable wrapper for peripheral nerve repair, *Journal of Neurosurgery*, 21, 737–750, 1964.
115. Li, S.T. et al., Peripheral nerve repair with collagen conduits, *Clinical Materials*, 9, 195–200, 1992.
116. Li, S.T. et al., Semi-permeable collagen nerve conduits for peripheral nerve regeneration, *Polymer and Materials Science Engineering*, 62, 575–582, 1990.
117. Archibald, S.J. et al., A collagen-based nerve guide conduit for peripheral nerve repair: an electrophysiological study of nerve regeneration in rodents and nonhuman primates, *Journal of Comparative Neurology*, 306, 685–696, 1991.
118. Archibald, S.J. et al., Monkey median nerve repaired by nerve graft or collagen nerve guide tube, *Journal of Neuroscience*, 15, 4109–4123, 1995.
119. Komiyama, T. et al., A novel technique to isolate adult Schwann cells for an artificial nerve conduit, *Journal of Neuroscience Methods*, 122, 195–200, 2003.
120. Wang, J. et al., Study *in vitro* of populating autogenous Schwann cells into chemical extracted allogenous nerve, *Chinese Journal of Traumatology,* 5, 326–328, 2002.
121. Timmer, M. et al., Axonal regeneration across long gaps in silicone chambers filled with Schwann cells overexpressing high molecular weight FGF–2, *Cell Transplant*, 12, 265–277, 2003.
122. Fukaya, K. et al., Oxidized galectin-1 stimulates the migration of Schwann cells from both proximal and distal stumps of transected nerves and promotes axonal regeneration after peripheral nerve injury, *Journal of Neuropathology and Experimental Neurology,* 62, 162–172, 2003.

123. Geuna, S. et al., Schwann-cell proliferation in muscle-vein combined conduits for bridging rat sciatic nerve defects, *Journal of Reconstructive Microsurgery*, 19, 119–123, 2003.
124. Ahmed, Z., S. Underwood, and R.A. Brown, Nerve guide material made from fibronectin: assessment of *in vitro* properties, *Tissue Engineering*, 9, 219–231, 2003.
125. Dvali, L. and S. Mackinnon, Nerve repair, grafting, and nerve transfers. *Clinical Plastic Surgery,* 30, 203–221, 2003.
126. Xu, X. et al., Peripheral nerve regeneration with sustained release of poly(phosphoester) microencapsulated nerve growth factor within nerve guide conduits, *Biomaterials*, 24, 2405–2412, 2003.
127. Zhang, F. and W.C. Lineaweaver, Gene transfer with DNA strand technique and peripheral nerve injuries, *Journal of Long Term Efficacy of Medical Implants*, 12, 85–96, 2002.
128. Zhu, J.Y. et al., Expression of adenovirus-mediated neurotrophin-3 gene in Schwann cells of sciatic nerve in rats, *Chinese Journal of Traumatology,* 6, 75–80, 2003.
129. Jubran, M. and J. Widenfalk, Repair of peripheral nerve transections with fibrin sealant containing neurotrophic factors, *Experimental Neurology,* 181, 204–212, 2003.
130. Karaismailoglu, T.N. et al., Histological and electrophysiological assessment of the results of primary and secondary neurorrhaphy in a rabbit model, *Journal of Orthopedic Science,* 8, 88–91, 2003.
131. Menovsky, T. and J.F. Beek, Carbon dioxide laser-assisted nerve repair: effect of solder and suture material on nerve regeneration in rat sciatic nerve, *Microsurgery,* 23, 109–116, 2003.
132. Scharpf, J. et al., A novel technique for peripheral nerve repair, *Laryngoscope*, 113, 95–101, 2003.
133. Wieken, K. et al., Nerve anastomosis with glue: comparative histologic study of fibrin and cyanoacrylate glue, *Journal of Reconstructive Microsurgery,* 19, 17–20, 2003.
134. Brushart, T.M., Motor axons preferentially reinnervate motor pathways, *Journal of Neuroscience,* 13, 2730–2738, 1993.
135. Madison, R.D., S.J. Archibald, and T.M. Brushart, Reinnervation accuracy of the rat femoral nerve by motor and sensory neurons, *Journal of Neuroscience*, 16, 5698–5703, 1996.
136. Brushart, T.M. et al., Contributions of pathway and neuron to preferential motor reinnervation, *Journal of Neuroscience,* 18, 8674–8681, 1998.
137. Martini, R., M. Schachner, and T.M. Brushart, The L2/HNK-1 carbohydrate is preferentially expressed by previously motor axon-associated Schwann cells in reinnervated peripheral nerves, *Journal of Neuroscience*, 14, 7180–7191, 1994.
138. Mears, S., M. Schachner, and T.M. Brushart, Antibodies to myelin-associated glycoprotein accelerate preferential motor reinnervation, *Journal of the Peripheral Nervous System*, 8, 91–99, 2003.
139. Nix, W.A. and H.C. Hopf, Electrical stimulation of regenerating nerve and its effect on motor recovery, *Brain Research*, 272, 21–25, 1983.
140. Al-Majed, A.A. et al., Brief electrical stimulation promotes the speed and accuracy of motor axonal regeneration, *Journal of Neuroscience*, 20, 2602–2608, 2000.
141. Al-Majed, A.A., T.M. Brushart, and T. Gordon, Electrical stimulation accelerates and increases expression of BDNF and trkB mRNA in regenerating rat femoral motoneurons, *European Journal of Neurosciemce,* 12, 4381–4390, 2000.

142. Kim, D.H. et al., Outcomes of surgery in 1019 brachial plexus lesions treated at Louisiana State University Health Sciences Center, *Journal of Neurosurgery*, 98, 1005–1016, 2003.
143. Kim, D.H. et al., Surgical outcomes of 654 ulnar nerve lesions, *Journal of Neurosurgery*, 98, 993–1004, 2003.

4 Cellular Brain Ischemia and Stroke: Neuroprotection, Metabolism, and New Strategies for Brain Recovery

Kelley A. Foster, Christopher J. Beaver, Larry B. Goldstein, and Dennis A. Turner

CONTENTS

0-8493-1482-8/05/$0.00+$1.50

4.1 INTRODUCTION

Each year 4.6 million people die from stroke worldwide and 75% of these cases occur in industrialized countries.[1] In the U.S., stroke is the third leading cause of mortality, with 4.7 million survivors, 15 to 30% of whom are left with permanent disabilities and 20% of whom require long-term institutional care.[2] Significant social, financial, and personal problems occur as a result of these disabilities.[3] *Stroke* is a generic term, encompassing a wide variety of vascular diseases affecting the nervous system. Treatment of these diverse disease processes necessarily involves several different approaches.

Because the brain relies completely on a constant supply of oxygen and glucose for normal function, ischemic injury can occur rapidly if the delivery of these substrates is impaired as a result of transient or permanent cessation of blood flow. Such ischemic injury occurs in nearly 80% of stroke cases due to occlusion of either a major proximal or cerebral artery, most commonly as a result of an embolus or local thrombus. The remaining causes of stroke relate primarily to bleeding in or around the brain.

Acute revascularization and neuroprotective strategies have been the two most extensively studied specific approaches to the treatment of acute ischemic stroke. Of the 178 controlled clinical trials of acute stroke therapies conducted in the past century, only trials of intravenous tissue plasminogen activator (tPA) have been sufficiently positive to lead to approval by the U.S. Food and Drug Administration.[3,4]

Despite showing promise in preclinical studies, none of the more than 114 stroke trials that examined more than 49 neuroprotective drugs have been positive.[4] This discrepancy between preclinical data and the results of clinical trials illustrates the significant challenge of translational neuroscience. These difficulties may have arisen from the use of unsuitable preclinical animal models, inappropriate extrapolation of preclinical data to human trials, or poor clinical trial design.[5,6] However, these multiple failures and experiences can provide useful information to help guide new translational approaches to stroke therapy.

4.2 TYPES OF STROKES AND CEREBRAL ISCHEMIA EVENTS

Causes of ischemic stroke include extracranial or intracranial steno-occlusive disease affecting large- or medium-sized arteries most frequently related to atherosclerosis, embolization from a cardiac or arterial source, and occlusion of small intracranial vessels.[7] In up to 40% of cases, the cause is unknown or the stroke is due to multiple possible etiologies. Atherosclerosis occurs as a result of a complex series of processes leading to arterial injury with cholesterol deposition. Atherosclerotic plaques can provide a nidus for platelet aggregation and thrombus formation, or they can rupture. They can then occlude the artery at the site of clot formation or lead to emboli that can block a distal vessel.

A variety of cardiac conditions can lead to embolization. They include arterial fibrillation, valvular heart disease, ventricular or septal aneurysm, and cardiomyopathies. Small vessel intracranial disease is most frequently associated with hypertension and leads to ischemia in the distribution of penetrating arteries, resulting in so-called "lacunar" syndromes. A large number of other less common conditions including arterial dissection, nonatherosclerotic vasculopathies, hypercoagulable states, and hematological disorders can also lead to ischemic stroke.

Temporary focal ischemia (transient ischemic attacks or TIAs) may also occur. TIAs are traditionally defined as producing neurological symptoms lasting less than 24 hours, but most are far shorter. They are not only harbingers of ischemic stroke, but may also reflect cerebral infarction with transient symptoms (i.e., stroke with rapid functional recovery).[8] Other less common causes of stroke include intracerebral hemorrhage and subarachnoid hemorrhage (SAH).

SAH usually results from rupture of saccular aneurysms most commonly located at branch points in major arteries at the base of the brain. SAH can cause the subarachnoid space to fill with blood at nearly arterial pressure, resulting in direct brain injury due to decreased perfusion of the brain. The presence of blood around major vessels also can lead to delayed cerebral vasospasm (see Chapter 11), then to delayed ischemic stroke due to vessel narrowing and lack of perfusion.[9]

In contrast to focal ischemia caused by arterial occlusion, global ischemia can result from other types of conditions, such as cardiac arrest, near-drowning, or hypotension. Depending on its severity and on other factors, less than 5 minutes of global ischemia can be tolerated before lasting damage occurs.[10] Cerebellar Purkinje cells, CA1 hippocampal pyramidal neurons, and layers 3 and 5 of the neocortex are relatively more vulnerable to global ischemia than other areas of the brain.[10]

4.3 CELLULAR CONSEQUENCES OF STROKE

Although the brain only comprises 2.5% of body weight, it accounts for nearly 25% of basal metabolism. Neuronal function and survival are highly dependent on aerobic metabolism.[11] When a cerebral artery becomes occluded, the lack of oxygen and glucose rapidly leads to neuronal death unless blood supply is restored. However, before this final stage takes place, a cascade of multiple biochemical events is initiated and includes the interactions of a number of different cells in the ischemic area, including neurons, mitochondria, astrocytes, fibroblasts, smooth muscle cells, endothelial cells, and blood components.[12–14] The process begins with the impairment of energetics required to maintain ionic gradients (see Figure 4.1).[15] With the loss of membrane potential, neurons and glia become depolarized,[16] which in turn activates voltage-dependent Ca^{2+} channels. This activation leads to the release of excitatory amino acids into the extracellular space.

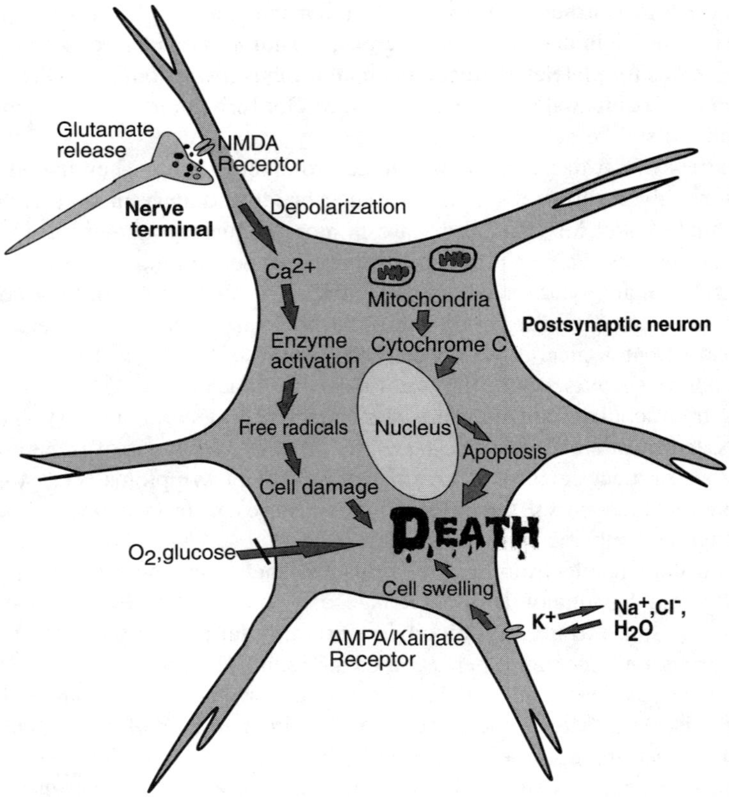

FIGURE 4.1 (See color insert following page 146.) Mechanisms of cell death. A typical neuron is represented indicating a variety of perturbed physiological mechanisms leading to cell death. These mechanisms include excess glutamate stimulation and secondary depolarization (excitotoxicity); loss of substrate (oxygen or glucose); free radical formation, particularly following reoxygenation; apoptosis initiated by cytochrome C release from mitochondria; and cell swelling induced by water influx.

Excitatory amino acids further accumulate because their presynaptic uptake is energy dependent. This can lead to further injury in ischemic neurons that otherwise might remain above the threshold of viability. Activation of excitatory amino acid receptors leads to further sodium and calcium entry.[17] Several different types of excitatory amino acid receptors have been identified pharmacologically. The N-methyl-D-aspartate (NMDA) receptors are gated channels that are highly permeable to Ca^{2+}. Ca^{2+} accumulation is also triggered secondarily by Na^+ influx through α-amino-ε-hydroxy-5-methyl-4-isoxazole-propionic acid (AMPA)-, kainate-, and NMDA-receptor gated channels through activation of voltage-gated Ca^{2+} channels and reverse operation of the Na^+/Ca^{2+} exchanger.[17,18]

Na^+ and Cl^- enter the neurons via monovalent ion channels (such as AMPA- or kainate-receptor gated channels) as a result of glutamate-mediated overactivation.

Inhibitory neurotransmitters (primarily gamma aminobutyric acid [GABA]) are important to slow the excitatory cascade; these neurotransmitters result in chloride flux into cells. The overall effect of these changes in Na and Cl ionic gradients is the passive influx of water leading to cellular edema.

At the same time, K^+ exits the neurons as part of the inhibitory currents following action potentials. In conjunction with the accumulation of extracellular glutamate, increased K^+ levels can promote repeated neuronal depolarizations in the penumbral regions (see later discussion). These "peri-infarct depolarizations" are closely related to hypoxic spreading depression or anoxic depolarization.[19] Frequent neuronal depolarizations result in increased metabolic demands because of severe neuronal membrane depolarization, further worsening ischemic injury and increasing the zone of frank infarction.[19,20]

The accumulation of calcium also initiates a cascade of processes leading to delayed tissue damage. For example, calcium induces proteolytic enzymes, which degrade cytoskeletal proteins. It activates enzymes that lead to the formation of free radical species causing lipid peroxidation and membrane damage. Oxygen free radicals can also promote inflammation and apoptosis. Mitochondria are significant sources of oxygen free radicals that can damage mitochondrial membranes. Oxidation impairs the function of mitochondrial proteins that participate in adenosine triphosphate (ATP) production, electron transport, and H^+ extrusion.[21] Accumulation of calcium and free radical formation in mitochondria favors the formation of permeability transition pores that induce cell death.[22]

Neurons permanently lose membrane potential when blood flow drops more than 20% below normal for more than a very short period.[23] This ischemic core region is surrounded by an area known as the penumbra that is characterized by more modest reductions in blood flow and associated with impaired neuronal function.[24] Although the core of infarcted tissue is not salvageable, the ischemic penumbra partially preserves energy metabolism[25] that may be either reversible or may proceed on to infarction as a result of the cascade of processes including those previously discussed.[26]

Programmed cell death (apoptosis) and excitotoxic necrosis can occur simultaneously in the ischemic brain.[27,28] Apoptosis is a genetically regulated program in which protein-cleaving enzymes known as caspases promote cell death. Caspases 1 and 3 seem to be the predominant proteins involved.[26] Mitochondria release cytochrome C, which induces apoptosis.[29] Several factors determine whether apoptotic or delayed excitotoxic cell death predominates. The factors include the maturity of the neurons, the extent of the injury, the availability of trophic support, and the intracellular free calcium concentration.[30]

In addition to these mechanisms, the accumulation of free radicals and calcium-activated intracellular second messenger systems produces inflammatory mediators such as platelet-activating factor, tumor necrosis factor-α, and interleukin-1β.[13] These inflammatory mediators activate microglia and result in leukocyte infiltration through an increase in endothelial adhesion molecules.[26] As a result of the interactions of complementary receptors on neutrophils and adhesion molecules, the neutrophils adhere to the endothelium, travel through the vascular wall, and enter the brain parenchyma. Post-ischemic inflammatory processes can also contribute to secondary neuronal injury and final infarct size[31,32] through a number of mechanisms including microvascular obstruction by neutrophils[33] and the production of toxic mediators.

Zinc may also be an important mediator of secondary neuronal injury. Under normal physiological conditions, zinc modulates the action of NMDA-receptor gated calcium channels and is critical for the action of several metalloenzymes and transcription factors. Zinc is variably released from excitatory nerve terminal vesicles upon normal synaptic functioning. During ischemia, zinc is thought to be excessively released across the plasma membrane through a number of mechanisms including: activation of voltage-gated calcium channels, NMDA-receptor gated channels, transport exchange for intracellular Na^+, and Ca^{2+}-permeable AMPA receptors.[34] Zinc accumulates in neuronal cell bodies after its release from synaptic terminals.[35] The release of zinc causes apoptosis or necrosis, depending on the extent of the exposure, possibly through direct inhibition of aerobic glycolysis and depletion of energy.[30]

Preclinical animal studies show that the time between initiation of ischemia and the delivery of a putative neuroprotective drug is critical.[36–39] Depending on experimental conditions, neurons occupying the bordering areas of the ischemic territory may be able to survive up to 48 hours following ischemic insult.[40] However, the therapeutic window is considerably shorter.

Understanding of the pathophysiological mechanisms involved in ischemic injury led numerous research groups to develop possible treatments targeting various steps of the cascade. Over the past 10 to 15 years, several animal models of both focal and global ischemia have been developed in attempts to simulate the neuropathological consequences of human stroke. The majority of the early treatments sought to modulate the initial metabolic events following ischemia, in particular excitotoxic mechanisms by using a variety of NMDA receptor antagonists and calcium channel blockers.[12,41] Free-radical scavengers, caspase inhibitors and GABA agonists have also been evaluated. Although results in animal models are promising, all attempts to translate these findings into an efficacious clinical treatment have failed.[42,43]

4.4 PROBLEMS IN TRANSLATION OF STROKE TREATMENTS FROM BENCH TO BEDSIDE

Aside from the neuroanatomical, pathophysiological, pharmacokinetic, and genetic differences among laboratory animals (in particular rodents) and humans, fundamental differences also exist in the designs of preclinical studies and clinical trials:[44] (1) treatment window following stroke, (2) target area of the brain (gray versus white matter), (3) duration of drug treatment, (4) pharmacokinetics, and (5) outcome measures. Laboratory studies are tightly controlled; whereas, human clinical trials involve heterogeneous subjects. Because no approach has yet been successful, the type of preclinical studies that are sufficient to warrant proceeding to clinical trials remains uncertain.

4.4.1 Animal Models of Ischemic Stroke

A variety of animal models have been used to study ischemic stroke to evaluate potential therapies.[45] The most frequently used models of global ischemia involve either bilateral carotid artery occlusion in gerbils or bilateral carotid occlusion with hypotension or four-vessel occlusion in rats. Models of focal ischemia have been developed in a number of animal species and can involve transient or permanent arterial occlusion.

The damage following permanent occlusion results in an ischemic core area surrounded by a penumbral region of varying size. The middle cerebral artery (MCA) occlusion model is among the most commonly employed.[46] An intraluminal thread is used to cause the vessel occlusion and can be withdrawn after 1 to 2 hours to mimic reperfusion or can be left in place to cause permanent occlusion. Experimental factors such as trauma, temperature regulation, stress, and anesthetic use (some of which can have neuroprotective effects alone or in combination with experimental drugs[47]) may complicate interpretation of the results.

Animal models have greatly aided our understanding of the ischemic penumbra and other pathophysiological mechanisms of stroke.[48] Although the results from animal experiments have provided the principles guiding the design of human clinical trials, the results should be used with caution.

Many laboratory animal models are intended to explain basic pathophysiological mechanisms of ischemia and have not been validated for predicting drug efficacy in humans. This is because of a number of important differences between animal models and human strokes. For example, the infarct volume resulting from occlusion in animal models is both uniform and reproducible, and therefore does not necessitate the need for large sample numbers. Experimental conditions such as body temperature, glucose levels, blood pressure, acid-base balance, and oxygenation are tightly regulated and may alter an animal's response to an ischemic insult. In contrast, human stroke is a highly variable clinical condition as a result of differences in location, cause, severity, and reversibility. Stroke types vary considerably in humans (cortical, mixed cortical–subcortical, pure subcortical, white matter, or ischemic and hemorrhagic strokes).

Most animal stroke models use lissencephalic species such as rodents (humans are gyrencephalic). Animal models do not generally consider co-morbid disease states such as diabetes, hypertension, and infections.[44] Humans typically receive a number of different drugs to treat co-morbid conditions that alter the underlying milieu as compared to experimental conditions. In addition, animals used in stroke models are most commonly young as compared to the typically aged human who is afflicted with stroke. All these variables limit extrapolation from the animal results, even when a study considers the same stroke type in preclinical and clinical situations.

Changes in some of the methodology used in laboratory models can make them more relevant to human stroke. First, the occlusion should be transient so as to enable entry of the drug to the site of injury. This would also better reflect the condition in humans in which varying degrees of perfusion are reestablished through collaterals or clot lysis. Drugs should be evaluated in a number of animal species and models to support the generalizability of their purported effects. Allocation of treatment and outcome assessment should be blinded or masked to avoid potential bias. Experiments should be carried out in aged animals to match the human ages commonly observed in stroke patients. Assessment of functional behavioral outcomes in addition to structural outcomes, such as volume of infarction, is essential because functional outcome is the basis of clinical trial assessment. Outcome assessment should be delayed as long as feasible based on animal species. Human trials generally conduct outcome assessments at least 3 months after stroke.

4.4.2 Time Windows of Treatment

With the advent of thrombolytic therapy, stroke is now considered an emergency condition with the same priority as acute myocardial infarction. Many hospitals have developed acute stroke teams, and communities are being organized to facilitate the rapid transportation of stroke patients to appropriate facilities. In addition, efforts have been made to increase public and professional awareness of stroke. People at risk of stroke and their families and friends should be alerted to the common symptoms of stroke.[49]

Unlike rigorously controlled preclinical studies, the time taken to arrive at a hospital following stroke and therefore the time at which the patient is available for treatment after the actual onset of ischemia varies. Between 1995 and 1999, the median time to entry into an acute stroke clinical trial was 14.3 hours, compared to a median permitted entry window of 12 hours.[4] Past studies suggest that irreversible focal injury takes place after only a few minutes and is complete after 6 hours.[10] Although individuals may have salvageable tissue up to 6 hours or longer following a stroke, the progression of damage varies among patients and depends on collateral circulation and other factors.[50]

The failures of past neuroprotectant trials may have in part been due to the administration of the putative neuroprotectants after irreversible injury had occurred. Therefore, potential neuroprotectants must be tested at realistic time points in preclinical studies, but at time intervals longer than minutes. A drug that is efficacious in animal models only if given immediately after arterial occlusion is unlikely to be of benefit. For example NXY-059, a novel nitrone, is effective when administered 3 to 6 hours following recirculation in transient focal MCA occlusion models[51] and at 4 hours in permanent focal MCA occlusion models.[52] Therefore, it would not be reasonable to initially test this drug in humans beyond 6 hours.

4.4.3 Early vs. Late Outcome Definition

Preclinical studies have commonly used histological endpoints to assess therapeutic efficacy. These histological outcomes (i.e., reduction in infarct size) have been generally assessed between 48 and 96 hours. However, ischemic injury can continue to develop for weeks or even months.[53] As a result, early histological endpoints can lead to erroneous conclusions. For example, MK-801 appeared to reduce infarct size at 3 days following ischemic insult but the benefit was not significant after 4 weeks.[54] A number of other drugs including SNX-111 (N-type calcium channel antagonist),

NBQX (AMPA antagonist), and flavopiridol (cyclin-dependent kinase inhibitor) showed potential neuroprotection 1 week following ischemia, but had no effect if the assessment was carried out 4 weeks post-insult.[55,56]

In comparison to preclinical animal studies in which injury is assessed histologically at early time points, clinical trials rely on behavioral and functional outcomes at later stages (generally at 3 months following stroke)[4,57] to assess the effectiveness of intervention. Early behavioral assessments are suggested to be more predictive than histological endpoints.[58] For example, some drugs may be effective in improving functional outcome but may not reduce the resulting infarct size, suggesting that the drugs are acting via other mechanisms.

Such mechanisms may include stimulation of neuronal sprouting and protection against retrograde neuronal death.[59,60] Therefore, in addition to infarct size assessments, preclinical studies should include functional measures of motor, sensory or cognitive deficits in order to gauge the therapeutic efficacy.[61,62] A large variety of tests have been developed for this purpose (see Gladstone et al.[44] for references). Recent preclinical studies have employed complex behavioral tasks as endpoints for determining whether the treatment in question will aid in the reduction of ischemia-related disability.[63,64]

4.4.4 Regional Differences in Target Areas of Brain

Preclinical neuroprotectant studies have targeted the ischemic penumbra. However, in some patients the penumbra may only account for a small percentage of the total infarct volume. To increase the likelihood of detecting a drug effect, clinical studies should target patients with sufficiently large penumbrae.[6,50] However, the optimal way of detecting the penumbra in the context of a clinical trial has not been fully established, and no treatment has been proven efficacious with the use of this approach.

Past clinical trials tended to treat stroke as a single disease entity. Only 62 (35%) of the 178 published stroke trials specified a particular stroke territory (e.g., carotid artery, MCA).[4] The majority of drug therapies tested in animal models targeted gray matter. In comparison to the rodent brain, the human brain contains a higher proportion of white matter (including axons) that may not be salvageable using therapeutic agents that only target gray matter.[65–67] Approximately a third of strokes involve deep white matter and may not respond to neuroprotective therapy. It is therefore possible that potentially successful neuroprotectants have failed due to the inclusion of patients with white matter injuries in clinical trials. Clinical trials should limit selection to patients most likely to benefit. This is particularly important for Phase II clinical trials that provide data critical for a decision to proceed with or defer a large Phase III efficacy study.[5]

4.4.5 Pharmacokinetics, Safety Issues, and Appropriate Dosing

In order for a drug to be effective as a neuroprotectant, it must satisfy a number of criteria. First, it must be able to reach the target region and cells within the brain. Because the blood–brain barrier (BBB) is often damaged by the ischemia to variable extents at different times, some drugs normally excluded from the brain may still be able to reach target tissue, but this is likely to be variable.

Drugs that can cross the BBB may in some conditions have preferred access, depending on diffusion and vascular stasis; to cross the BBB, a drug must be lipid-soluble and have a molecular weight below 500 Da.[68] Recombinant proteins, monoclonal antibodies, gene therapy, and antisense drugs, all of which could serve as potential neuroprotectants, are too large to cross from the systemic circulation into the brain and may require direct infusion into the brain for effective delivery. Because no gene or drug targeting strategies are clinically available, drug testing is now limited to small lipid-soluble drugs that represent only 2% of all potential candidates for drug development.[68]

However, even this limited class of drugs may still exhibit poor access to the critical brain regions targeted. Because of the vascular occlusion, access to the ischemic region is likely to be reduced. Thus, directly sampling the area of brain targeted for drug levels may be a critical control to evaluate whether access into the critical region is possible. After this critical initial point is established, mechanisms of action may be thoroughly assessed by histological and behavioral outcomes.

Dosing is another important consideration. The neuroprotectant dose of a drug in animal models may result in intolerable toxicity in humans. Testing these drugs at doses below those required for efficacy in animal models is less likely to be successful in humans. Duration of treatment must also be considered. Although a drug exhibits neuroprotective properties after a single dose, multiple doses over the period in which the infarct is evolving may or may not increase its clinical efficacy.[69,70] The lengths of treatment have varied from a single injection, to continuous infusions, to several doses extending 3 months after a stroke.[69] Full dose response studies must be performed to avoid problems associated with an inverted U-shaped dose response curve.[71]

Several factors may influence the doses required for therapeutic treatment. If given by constant intravenous infusion, a lipid-soluble drug will accumulate in the cerebral tissues faster than a hydrophilic drug and will take longer to clear from the tissues. This delay may lead to increased toxicity. Therefore, plasma-level calculations may overestimate the levels needed for *in vivo* activity. Other considerations include the receptor-binding properties of the drug that will determine the loading dose and the need for maintenance infusion, the clearance and volume of distribution of the drug, and its therapeutic index.[69]

The duration of therapy is also influenced by side effects the drug might produce and pathophysiological changes following the stroke. For example, the more potent NMDA antagonists produce psychomimetic effects that might preclude the drug from administration over days to weeks.[69] As a result of the loss of autoregulation in acute stroke, drugs that produce hemodynamic effects may increase or decrease cerebral blood flow which in turn could exacerbate edema or worsen ischemia. Moderate increases in blood pressure, however, could be beneficial in improving blood flow and local perfusion.[69] In addition, these drugs may be used to increase the initial time window during which longer lasting drugs may be administered.[56]

4.4.6 Clinical Outcome Measures and Statistical Issues with Clinical Trials

A variety of measures have been used to assess outcomes in clinical stroke trials. Less than half of the published clinical trials utilized validated outcome measures and only 17% indicated primary endpoint.[4] The choice of outcome measures can play a fundamental role in whether a therapeutic agent is deemed successful.[6,57,72] Outcome can be assessed at the level of impairment (NIH Stroke Scale), disability (Barthel Index), or social handicap (Rankin Index). In addition, although they are not yet widely employed, scales that incorporate quality-of-life assessment and pharmacoeconomic analysis can be used as secondary outcome measures.[73] Final outcome assessments are generally carried out at least 3 months after stroke as recovery has usually reached a plateau by that time.

The use of a dichotomous division of a continuous scale may help determine whether a patient has achieved a clinically significant benefit.[73] Global statistics can take into account multiple assessment scales and can be used to provide overall assessments of benefits.[74] The National Institute of Neurological Disorders and Stroke's rtPA trial[75] utilized this approach.

Statistical power is an important consideration for clinical trial design. Studies must have sufficient statistical power to ensure that a lack of a treatment effect results from a lack of biological effect of the intervention and is not due to insufficient sample size. Kidwell et al.[4] calculated the sample sizes required for a 5% reduction in the proportion of patients dead or disabled at 6 months as 3148 (reduction from 60% to 55%; 80% power, alpha = 0.05). The mean sample size per trial of the 178 controlled clinical trials for acute ischemic stroke performed up until 1999 was 415 patients. The mean sample size for neuroprotective trials was 186 patients (median 69). Potentially efficacious drugs might have been abandoned because of a type II statistical error.[4] The Stroke Therapy Academic Industry Roundtable has developed a series of recommendations for translating preclinical studies into clinical trials based on these reviews and other considerations.[5,6,76]

4.4.7 Clinical Trials

Following successful outcomes in animal models, a drug may be assessed further in human clinical trials that consist of three phases. Phase I trials are conducted in healthy volunteers to determine whether untoward toxicity is present and to evaluate the maximal tolerated dose. Phase II studies are performed in persons who have the disease and include questions focused on dose finding, safety, and potential efficacy. Phase III trials are large-scale studies with sufficient statistical power to assess efficacy.[73]

Phase I trials are often conducted in young healthy volunteers. In contrast, stroke patients are most frequently elderly, where age-related changes in cerebral dynamics and vasculature can significantly affect toxicity as well as pharmacokinetics and regional cerebral blood flow. Therefore, the inclusion of healthy elderly patients in Phase I trials may help avoid under-recognition of potential side effects in the eventual target population.[73]

Phase II and III trials are sometimes combined to reduce the numbers of patients who need to be included and save time.[77] This results in having to use preclinical and Phase I data to develop a protocol for clinical efficacy.[73] Phase II trials are sometimes divided into IIa and IIb studies.[6] Phase IIa studies often focus on providing initial toxicity data and exploring dosing and pharmacokinetic issues. Phase IIb trials are important for refining patient selection, dose, route, timing, duration of therapy, and for better understanding of side effects, pharmacokinetics, and drug interactions.[6]

4.5 FUTURE POTENTIAL TREATMENTS AND OPPORTUNITY TIME WINDOWS

The consequences of acute ischemic injury evolve over time, and drug treatments corresponding to various successive events within the ischemic cascade may be developed in the future. A "cocktail" of therapies may need to be developed and tested to address these potential overlapping therapeutic windows.[6] Combination therapy may also reduce the dose-limiting toxicity encountered in the use of single agents if multiple agents can be administered at lower doses. In some cases, the administration of a second drug may improve the action of the first. For example, the combination of a thrombolytic agent and a neuroprotectant may increase the chances of the latter drug reaching the site of injury within the required time window.[73]

Combination therapy may also offer synergistic effects. For example, the administration of insulin with the noncompetitive NMDA antagonist, dizocilpine, in diabetic rats following ischemia resulted in additive neuroprotective effects.[78] However, testing combined administration of unproven drugs provides additional challenges for clinical trial design.[6]

Some stroke treatments are more appropriate at one time period in the evolution of stroke than at others. Three time periods will be considered here. The first is minutes to hours, the second is hours to days, and the third is days to months.

4.5.1 Post-Stroke (Time Frame of Minutes to Hours)

Neuroprotective therapies in the ischemic core will be helpful only if the blood supply to the ischemic brain can be reestablished. Hypoperfusion in the core and penumbra accounts for a greater proportion of the resulting injury than the subsequent degradative processes that occur in the penumbral region.[79] In addition to avoiding relative hypotension, the primary treatment for hypoperfusion is the use of interventions with the potential to restore flow, such as the use of a clot lysing drug (tPA, for example).

Other approaches include mechanical clot disruption and the use of suction devices, lasers, and ultrasound.[80] Although intravenous rt-PA remains the only FDA-approved thrombolytic drug therapy for stroke,[75] other drugs including both long-used and novel thrombolytics and glycoprotein IIb/IIIa receptor antagonists are being evaluated.[81] Even if the blood vessels can be reopened, there is a risk of hemorrhagic infarction following reperfusion into ischemic areas. Thus, early reestablishment of blood supply is critical if possible.

4.5.2 Post-Stroke (Time Frame of Hours to Days)

4.5.2.1 Neuroimaging Techniques

Positron emission tomography (PET), diffusion–perfusion magnetic resonance imaging (MRI), and computerized tomography (CT) perfusion are now used to identify patients with potentially salvageable penumbral regions. In one study, the penumbral region accounted for 18% of the final infarct volume; the remaining 82% of the affected brain tissue was critically hypoperfused (70%) or sufficiently perfused (12%).[79] PET, the "gold standard" is not logistically feasible to guide urgent clinical treatment because it is not widely available and requires considerable set-up time. MRI techniques such as diffusion–perfusion weighted imaging, MR spectroscopy, and CT perfusion may prove more useful in detecting salvageable brain as part of routine clinical practice.[7,82] Because of person-to-person variations in collateral blood supplies, the use of neuroimaging may also allow the use of treatments that are not based solely on time since stroke onset.[82–84]

4.5.2.2 Receptor Antagonists, Calpain Inhibitors, and Free Radical Scavengers

Several other drugs intended to limit ischemic injury are being developed. The combination of NMDA antagonists with AMPA or kainate receptor antagonists may confer protection to oligodendrocytes and GABAergic neurons with Ca^{2+}-permeable AMPA receptors.[30] Toxicity can severely limit the clinical efficacy of otherwise useful treatment approaches. This applies to many drugs aimed at blocking excitotoxicity. Developing more targeted drugs may limit side effects. For example, ifenprodil acts on NR2B-containing NMDA receptors and they are expressed in greater proportions in the forebrain compared to the hindbrain.[30,85] Therefore, it is anticipated that its psychometric side effects might be reduced as compared to other drugs of this class.

Calpains are also receiving attention because they are proteolytic enzymes activated by calcium and may be potential targets for therapeutic agents. Calpains are activated following ischemia and break down cytoskeletal proteins such as spectrin. Calpain inhibitors including AK275, AK295, and MDL 28,170 are neuroprotective following ischemia in rats.[86–88] MDL 28,170 reduced infarct volume when administreed up to 6 hours following MCA occlusion.[88]

A number of potential therapeutic agents have been developed to reduce reperfusion-related injuries that involve the accumulation of oxygen free-radicals and inflammatory cells. The agents include superoxide dismutase, catalase, glutathione, iron chelators, vitamin E, alphaphenyl nitrogen (PBN), dimethylthiourea, oxypurinol, and tirilazad mesylate. They may act by reducing cytotoxic and vasogenic brain edema, aiding in Ca^{2+} homeostasis reestablishment, and antagonizing glutamate excitotoxicity.[80] Some have been tested in clinical trials, but none has yet proven efficacious.

One of the consequences of oxygen free radical formation is the destruction of single-strand DNA. This leads to the activation of poly (ADP-ribose) polymerase (PARP), a repair enzyme that depletes cellular nicotinamide adenine dinucleotide (NAD^+) and ATP.[89] Studies that eliminated the PARP gene or administered PARP inhibitors showed reduced infarction following ischemia.[89] However, the PARP enzyme may be important in DNA repair and genomic stability, particularly after partial DNA disruption from ischemia. It has also been hypothesized that because PARP activation involves NAD^+ that then depletes the metabolic pool of NADH, enhancing the pool of NAD^+ may contribute to enhanced cell functioning. Several papers have suggested that direct nicotinamide treatment may be effective at repleting the pool of metabolic NADH and also facilitating the repair processes of PARP.

4.5.2.3 Anti-Apoptosis/Necrosis Agents

Drug therapies (cycloheximide and anisomycin) have also been developed to inhibit apoptosis and have demonstrated neuroprotection in focal and global models of ischemia.[90,91] Thought to act through this mechanism, a caspase-3 inhibitor [N-benzyloxycarbonyl-Asp(Ome)-Glu(Ome)-Val-Asp(Ome)-fluoromethylketone or z-DEVD.FMK] reduced infarct size following transient ischemia.[11]

Neuronal apoptosis inhibitor protein (NAIP), a novel anti-apoptotic gene, is a group II (3 and 7) caspase inhibitor that may be able to reduce apoptosis.[92] Other inhibitors such as *N*-benzyloxycarbonyl-Val-Ala-Asp-fluoromethyl ketone (Z-VAD.FMK) and *N*-benzyloxycarbonyl-Asp-Glu-Val-Asp-fluoromethyl ketone (z-DEVD.FMK) that are not caspase selective and also block cathepsins reduce behavioral and cellular deficits as well as infarct volume following focal ischemia.[33,93,94]

A combination of anti-apoptotic and antinecrotic therapies may be advantageous. For example, the combined administration of dextrorphan and cycloheximide reduced infarct volume following transient ischemia (MCAO) in rats by 87%, which was greater than the reduction resulting from the use of either agent alone (~65%).[95] Another example of combination therapy is the use of MK-801 and z-VAD.fmk, which also reduced infarct size following ischemia.[96] It has been suggested, however, that if necrosis is reduced, apoptosis may become unmasked or promoted.[44]

4.5.2.4 Zinc Toxicity Treatment

Zinc toxicity following stroke is another potential area of therapeutic application. One consequence of zinc exposure is an increase in dihydroxy-acetone phosphate, a glycolytic intermediate, that in turn causes a decrease in neuronal ATP levels. It has been suggested that the administration of pyruvate, an energy substrate, can help ease the ATP loss. It has been postulated that the failure of calcium channel antagonists may in part be due to perturbations in zinc levels following ischemic injury. The reduction of zinc release from nerve terminals may be accomplished by a dietary restriction of zinc.[11] Other approaches to lessening the toxic effects of zinc could include the upregulation of both metallothioneins and cellular zinc extrusion transporters and the implementation of mechanisms that prevent energy metabolism interference.[30] Unfortunately, clinical trials targeting zinc have also failed.

4.5.2.5 Anti-Inflammatory Treatments

Neurons have been the primary targets of neuroprotective strategies. However, white matter and axons are also damaged following ischemia. Astrocytes may be injured as a result of the release of inflammatory mediators following ischemic insult as well as zinc toxicity.[97] Axons and oligodendrocytes are thought to incur damage as a consequence of calcium influx through the Na^+/Ca^{2+} exchanger and AMPA receptor over-stimulation, respectively.[98] The release of glutamate via reverse Na^+-glutamate transport may also contribute to oligodendrocyte damage.[65]

Potential therapeutic targets may involve interfering with various steps of the inflammatory cascades. For example, microvascular occlusion may be reduced by the inhibition of leukocyte adherence to blood vessels in the ischemic area. Other strategies include directing antibodies toward molecules such as intercellular adhesion molecule-1 (ICAM-1)[99] and inhibiting the release of proinflammatory cytokines from astrocytes and microglia such as interleukin-1β (IL-1β) or tumor necrosis factor-α (TNF-α). The use of statins and estrogens may also have the potential to reduce injury following ischemic insult through upregulation of endothelial nitric oxide synthase[100] and antioxidant and trophic mechanisms,[101] respectively.

The use of growth factors may also be beneficial in treating ischemic injury and promoting functional recovery. Exogenous compounds such as nerve growth factor (NGF), brain-derived neurotrophic factor (BDNF), neurotrophins 4/5 (NT-4/5), basic fibroblast growth factor, and insulin-like growth factor-1 (IGF-1) can all reduce injuries in rats subjected to cerebral ischemia.[30] One clinical trial of fibroblast growth factor (FGF) was stopped because of toxicity.

The inflammatory response is initiated and regulated by the complement system that consists of a number of cascades. The complement system causes injury in animal models of ischemia through the production of anaphylotoxins C3a and C5a and endothelial cell adhesion molecule upregulation.[102] The complement cascade offers several sites of potential therapeutic intervention. For example, soluble complement receptor-1 (sCR1), a strong inhibitor of complement activation, reduced neurological deficits and decreased platelet and polymorphonuclear leukocytes (PMN) accumulations following MCAO and reperfusion in mice.[103]

Protein kinase C increases both the vesicular release of glutamate and neuronal excitability.[104] Pretreatment with agents such as staurosporine, a broad-spectrum protein kinase inhibitor[105] and 1-(5-isoquinolinesulfonyl)-2-methylpiperazine dihydrochloride (H-7)[106] decreased neuronal cell death following global cerebral ischemia and the accumulation of extracellular glutamate, respectively.

The mitogen-activated protein (MAP) kinase pathways may also be activated during ischemia. These include the c-Jun NH_2-terminal kinases (JNKs), p38 kinases, and extracellular signal-regulated kinases (ERKs). The p38 inhibitor, SB203580, administered as a pretreatment reduced neuronal death in a global model of ischemia.[107] However, following transient focal ischemia, it was not effective.[108] The PD98059 ERK inhibitor also given as a pretreatment decreased the volume of infarction following transient focal ischemia.[108]

4.5.2.6 Hypothermia

In addition to the plethora of pharmacological agents that may provide therapeutic benefits following stroke, physiological variables can be manipulated to confer protection. Hypothermia has been studied for the past 50 years because of its protective benefits[109] and has been used to protect organs during cardiovascular and neurosurgical procedures. In the case of stroke, reduced body temperatures in patients admitted to hospitals resulted in both lower mortality rates and improved functional outcomes.[110,111] Among the potential mechanisms by which hypothermia offers protection are: reduction in metabolic rate, thereby delaying the depletion of high-energy phosphates, inhibition of excitatory neurotransmitter release and oxygen radical production, decrease in intracranial pressure and anti-convulsant activity, and suppression of initiation of spreading depression (see Clifton et al. and Bernard and Buist[112,113] for references).

Temperatures of 32 to 34°C have been demonstrated in animal models to be safe and produce a minimum number of side effects.[112,114,115] In patients who suffer hypothermic side effects such as platelet dysfunction, rebound hyperthermia, and pneumonia,[116] a combination of more modest reductions in temperature (2 to 3°C) with neuroprotective drug therapies may be more effective. The efficacy of hypothermia seems to depend on the length of its application following ischemic injury. Hippocampal CA1 cells are protected when the duration of hypothermia is increased from 12 to 24 hours.[117–119] The time window between the onset of ischemia and irreversible cell injury increases as the duration of hypothermia is increased.[120] Preliminary clinical trials of hypothermia are promising.[121,122]

In contrast to hypothermia, any increases in brain temperature above normal (37°C) following stroke can exacerbate ischemic injury.[123,124] Hyperthermia in stroke patients is associated with increases in morbidity and mortality rates.[125,126] Hyperthermia has also been shown to interfere with the actions of therapeutic agents such as MK-801 and thrombolytic treatments.[127] Fevers must be treated aggressively in patients with ischemic stroke.[7]

4.5.2.7 Hyperglycemia Treatment

Hyperglycemia, another physiological variable that can be manipulated in the clinical environment, has been associated with poor outcomes following strokes in animal studies and clinical trials. The multicenter Trial of ORG 10172 in Acute Stroke Treatment (TOAST) found that higher blood glucose levels resulted in worse outcomes (odds ratio: 0.82 for every 100 mg/dL increase in glucose; $p = 0.03$).[128] Hyperglycemia increases cerebral lactate concentrations, causes neuronal and glial damage, and increases infarct volume.[129–131] Pre-ischemic hyperglycemia also increases extracellular glutamate concentrations during ischemia, which results in exacerbated cell damage in the neocortex.[132] In contrast, relative hypoglycemia in the presence of permanent focal ischemia results in a smaller infarct volume as compared to severe hyperglycemic conditions.[130] Insulin was neuroprotective in a number of animal studies following global and focal ischemia (see Kagansky et al.[133] for references). IGF-1 has also demonstrated neuroprotective properties.[134]

The Glucose Insulin in Stroke Trial is examining the potential protective effects of the combined administration of glucose, potassium, and insulin (GKI) in stroke patients with mild to moderate hyperglycemia. Results from the pilot study indicate a slightly lower mortality rate in GKI patients compared to controls (28 versus 32%).* During the first 24 hours of hospitalization following stroke, hyperglycemia should be avoided by excluding the administration of dextrose-containing solutions. By consensus, the upper limit of glucose concentration range in all patients should be maintained at ≤300 mg/dL.[7]

4.5.3 Post-Stroke (Time Frame of Days to Months)

A number of therapeutic approaches can be employed in the days or months after stroke.[49] In addition to drugs aimed at secondary prevention, orally active drugs may eventually be developed to confer long-lasting neuroprotection in persons at risk for recurrent stroke.[69] Pharmacological strategies designed to facilitate the recovery process are also under investigation. For example, amphetamine enhances sensory and motor function following ischemia.[135] Other drugs such as yohimbine,[136] phenylpropanolamine,[137] and methylphenidate[138] enhance motor recovery following brain injury as a result of their effects on norepinephrine.

However, drugs that decrease norepinephrine release such as clonidine hydrochloride (α2-adrenergic receptor agonist), prazosin, and phenoxybenzamine (α1-adrenergic receptor antagonists) interfere with motor recovery following brain injury.[139] Therefore, the use of certain drugs given for nonstroke morbidities should be avoided because they may interfere with long-term stroke outcomes.[139]

Other novel approaches aimed at improving post-stroke recovery include stem cell transplantation and gene therapy (see Chapter 2). In rat models of stroke, the transplantation of cultured neuronal cells improved motor and cognitive deficits and was safe.[140,141] An initial trial was conducted in humans. Cultured neurons (human precursor cell lines differentiated into neurons) were injected into the area of infarction. No major adverse consequences appeared as long as 12 to 18 months following transplantation, but clinical benefit remains uncertain.[142]

Physiotherapeutic approaches are central in reducing mortality and improving long-term outcomes. Patient mobilization can also help to reduce the occurrence of pneumonia and secondary thromboembolic events.[143] A variety of new physiotherapeutic approaches are under investigation including constraint-induced therapy, robot-assistive training, and supported treadmill training.[144]

4.5.4 Surgical Treatment Options

Although this discussion has focused on medical interventions, surgical treatments are also being explored for stroke treatment. For example, hemicraniectomy (removal of the skull and dura on one side of the head for decompression of the brain) may be useful in patients at risk for herniation after nondominant hemisphere stroke.[121] Intensive care management of patients with acute ischemic stroke is also evolving (see Chapter 13). For example, the potential impact of monitoring physiological

* Internet Stroke Center: www.strokecenter.org.

parameters such as brain tissue oxygenation are being explored, together with aggressive hyperdynamic therapy to enhance blood flow into ischemic regions and collateral formation.

4.6 CONCLUSIONS

Stroke remains one of the leading causes of death and disability worldwide. Attempts to develop effective drug therapies for stroke-related brain damage have been fraught with difficulties. A number of issues must be addressed before successful results from preclinical studies can be translated to the treatment of stroke patients. Valuable lessons learned from past failures can be used to increase the chances of producing efficacious drug therapies for stroke. Many promising avenues of stroke treatment remain, but enhancing their delivery to the vascular-compromised brain remains a further challenge for the future.

REFERENCES

1. Bonita, R. and Beaglehole, R., Monitoring stroke: an international challenge, *Stroke*, 26 (4), 541, 1995.
2. American Heart Association, *Heart Disease and Stroke Statistics: 2003 Update*, Dallas, TX, 2003.
3. Bonita, R., Epidemiology of stroke [comment], *Lancet*, 339 (8789), 342, 1992.
4. Kidwell, C.S. et al., Trends in acute ischemic stroke trials through the 20th century, *Stroke*, 32 (6), 1349, 2001.
5. Finklestein, S.P. et al., Stroke Therapy Academic Industry Roundtable I: recommendations for standards regarding preclinical neuroprotective and restorative drug development, *Stroke*, 30 (12), 2752, 1999.
6. Albers, G.W. et al., Stroke Therapy Academic Industry Roundtable II: recommendations for clinical trial evaluation of acute stroke therapies [comment], *Stroke*, 32 (7), 1598, 2001.
7. Adams, H.P., Jr. et al., Guidelines for the early management of patients with ischemic stroke: scientific statement from the Stroke Council of the American Stroke Association, *Stroke,* 34 (4), 1056, 2003.
8. Bogousslavsky, J. and Regli, F., Cerebral infarction with transient signs (CITS): do TIAs correspond to small deep infarcts in internal carotid artery occlusion? *Stroke,* 15 (3), 536, 1984.
9. Mayberg, M.R. et al., Guidelines for the management of aneurysmal subarachnoid hemorrhage: a statement for healthcare professionals from a special writing group of the Stroke Council, American Heart Association, *Stroke,* 25 (11), 2315, 1994.
10. Zivin, J.A., Factors determining the therapeutic window for stroke, *Neurology*, 50 (3), 599, 1998.
11. Lee, J.M. et al., Brain tissue responses to ischemia, *Journal of Clinical Investigations*, 106 (6), 723, 2000.
12. Choi, D.W., Glutamate neurotoxicity and diseases of the nervous system, *Neuron*, 1 (8), 623, 1988.
13. Rothwell, N.J. and Hopkins, S.J., Cytokines and the nervous system II: actions and mechanisms of action [comment], *Trends in Neurosciences*, 18 (3), 130, 1995.

14. Kristian, T. and Siesjo, B.K., Calcium–related damage in ischemia, *Life Sciences*, 59 (5–6), 357, 1996.
15. Martin, R.L., Lloyd, H.G., and Cowan, A.I., The early events of oxygen and glucose deprivation: setting the scene for neuronal death? *Trends in Neurosciences*, 17 (6), 251, 1994.
16. Katsura, K., Kristian, T., and Siesjo, B.K., Energy metabolism, ion homeostasis, and cell damage in the brain, *Biochemical Society Transactions*, 22 (4), 991, 1994.
17. Choi, D.W., Calcium-mediated neurotoxicity: relationship to specific channel types and role in ischemic damage, *Trends in Neurosciences*, 11 (10), 465, 1988.
18. Choi, D.W., Calcium: still center-stage in hypoxic–ischemic neuronal death, *Trends in Neurosciences,* 18 (2), 58, 1995.
19. Hossmann, K.A., Peri-infarct depolarizations, *Cerebrovascular and Brain Metabolism Reviews*, 8 (3), 195, 1996.
20. Mies, G., Iijima, T., and Hossmann, K.A., Correlation between peri-infarct DC shifts and ischaemic neuronal damage in rat, *Neuroreport*, 4 (6), 709, 1993.
21. Zhang, Y. et al., The oxidative inactivation of mitochondrial electron transport chain components and ATPase, *Journal of Biological Chemistry*, 265 (27), 16330, 1990.
22. Matsumoto, S. et al., Blockade of the mitochondrial permeability transition pore diminishes infarct size in the rat after transient middle cerebral artery occlusion, *Journal of Cerebral Blood Flow and Metabolism*, 19 (7), 736, 1999.
23. Hossmann, K.A., Viability thresholds and the penumbra of focal ischemia [comment], *Annals of Neurology*, 36 (4), 557, 1994.
24. Kato, H. and Kogure, K., Biochemical and molecular characteristics of the brain with developing cerebral infarction, *Cellular and Molecular Neurobiology.* 19 (1), 93, 1999. Erratum in *Cellular and Molecular Neurobiology*, 20 (3), 417, 2000.
25. Kuroda, S. and Siesjo, B.K., Reperfusion damage following focal ischemia: pathophysiology and therapeutic windows, *Clinical Neuroscience*, 4 (4), 199, 1997.
26. Dirnagl, U., Iadecola, C., and Moskowitz, M.A., Pathobiology of ischaemic stroke: an integrated view, *Trends in Neurosciences,* 22 (9), 391, 1999.
27. Martin, L.J. et al., Neurodegeneration in excitotoxicity, global cerebral ischemia, and target deprivation: perspective on the contributions of apoptosis and necrosis, *Brain Research Bulletin*, 46 (4), 281, 1998.
28. Choi, D.W., Ischemia-induced neuronal apoptosis, *Current Opinions in Neurobiology*, 6 (5), 667, 1996.
29. Fujimura, M. et al., Cytosolic redistribution of cytochrome C after transient focal cerebral ischemia in rats, *Journal of Cerebral Blood Flow and Metabolism*, 18 (11), 1239, 1998.
30. Lee, J.M., Zipfel, G.J., and Choi, D.W., The changing landscape of ischaemic brain injury mechanisms, *Nature*, 399 (6738; Suppl.), A7, 1999.
31. Degraba, T.J., The role of inflammation after acute stroke: utility of pursuing anti-adhesion molecule therapy, *Neurology*, 51 (Suppl, 3), S62, 1998.
32. Hallenbeck, J.M., Inflammatory reactions at the blood–endothelial interface in acute stroke, *Advances in Neurology*, 71, 281, 1996.
33. Hara, H. et al., Inhibition of interleukin 1-beta converting enzyme family proteases reduces ischemic and excitotoxic neuronal damage, *Proceedings of the National Academy of Sciences of the United States of America*, 94 (5), 2007, 1997.
34. Sensi, S.L. et al., Measurement of intracellular free zinc in living cortical neurons: routes of entry, *Journal of Neuroscience*, 17 (24), 9554, 1997.
35. Sorensen, J.C. et al., Rapid disappearance of zinc positive terminals in focal brain ischemia, *Brain Research*, 812 (1–2), 265, 1998.

36. Astrup, J., Siesjo, B.K., and Symon, L., Thresholds in cerebral ischemia: the ischemic penumbra, *Stroke,* 12 (6), 723, 1981.
37. Pulsinelli, W., Pathophysiology of acute ischaemic stroke, *Lancet,* 339 (8792), 533, 1992.
38. Grotta, J., Rodent models of stroke limitations: what can we learn from recent clinical trials of thrombolysis? [comment], *Archives of Neurology,* 53 (10), 1067, 1996.
39. Ginsberg, M.D., The concept of the therapeutic window: a synthesis of critical issues, in *Nineteenth Princeton Stroke Conference*, Moskowitz, M.A. and Caplan, L.R., Eds., Butterworth–Heinemann, Boston, 1995, p. 331.
40. Heiss, W.D. et al., Progressive derangement of peri-infarct viable tissue in ischemic stroke, *Journal of Cerebral Blood Flow and Metabolism,* 12 (2), 193, 1992.
41. Rothman, S.M. and Olney, J.W., Glutamate and the pathophysiology of hypoxic–ischemic brain damage, *Annals of Neurology,* 19 (2), 105, 1986.
42. Martinez-Vila, E. and Sieira, P.I., Current status and perspectives of neuroprotection in ischemic stroke treatment, *Cerebrovascular Diseases,* 11 (Suppl. 1), 60, 2001.
43. Fisher, M. and Schaebitz, W., An overview of acute stroke therapy: past, present, and future, *Archives of Internal Medicine,* 160 (21), 3196, 2000.
44. Gladstone, D.J. et al., Toward wisdom from failure: lessons from neuroprotective stroke trials and new therapeutic directions, *Stroke,* 33 (8), 2123, 2002.
45. Hossmann, K.A., Animal models of cerebral ischemia 1: review of literature, *Cerebrovascular Diseases,* 1, 2, 1991.
46. Mohr, J.P. and Barnett, H.J.M., Classification of ischemic strokes, in *Stroke: Pathophysiology, Diagnosis, and Management,* Barnett, H.J.M. et al., Eds., Churchill Livingstone, New York, 1986, p. 281.
47. Lightfoote, W.E., II, Molinari, G.F., and Chase, T.N., Modification of cerebral ischemic damage by anesthetics, *Stroke,* 8 (5), 627, 1977.
48. Ginsberg, M.D., The validity of rodent brain ischemia models is self-evident [comment], *Archives of Neurology,* 53 (10), 1065, 1996.
49. Adams, H.P., Jr., Treating ischemic stroke as an emergency, *Archives of Neurology,* 55 (4), 457, 1998.
50. Fisher, M., Characterizing the target of acute stroke therapy, *Stroke,* 28 (4), 866, 1997.
51. Kuroda, S. et al., Neuroprotective effects of a novel nitrone, NXY-059, after transient focal cerebral ischemia in the rat, *Journal of Cerebral Blood Flow and Metabolism,* 19 (7), 778, 1999.
52. Sydserff, S.G. et al., Effect of NXY-059 on infarct volume after transient or permanent middle cerebral artery occlusion in the rat: studies on dose, plasma concentration and therapeutic time window, *British Journal of Pharmacology,* 135 (1), 103, 2002.
53. Drummond, J.C., Piyash, P.M., and Kimbro, J.R., Neuroprotection failure in stroke [comment], *Lancet,* 356 (9234), 1032, 2000.
54. Valtysson, J. et al., Neuropathological endpoints in experimental stroke pharmacotherapy: the importance of both early and late evaluation, *Acta Neurochirurgica,* 129 (1–2), 58, 1994.
55. Colbourne, F. et al., Continuing postischemic neuronal death in CA1: influence of ischemia duration and cytoprotective doses of NBQX and SNX-111 in rats, *Stroke,* 30 (3), 662, 1999.
56. Wang, F. et al., Inhibition of cyclin-dependent kinases improves CA1 neuronal survival and behavioral performance after global ischemia in the rat, *Journal of Cerebral Blood Flow and Metabolism,* 22 (2), 171, 2002.

57. Duncan, P.W., Jorgensen, H.S., and Wade, D.T., Outcome measures in acute stroke trials: a systematic review and some recommendations to improve practice [comment], *Stroke,* 31 (6), 1429, 2000.
58. Corbett, D. and Nurse, S., The problem of assessing effective neuroprotection in experimental cerebral ischemia, *Progress in Neurobiology,* 54 (5), 531, 1998.
59. Kawamata, T. et al., Intracisternal basic fibroblast growth factor (bFGF) enhances behavioral recovery following focal cerebral infarction in the rat, *Journal of Cerebral Blood Flow and Metabolism,* 16 (4), 542, 1996.
60. Kawamata, T. et al., Intracisternal osteogenic protein-1 enhances functional recovery following focal stroke, *Neuroreport,* 9 (7), 1441, 1998.
61. Hunter, A.J., Mackay, K.B., and Rogers, D.C., To what extent have functional studies of ischaemia in animals been useful in the assessment of potential neuroprotective agents? *Trends in Pharmacological Sciences,* 19 (2), 59, 1998.
62. Hudzik, T.J. et al., Long-term functional end points following middle cerebral artery occlusion in the rat, *Pharmacology, Biochemistry and Behavior,* 65 (3), 553, 2000.
63. Lyden, P.D. et al., Quantitative effects of cerebral infarction on spatial learning in rats, *Experimental Neurology,* 116 (2), 122, 1992.
64. Grotta, J.C. et al., CGS-19755, a competitive NMDA receptor antagonist, reduces calcium-calmodulin binding and improves outcome after global cerebral ischemia, *Annals of Neurology,* 27 (6), 612, 1990.
65. Stys, P.K., Anoxic and ischemic injury of myelinated axons in CNS white matter: from mechanistic concepts to therapeutics, *Journal of Cerebral Blood Flow and Metabolism,* 18 (1), 2, 1998.
66. Muir, K.W. and Grosset, D.G., Neuroprotection for acute stroke: making clinical trials work, *Stroke,* 30 (1), 180, 1999.
67. Small, D.L. and Buchan, A.M., Animal models, *British Medical Bulletin,* 56 (2), 307, 2000.
68. Pardridge, W.M., Why is the global CNS pharmaceutical market so under-penetrated? *Drug Discovery Today,* 7 (1), 5, 2002.
69. Dyker, A.G. and Lees, K.R., Duration of neuroprotective treatment for ischemic stroke, *Stroke,* 29 (2), 535, 1998.
70. Coimbra, C. et al., Long-lasting neuroprotective effect of postischemic hypothermia and treatment with an anti-inflammatory/antipyretic drug: evidence for chronic encephalopathic processes following ischemia, *Stroke,* 27 (9), 1578, 1996.
71. Muir, K.W. and Lees, K.R., Clinical experience with excitatory amino acid antagonist drugs, *Stroke,* 26 (3), 503, 1995.
72. Lees, K.R., Advances in neuroprotection trials, *European Neurology,* 45 (1), 6, 2001.
73. Degraba, T.J. and Pettigrew, L.C., Why do neuroprotective drugs work in animals but not humans? *Neurologic Clinics,* 18 (2), 475, 2000.
74. Tilley, B.C. et al., Use of a global test for multiple outcomes in stroke trials with application to the National Institute of Neurological Disorders and Stroke t-PA Stroke Trial, *Stroke,* 27 (11), 2136, 1996.
75. National Institute of Neurological Disorders and Stroke rt-PA Stroke Study Group, Tissue plasminogen activator for acute ischemic stroke [comment], *New England Journal of Medicine,* 333 (24), 1581, 1995.
76. Fisher, M., Recommendations for advancing development of acute stroke therapies: Stroke Therapy Academic Industry Roundtable 3, *Stroke,* 34 (6), 1539, 2003.
77. Storer, B.E., A sequential Phase II/III trial for binary outcomes, *Statistics in Medicine,* 9, 229, 1990.

78. Bomont, L. and Mackenzie, E.T., Neuroprotection after focal cerebral ischaemia in hyperglycaemic and diabetic rats, *Neuroscience Letter*, 197 (1), 53, 1995.
79. Heiss, W.D. et al., Which targets are relevant for therapy of acute ischemic stroke? [comment], *Stroke,* 30 (7), 1486, 1999.
80. Lindsberg, P.J. et al., The future of stroke treatment, *Neurologic Clinics*, 18 (2), 495, 2000.
81. Verheugt, F.W., Acute coronary syndromes: drug treatments, *Lancet*, 353 (Suppl. 2), SII20, 1999.
82. Baird, A.E. and Warach, S., Magnetic resonance imaging of acute stroke, *Journal of Cerebral Blood Flow and Metabolism*, 18 (6), 583, 1998. Erratum in *Journal of Cerebral Blood Flow and Metabolism*, 18 (10), preceding 19047, 1998.
83. Barber, P.A. et al., Absent middle cerebral artery flow predicts the presence and evolution of the ischemic penumbra, *Neurology*, 52 (6), 1125, 1999.
84. Nabavi, D.G. et al., Perfusion mapping using computed tomography allows accurate prediction of cerebral infarction in experimental brain ischemia, *Stroke,* 32 (1), 175, 2001.
85. Portera-Cailliau, C., Price, D.L., and Martin, L.J., N-methyl-D-aspartate receptor proteins NR2A and NR2B are differentially distributed in the developing rat central nervous system as revealed by subunit-specific antibodies, *Journal of Neurochemistry*, 66 (2), 692, 1996.
86. Bartus, R.T. et al., Calpain inhibitor AK295 protects neurons from focal brain ischemia: effects of post-occlusion intra-arterial administration, *Stroke,* 25 (11), 2265, 1994.
87. Bartus, R.T. et al., Postischemic administration of AK275, a calpain inhibitor, provides substantial protection against focal ischemic brain damage, *Journal of Cerebral Blood Flow and Metabolism*, 14 (4), 537, 1994.
88. Markgraf, C.G. et al., Six-hour window of opportunity for calpain inhibition in focal cerebral ischemia in rats, *Stroke,* 29 (1), 152, 1998.
89. Szabo, C. and Dawson, V.L., Role of poly(ADP-ribose) synthetase in inflammation and ischaemia-reperfusion, *Trends in Pharmacological Sciences*, 19 (7), 287, 1998.
90. Goto, K. et al., Effects of cycloheximide on delayed neuronal death in rat hippocampus, *Brain Research*, 534 (1–2), 299, 1990.
91. Shigeno, T. et al., Reduction of delayed neuronal death by inhibition of protein synthesis, *Neuroscience Letter*, 120 (1), 117, 1990.
92. Robertson, G.S. et al., Neuroprotection by the inhibition of apoptosis, *Brain Pathology*, 10 (2), 283, 2000.
93. Endres, M. et al., Attenuation of delayed neuronal death after mild focal ischemia in mice by inhibition of the caspase family, *Journal of Cerebral Blood Flow and Metabolism*, 18 (3), 238, 1998.
94. Schotte, P. et al., Non–specific effects of methyl ketone peptide inhibitors of caspases, *FEBS Letters*, 442 (1), 117, 1999.
95. Du, C. et al., Additive neuroprotective effects of dextrorphan and cycloheximide in rats subjected to transient focal cerebral ischemia, *Brain Research*, 718 (1–2), 233, 1996.
96. Ma, J., Endres, M., and Moskowitz, M.A., Synergistic effects of caspase inhibitors and MK-801 in brain injury after transient focal cerebral ischaemia in mice, *British Journal of Pharmacology*, 124 (4), 756, 1998.
97. Choi, D.W. and Koh, J.Y., Zinc and brain injury, *Annual Review of Neuroscience*, 21, 347, 1998.

98. Waxman, S.G. et al., Anoxic injury of rat optic nerve: ultrastructural evidence for coupling between Na^+ influx and Ca^{2+}-mediated injury in myelinated CNS axons, *Brain Research*, 644 (2), 197, 1994.
99. Zhang, R.L. et al., Anti-ICAM-1 antibody reduces ischemic cell damage after transient middle cerebral artery occlusion in the rat, *Neurology*, 44 (9), 1747, 1994.
100. Endres, M. et al., Stroke protection by 3-hydroxy-3-methylglutaryl (HMG)-CoA reductase inhibitors mediated by endothelial nitric oxide synthase, *Proceedings of the National Academy of Sciences of the United States of America*, 95 (15), 8880, 1998.
101. Toung, T.J., Traystman, R.J., and Hurn, P.D., Estrogen-mediated neuroprotection after experimental stroke in male rats, *Stroke*, 29 (8), 1666, 1998.
102. De Keyser, J., Sulter, G., and Luiten, P.G., Clinical trials with neuroprotective drugs in acute ischaemic stroke: are we doing the right thing? [comment], *Trends in Neurosciences*, 22 (12), 535, 1999.
103. Huang, J. et al., Neuronal protection in stroke by an sLex-glycosylated complement inhibitory protein, *Science*, 285 (5427), 595, 1999.
104. Kaczmarek, L.K., The role of protein kinase C in the regulation of ion channels and neurotransmitter release, *Trends in Neurosciences*, 10, 30, 1987.
105. Hara, H. et al., Staurosporine, a novel protein kinase C inhibitor, prevents postischemic neuronal damage in the gerbil and rat, *Journal of Cerebral Blood Flow and Metabolism*, 10 (5), 646, 1990.
106. Nakane, H. et al., Protein kinase C modulates ischemia-induced amino acids release in the striatum of hypertensive rats, *Brain Research*, 782 (1–2), 290, 1998.
107. Sugino, T., Nozaki, K., and Hashimoto, N., Activation of mitogen-activated protein kinases in gerbil hippocampus with ischemic tolerance induced by 3-nitropropionic acid, *Neuroscience Letters*, 278 (1–2), 101, 2000.
108. Alessandrini, A. et al., MEK1 protein kinase inhibition protects against damage resulting from focal cerebral ischemia, *Proceedings of the National Academy of Sciences of the United States of America*, 96 (22), 12866, 1999.
109. Bigelow, W.G., Lindsay, W.K., and Greenwood, W.F., Hypothermia: its possible role in cardiac surgery, *Annals of Surgery*, 132, 849, 1950.
110. Reith, J. et al., Body temperature in acute stroke: relation to stroke severity, infarct size, mortality, and outcome [comment], *Lancet*, 347 (8999), 422, 1996.
111. Jorgensen, H.S. et al., What determines good recovery in patients with the most severe strokes? The Copenhagen Stroke Study, *Stroke*, 30 (10), 2008, 1999.
112. Clifton, G.L. et al., Systemic hypothermia in treatment of brain injury, *Journal of Neurotrauma*, 9 (Suppl. 2), S487, 1992.
113. Bernard, S.A. and Buist, M., Induced hypothermia in critical care medicine: a review, *Critical Care Medicine*, 31 (7), 2041, 2003.
114. Marion, D.W. et al., The use of moderate therapeutic hypothermia for patients with severe head injuries: a preliminary report [comment], *Journal of Neurosurgery*, 79 (3), 354, 1993.
115. Piepgras, A. et al., Rapid active internal core cooling for induction of moderate hypothermia in head injury by use of an extracorporeal heat exchanger, *Neurosurgery*, 42 (2), 311, 1998.
116. Sessler, D.I., Mild perioperative hypothermia, *New England Journal of Medicine*, 336 (24), 1730, 1997.
117. Colbourne, F. and Corbett, D., Delayed and prolonged post-ischemic hypothermia is neuroprotective in the gerbil, *Brain Research*, 654 (2), 265, 1994.

118. Colbourne, F. and Corbett, D., Delayed postischemic hypothermia: a six-month survival study using behavioral and histological assessments of neuroprotection, *Journal of Neuroscience*, 15 (11), 7250, 1995.
119. Colbourne, F., Sutherland, G., and Corbett, D., Postischemic hypothermia: a critical appraisal with implications for clinical treatment, *Molecular Neurobiology*, 14 (3), 171, 1997.
120. Corbett, D. and Thornhill, J., Temperature modulation (hypothermic and hyperthermic conditions) and its influence on histological and behavioral outcomes following cerebral ischemia, *Brain Pathology*, 10 (1), 145, 2000.
121. Schwab, S. et al., Early hemicraniectomy in patients with complete middle cerebral artery infarction, *Stroke,* 29 (9), 1888, 1998.
122. Schwab, S. et al., Feasibility and safety of moderate hypothermia after massive hemispheric infarction [comment], *Stroke,* 32 (9), 2033, 2001.
123. Busto, R. et al., Small differences in intraischemic brain temperature critically determine the extent of ischemic neuronal injury, *Journal of Cerebral Blood Flow and Metabolism*, 7 (6), 729, 1987.
124. Morikawa, E. et al., The significance of brain temperature in focal cerebral ischemia: histopathological consequences of middle cerebral artery occlusion in the rat, *Journal of Cerebral Blood Flow and Metabolism*, 12 (3), 380, 1992.
125. Hindfelt, B., The prognostic significance of subfebrility and fever in ischaemic cerebral infarction, *Acta Neurologica Scandinavica*, 53 (1), 72, 1976.
126. Castillo, J. et al., Timing for fever-related brain damage in acute ischemic stroke [comment], *Stroke,* 29 (12), 2455, 1998.
127. Memezawa, H. et al., Hyperthermia nullifies the ameliorating effect of dizocilpine maleate (MK-801) in focal cerebral ischemia, *Brain Research*, 670 (1), 48, 1995.
128. Bruno, A. et al., Acute blood glucose level and outcome from ischemic stroke: Trial of ORG 10172 in Acute Stroke Treatment (TOAST) Investigators, *Neurology*, 52 (2), 280, 1999.
129. Siesjo, B.K. et al., Molecular mechanisms of acidosis-mediated damage, *Acta Neurochirurgica Supplementum*, 66, 8, 1996.
130. Anderson, R.E. et al., Effects of glucose and PaO2 modulation on cortical intracellular acidosis, NADH redox state, and infarction in the ischemic penumbra, *Stroke,* 30 (1), 160, 1999.
131. Hoxworth, J.M. et al., Cerebral metabolic profile, selective neuron loss, and survival of acute and chronic hyperglycemic rats following cardiac arrest and resuscitation, *Brain Research*, 821 (2), 467, 1999.
132. Li, P.A. et al., Hyperglycemia enhances extracellular glutamate accumulation in rats subjected to forebrain ischemia [comment], *Stroke,* 31 (1), 183, 2000.
133. Kagansky, N., Levy, S., and Knobler, H., The role of hyperglycemia in acute stroke, *Archives of Neurology*, 58 (8), 1209, 2001.
134. Zhu, C.Z. and Auer, R.N., Intraventricular administration of insulin and IGF-1 in transient forebrain ischemia, *Journal of Cerebral Blood Flow and Metabolism*, 14 (2), 237, 1994.
135. Aichner, F., Adelwohrer, C., and Haring, H.P., Rehabilitation approaches to stroke, *Journal of Neural Transmission Supplementum*, (63), 59, 2002.
136. Feeney, D.M. and Westerberg, V.S., Norepinephrine and brain damage: alpha noradrenergic pharmacology alters functional recovery after cortical trauma, *Canadian Journal of Psychology*, 44 (2), 233, 1990.
137. Feeney, D.M. and Sutton, R.L., Pharmacotherapy for recovery of function after brain injury, *Critical Reviews in Neurobiology*, 3 (2), 135, 1987.

138. Kline, A.E. et al., Methylphenidate treatment following ablation-induced hemiplegia in rat: experience during drug action alters effects on recovery of function, *Pharmacology, Biochemistry and Behavior*, 48 (3), 773, 1994.
139. Goldstein, L.B., Potential effects of common drugs on stroke recovery, *Archives of Neurology*, 55 (4), 454, 1998.
140. Kleppner, S.R. et al., Transplanted human neurons derived from a teratocarcinoma cell line (NTera-2) mature, integrate, and survive for over one year in the nude mouse brain, *Journal of Comparative Neurology*, 357 (4), 618, 1995.
141. Borlongan, C.V. et al., Transplantation of cryopreserved human embryonal carcinoma-derived neurons (NT2N cells) promotes functional recovery in ischemic rats, *Experimental Neurology*, 149 (2), 310, 1998.
142. Kondziolka, D. et al., Transplantation of cultured human neuronal cells for patients with stroke [comment], *Neurology*, 55 (4), 565, 2000.
143. Johansson, B.B., Brain plasticity and stroke rehabilitation: the Willis lecture, *Stroke*, 31 (1), 223, 2000.
144. Goldstein, L.B., New approaches to poststroke rehabilitation, in *Pharmacology of Cerebral Ischemia*, Krieglstein, J. and Klumpp, S., Eds., MedPharm, Germany, 2002, p. 487.

5 Imaging and Functional Mapping of Local Circuits and Epilepsy

Kenneth M. Little and Michael M. Haglund

CONTENTS

5.1 INTRODUCTION

Optical imaging (OI) is a functional imaging method that measures changes in nervous tissue light reflectance or transmission. OI has been used to study brain functions in both normal and pathophysiological states. In addition to identifying brain function in a way that is not possible with single photon emission computed

0-8493-1482-8/05/$0.00+$1.50

tomography (SPECT), positron emission tomography (PET), and functional magnetic resonance imaging (fMRI), OI has provided a greater understanding of the physiologic mechanisms underlying these imaging methods. This chapter will present overviews of different imaging techniques, principles of optical imaging (see Figure 5.1), various studies performed at different levels of brain analysis, and finally, human open optical imaging studies.

5.2 OVERVIEW OF FUNCTIONAL IMAGING TECHNIQUES

Conventional anatomical imaging techniques such as x-ray radiography and computer assisted tomography (CAT) rely upon a photon source that is generated as electrons strike an anode (Bremstrahlung effect). The attenuation of these photons as they pass through tissue and strike silver nitrate film or a fluorescent screen reveals underlying anatomic structures within the interrogated tissue volume. To detect function rather than simply reflecting anatomic structures, functional imaging techniques in general rely upon similar principles and utilize various energy sources and detectors. Functional related changes in nervous tissue trigger water movement into or out of cells or alter the uptake of glucose or other tracers that are indicative of cellular metabolism.

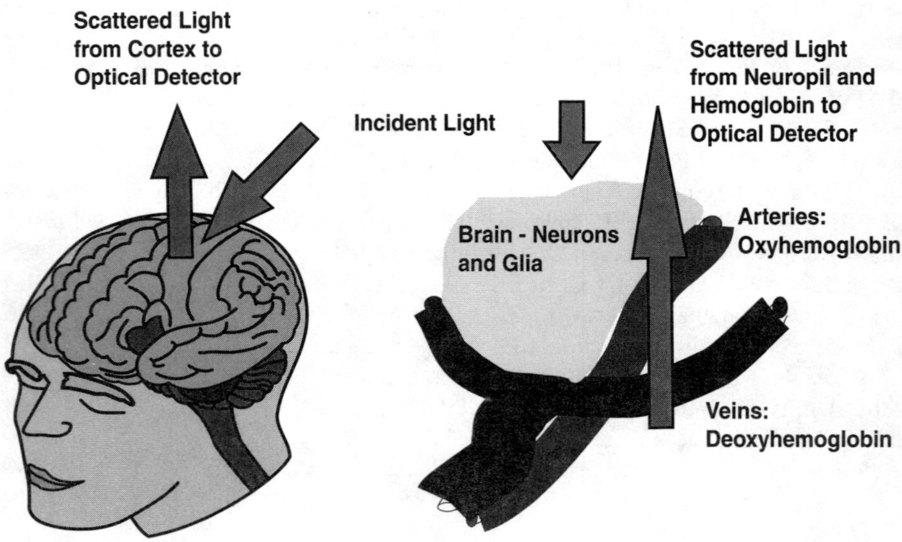

FIGURE 5.1 (See color insert following page 146.) Optical imaging at brain and neuropil levels. Two types of optical imaging approaches: at macro (whole brain) level (left) and at neuropil (micro) level (right). In intact brain, incident light is more or less scattered or absorbed by tissue and its contents, with the resulting scattered light detected by an optical system. At the micro level, scattering occurs in distinct elements that include the neuropil (that has its own intrinsic optical signal), and either oxyhemoglobin or deoxyhemoglobin within the blood vessels. Separate absorption and scattering characteristics appear, depending on the relative content of oxyhemoglobin and deoxyhemoglobin.

5.2.1 Functional MRI

Magnetic resonance imaging (MRI) relies upon an externally generated magnetic field gradient with local radiofrequency-induced disruption. The electromagnetic radiation emitted from the hydrogen dipoles as they reorient from high energy disorganized states to lower energy organized states within the gradient are measured by the detector. Functional MRI (fMRI) uses fast imaging techniques to indirectly detect active neuronal circuits based on relative increases in oxyhemoglobin. This physiological phenomenon results from a local increase in oxygenated blood delivery during neuronal activity. The local increase in oxygenated blood outstrips the generation of deoxyhemoglobin by active tissue, indicating a drop in oxygen extraction by the tissue. The relative decrease in deoxyhemoglobin is detected by blood oxygen level-dependent contrast magnetic resonance imaging (BOLD MRI) "downstream" from the metabolically active tissue.[1] An increasing amount of attention has focused on the initial negative "dip" in the BOLD MRI signal that may indicate transient relative tissue hypoxia before blood flow to the active tissue is increased.

Some studies have compared fMRI directly to cortical electrical stimulation mapping (ESM) performed via open craniotomy or grid stimulation for motor, somatosensory, and language mapping and have demonstrated a correlation between the two methods.[1–7] Significant discrepancies and sources of error, however, mitigate optimistic conclusions that these two modalities are highly correlative. For example, compared to sites identified by ESM, sites of increased activity on fMRI are considerably larger. The radial cortical projections of subsurface fMRI signals used to create functional cortical maps for computer-assisted surgical navigation may not correspond to cortical surface ESM-identified sites.[8] Additionally, due to brain shift during the craniotomy, precise fMRI localization may be prohibitively difficult.

The fMRI areas activated by motor tasks may identify nonessential motor cortex.[9] Differences in language tasks, imaging techniques, data analysis, and brain shift associated with craniotomy have made language mapping particularly difficult to corroborate with ESM results.[4,6,7,10,11] Language tasks that prove to be essential based on intraoperative ESM seem to be best activated on fMRI with semantic decision and verbal fluency tasks; multiple language tasks seem to provide greater sensitivity than any single task.[2,4,10,11] However, limiting fMRI language activation to only essential sites is heavily dependent upon the method of statistical data analysis.[7] According to available evidence, fMRI is subject to errors by (1) identifying areas that are not essential to neurological function, thus potentially limiting the resection unnecessarily; and (2) failing to identify areas that could cause postoperative deficits if resected. In its current state of development, fMRI should be used only as an adjunct to ESM for functional mapping.

Regarding its ability to predict postoperative deficits, one study of sensorimotor cortex in patients undergoing lesion resections demonstrated a correlation between the size of the margin between the lesion (not the resection margin) and area of fMRI activation and the presence of postoperative neurological deficits.[12] However, detailed analysis correlating postoperative deficits to the margin between the *resection cavity* and the area of functional activity is necessary.

Functional MRI has also been used to lateralize language function and several authors have compared fMRI directly to the Wada test (intra-arterial pentobarbital or IAP). Although many studies show promising results, no consensus has yet been reached about which language tasks best correlate with language measures used during IAP or methods of image acquisition and data analysis. Thus, fMRI may be limited in its predictive capability for postoperative deficits.

5.2.2 Positron Emission Tomography

Positron emission tomography (PET) scanning detects photons generated after nuclear decay from the annihilation of positrons with electrons. The photon detectors often consist of bismuth germinate or scintillators coupled to photomultiplier tubes to convert the photons into an electrical signal. PET can be used to detect neuronal activity based on metabolically dependent increased glucose utilization or associated increases in regional cerebral blood flow. The increase in metabolism is detected by fluoro-deoxyglucose (FDG-PET) and blood flow changes are detected by ^{15}O water PET.[13] Due to the limits of radiation dosing, the use of FDG-PET has been limited to mapping primary sensory and motor areas whereas ^{15}O water PET can be infused several times, making it suitable for mapping higher cognitive functions.[13–16]

PET scanning has undergone comparisons to ESM. As with fMRI, studies have shown that compared to essential language sites identified by ESM, sites of language-associated increased PET activity are considerably larger and they may identify language sites where ESM does not disrupt language. The PET-activated sites may be up to 1 cm from the site identified by ESM. Depending upon which tasks are combined to produce activation maps and the method of statistical analysis, PET may fail to identify essential language cortex. Although PET seems inaccurate and unreliable for language localization, it may be adequate for language lateralization. When PET was directly compared to IAP, a study demonstrated a positive predictive value for language lateralization in 80 to 91% of patients, depending on the method of image analysis.[16]

5.2.3 Single Photon Emission Computed Tomography

Single photon emission computed tomography (SPECT) is related to PET but does not require short-lived isotopes. It most often utilizes Technetium-99m (Tc-99m)-labeled pharmaceuticals. The radiolabeled drugs deposit in neural tissue where positrons are emitted. The emitted positrons release photons (gamma rays) as they annihilate with electrons. The gamma rays are detected by a scintillation counter rotating about the patient's head. SPECT scanning has been used to detect pathologic increases and decreases in regional blood flow. Compared to PET, the spatial resolution of SPECT is inferior, but it provides a convenient method of assessing regional cerebral perfusion. This estimate of regional perfusion is highly limited by the large voxel sizes and the static nature of the imaging process, so generally it has been limited to studies of brain death when it is used to detect an absence of cerebral blood flow.

5.2.4 Optical Imaging

Optical imaging (OI) is one of the most recently developed functional mapping techniques for the identification of epileptic foci and eloquent cortical regions.[17–20] Similar to most conventional imaging techniques, OI relies upon photons and their interactions with tissues to measure changes in a tissue's optical scattering and absorption from one physical state to another (see Figure 5.1). Unlike the higher energy photons generated by Bremstrahlung and positron annihilation, the photons used in OI are generated by light within the near-infrared (low frequency) to ultraviolet range, and are therefore of considerably lower energy. They may be captured by various forms of optical detectors and lens systems including a video camera, charge-coupled device (CCD) camera, single photodiode, photodiode array, or the observer's retina.

When the light frequency is changed through filtering, specific physiological properties coupled to neuronal activity can be measured, including regional cerebral blood volume changes at the capillary and venous levels, blood oxygenation changes, cytochrome redox states, and cellular, extracellular, or organelle swelling due to ion gradient changes. These properties can be measured across a broad range of spatial resolution from microscopic neuronal populations to macroscopic cortical regions, depending on the type of microscope and lens system used. In addition to excellent spatial resolution versatility, the temporal resolution is superior to fMRI and PET scanning, making it ideal for imaging epileptiform and functional activity. At near-infrared wavelengths, this technique allows for the added benefit of noninvasive monitoring.[18–21]

5.3 PRINCIPLES OF OPTICAL IMAGING

Optical imaging encompasses several subsets of analysis, including direct reflectance or transmittance imaging (at the same wavelength of light, to detect scattering or absorption), fluorescence imaging, and near-infrared imaging of hemoglobin. Additionally, the scale of analysis varies considerably, from single cell resolution to imaging larger regions of brain for functional activation.

5.3.1 Optical Properties of Neuronal Tissue

Electromagnetic radiation (i.e., photons) interacts with targeted tissue substrates in different ways. The principal interactions between photons and neural tissue relevant to optical imaging are *scattering* and *absorption*.

Scattering: Light scattering in neural tissue is a combination of phenomena. *Rayleigh scattering* is a scattering of light by objects that are small in comparison to the incident light wavelength; the scattering flux density is proportional to the fourth power of the incident light frequency. Thus, high frequency, short wavelength light at the blue end of the visible spectrum is scattered about 10 times more intensely than low frequency, long wavelength light at the red end of the spectrum. This makes near-infrared light more suitable for noninvasive optical imaging through skin and bone, compared to higher frequency light, as less energy is lost to scattering.

When light travels from one medium to another, a portion of the light will be *reflected* away and part will pass through at a *refracted* vector, as described by Snell's law. This submicroscopic process is due to the compositions and densities of scattering molecules at, for example, lipid membrane–cytoplasm interfaces where light reflection and refraction are altered by physiological processes that change the chemical composition or geometry of the interface. Scattering also occurs intracellularly at the organelle/cytoplasm boundaries (mitochondria, nuclei, etc.).

Light may be reflected or refracted from a moving particle (e.g., erythrocytes, albumin) and undergo a *frequency shift* (Doppler shift). By focusing light on blood vessels, Doppler shifts may be used to calculate changes in blood flow. When the incident light is polarized, intracellular, membrane, and extracellular macromolecules may exhibit different refractive indices depending upon the orientation of polarized light. This effect is referred to as *birefringence*. Birefringence can provide information about intermolecular associations within cell membranes and nerve fibers. In an electric field, molecules may become birefringent. This phenomenon, known as the "Kerr effect," can be seen during changes in transmembrane electric potential gradients. For example, axonal membrane molecules with dipole moments can change orientation in a highly ordered fashion during depolarization. This reorientation results in transient birefringence until electric potential gradients are reestablished.

Absorption: Light entering tissue may be either reflected or may pass through after refraction. However, some of the light entering will not pass through the tissue and is therefore said to be *absorbed*. After the energy is transferred to other molecules, the excited recipient molecules eventually return to their more stable ground states. Because energy is conserved, the energy from the absorbed light must be converted and may be dissipated as *fluorescent emission*, *thermal energy*, or *phosphorescence*. For example, the reduced form of nicotinamide adenine dinucleotide (NADH) exhibits intrinsic fluorescence whereas NAD^+ does not. Investigators have taken advantage of this difference to use fluorescence as a sensitive measure of intracellular oxidation states.

In addition to the light scattering and absorption properties intrinsic to neural tissue, photons may interact with a variety of structurally or functionally partitioned dyes to produce phosphorescence or fluorescence. While this can augment imaging based on a tissue's intrinsic optical properties, the advantage of imaging the intrinsic optical signal alone is that it obviates the need for administrating optically active chemicals. A tissue's optical properties in the absence of optical dyes are referred to collectively as intrinsic optical properties. The resulting signal is referred to as the intrinsic optical signal (IOS).

5.3.2 Physiologic Processes Underlying the Intrinsic Optical Signal

Several physiologic processes have been shown to alter the intrinsic optical properties of neural tissue and may be divided into *blood-independent* and *blood-dependent* events or processes.

Blood-independent processes: A blood-independent IOS is generated primarily by cellular processes (the interface between outer membrane and extracellular space) and intracellular interfaces. Cellular events contributing to the IOS have been investigated in neurons and axons in isolation and in brain slices, free from the influences of blood volume and blood oxygenation changes.[22,23] Evidence suggests that neurons and their subcellular organelles including the nuclei and mitochondria swell to varying degrees at different levels of brain activity or injury.[23]

Cellular and organelle swelling in response to transmembrane ion gradient changes may lead to intrinsic optical changes in more than one way. A decrease in the concentration of light-scattering particles results in decreased light scattering and increased light transmittance. Changes in the cellular membrane geometries during swelling lead to changes in light reflectance relative to the angle of the recording device (i.e., CCD camera). Finally, changes in the refractive indices across the membranes that accompany changes in concentration gradients lead to changes in light refraction. Non-neuronal cells in brain tissue also show IOS changes. For example, MacVicar et al. (2002) demonstrated that IOS changes in glia during stimulation are also in part explained by NA/K-2CL cotransporter-induced astrocyte swelling.[24]

Intracellular events other than those attributable to volume changes also contribute to the IOS. Cytochrome C oxidase (Cyt-Ox) is the terminal electron acceptor of the mitochondrial electron transport chain and the energy generated in this process is used to synthesize adenosine triphosphate (ATP). Early studies of Cyt-Ox and NADH redox states in isolated mitochondria demonstrated transient oxidation during increased cellular activity.[25] Using a modified Lambert–Beer law, the light absorption changes accompanying changing Cyt-Ox redox states may be used *in vitro and in vivo* to measure transient cellular energy metabolism changes associated with brain activity.[26]

Blood-dependent processes: Optical imaging *in vivo* adds the additional dimension of vascular changes, making the interpretation of IOS changes more complex. Frostig et al. postulated that activity-related vascular IOS changes represent changes in blood oxygenation and volume.[27] Similar to Cyt-Ox, oxygenated hemoglobin (Oxy-Hb) and deoxygenated hemoglobin (Deoxy-Hb) have characteristic absorption spectra and, based on a modified Lambert–Beer law, changes in their concentrations contribute to the IOS. Blood volume changes have been postulated as contributing to the IOS in that changes in total hemoglobin (Oxy-Hb + Deoxy-Hb) reflect changes in corpuscular blood volume.[27,28]

Finally, changes in a tissue's optical parameters occur with differing latencies. A fast component (onset of 2 to 3 seconds) correlates with neuronal membrane electrical potential changes and a slower component (onset of 3 to 6 seconds and resolving about 20 seconds after the stimulus is removed) that may be associated with cellular and organelle volume changes.[28]

5.4 OPTICAL IMAGING METHODS

Since their initial use by Hill and Keynes, several investigators have developed OI methods for *in vitro* and *in vivo* animal models. Over the past decades, further technical advances have led to OI use in humans. Current methods include specific techniques for brain slices, open cortical mapping, stereotactic surgery, and transcranial imaging. The following is a brief overview.

5.4.1 Brain Slices

Lipton was the first to investigate the effects of membrane depolarization on light scattering in cerebral cortex slices.[29] Lipton observed that optical reflectance increased when the superfusate osmolarity was increased, and reflectance decreased when the osmolarity decreased below baseline conditions. Assuming that cell volumes increase with decreasing extracellular osmolarity, it was concluded that cell volume changes were inversely related to reflectance changes. Electrical stimulation across the tissue or exposure to high potassium concentrations caused decreases in reflectance, indicating that stimulation also led to increases in cell volumes.

MacVicar and Hochman were the first to apply digital imaging methods to obtain high-resolution synaptic-evoked changes in light transmission through hippocampal brain slices.[30] Specifically, pyramidal CA1 neurons were imaged concurrently with microelectrode recordings during Schaffer collateral bipolar stimulation in the CA3 region. The authors used this method to conduct a sequence of experiments designed to determine the physiological mechanisms underlying optical changes. Aitken et al.[31] identified four subtypes of physiologic processes leading to IOS responses consisting of:

1. Synaptic activation
2. Hypoxia
3. Spreading depression in the presence of normoxia or hypoxia
4. Extracellular osmolarity changes

In addition to studying normal physiology, brain slice OI holds promise in uncovering mechanisms underlying epileptiform activity. Hochman et al. demonstrated that optical changes are associated with epileptiform burst discharges.[32] Using furosemide, the authors were able to block the optical changes and epileptiform activity without blocking synaptic transmission or reducing the hyperexcitable response to electrical stimulation. It may be inferred that optical changes are more closely related to epileptiform activity through synaptic hypersynchronization, rather than hyperexcitability. Also, the intrinsic optical signal changes seen in brain slice preparations may reflect mechanisms critical for the generation of synchronized activity (i.e., seizure activity). This last hypothesis suggests the possibility of applying brain slice OI to screening anti-epileptic drugs.[19]

5.4.2 Open Brain Mapping

The open OI mapping technique became feasible when Blasdel and Salama employed a television camera (120 × 120) to improve the spatial resolution previously achieved with standard photodiode arrays (12 × 12).[33] Blasdel went on to apply OI for *in vivo* functional mapping by augmenting the IOS with voltage-sensitive dyes to image ocular dominance columns and orientation preferences in nonhuman primate visual cortex. Grinvald, Frostig, Ts'o, Lieke, and colleagues were later able to identify similar functional regions in primate visual cortex without using voltage-sensitive dyes by directly imaging the IOS changes associated with cortical surface optical reflectance.[27,34–36] Haglund and colleagues were the first to employ OI intraoperatively in humans. They obtained OI maps of somatosensory, motor, and language cortex in patients undergoing awake craniotomies for intractable epilepsy.[37] Subsequently, Haglund and a handful of investigators employed OI to study sensorimotor,[37–40] language,[37,41,42] and cortical regions subserving higher cognitive functions such as face matching and short-term memory.[40]

Optical imaging represents a significant breakthrough for the study of functional and epileptiform cortical activation. The author's current intraoperative setup allows investigation of microscopic neuronal populations with a spatial resolution of about 60 μm to cortical regions as large as 5 × 5 cm. We image with a temporal resolution of 200 milliseconds. These benefits allow greater accuracy in intraoperatively delineating Rolandic and language cortex, identifying interictal epileptiform discharges, pinpointing the onsets of ictal events with precise localization, and directly observing the pathways by which seizure activity spreads. Because OI relies upon physiological cortical activation rather than direct stimulation from external electrical currents, it can facilitate and enhance the intraoperative identification of cortical regions subserving cognitive functions.

Several sources of artifacts can make the IOS difficult to discern; successful intraoperative OI requires minimizing patient movement, dampening physiologic brain pulsations, and uniform cortical surface illumination. One of the most critical strategies during OI is to minimize movement. Brain pulsations associated with hemodynamic and respiratory patterns cause spatial shifts during image acquisition making frame-to-frame IOS analysis difficult.[38] This artifact can be overcome during image acquisition with mechanical dampening and during image analysis with image "warping" algorithms. Mechanical dampening is achieved by placing a glass plate (4, 9, 16, or 25 cm^2) over the cortical surface in the area of interest. The glass plate is mounted to an adjustable mechanical arm mounted to a skull clamp. This rigid construct has become particularly important during awake craniotomies and during imaging of seizure activity when image acquisition is continuous over 1 or 2 minutes.

The brain surface is uniformly illuminated using a stable tungsten halogen light source. The incident light is filtered to the desired wavelength (typically using a 695-nm long-pass filter) with the operating theater darkened to minimize artifacts from ambient light sources. By selecting different wavelengths of light through filtering, differential aspects of the IOS, and therefore specific physiological processes, may be emphasized. For example, imaging through a 610- or 695-nm filter, as reported in human studies, emphasizes changes in hemoglobin oxygenation.[38,39,41–43]

A CCD camera is mounted on the operating microscope. To further minimize movement artifacts, the microscope is mounted to the operating table by a modified microscope base. Initially, a low-power objective is selected to allow for visualization of a relatively large cortical area (25 cm^2). During seizure focus localization or functional mapping, images are collected at a rate of about 50 Hz over a period of 1 minute. Using software developed by Daryl Hochman, image analysis can then be performed intraoperatively within 2 minutes.

Images are analyzed by subtracting a baseline image (i.e., prior to cortical activation) from all subsequent images, yielding data that reflects changes in the IOS from baseline. During analysis, statistical algorithms are applied to align successive images in order to compensate for residual microscopic movement artifacts. This is particularly important when images are acquired through a high-power objective where small movements are magnified. Each series is viewed intraoperatively to evaluate epileptiform or functional activity. Despite efforts to minimize movement, ambient light, and thermal artifacts, a small amount of noise in the processed images is difficult to avoid. While open OI has been successfully performed in humans, it remains a research tool and will require further reliability testing and technical modifications before it can become feasible for routine clinical use.

5.4.3 Stereotactic Approaches

In 2000, Giller et al. introduced an optical imaging method to aid the identification of border zones between deep nuclei and their surrounding white matter tracts during stereotactic pallidotomy, thalamotomy, and placement of deep brain stimulators.[44] The stereotactically implanted fiberoptic probe consisted of a central light-delivering fiber (tungsten light source) surrounded by six light returning fibers. Light emitted from the probe was scattered by the surrounding brain tissue (target tissue morphology and volume estimated as a π sphere with a 100-micrometer diameter) and delivered to a spectrometer through the six light detecting fibers.

Their earlier investigations utilized light in the near-infrared range between 500 and 1000 nm. The range was later changed to 350 to 850 nm to match hemoglobin's absorption peaks.[45] With each incremental (1 mm) advance of the probe, reflectance was recorded and plotted with respect to wavelength. The normalized data obtained from each resulting graph were validated using postoperative MRI or CT scans merged with preoperative MRIs to assess the probe's trajectory and the structures it passed through at each depth.

The measured slopes obtained from reflectance wavelength plots were significantly greater in white matter (mean of 2.5) compared to those of gray matter (mean of 0.82). The authors demonstrated that stereotactic near-infrared imaging could detect subcortical white matter–gray matter interfaces during stereotactic localization of deep brain nuclei. This may prove a reliable and technically simple alternative to currently used localization methods, including microelectrode recording.

Optical coherence tomography (OCT) was developed to provide high-resolution tomographic images of the retina and anterior eye.[46] This technique has been used in combination with endovascular catheters and endoscopes to image internal organ systems, including cardiac vessel lumina, gastrointestinal lumina, and genitourinary lumina. In principle, it is similar to Giller's fiberoptic probe, with the exception that the tip of the probe rotates at various frequencies through 360 degrees to provide a cross-sectional view. More recently, its utility in detecting functional activity by measuring light scattering changes during propagation of action potentials has been demonstrated in the sea slug abdominal ganglion. With stereotactic techniques, it may be possible to adapt OCT to study functional activity in deep brain nuclei.

5.4.4 Transcranial Techniques

Near-infrared spectroscopy (NIRS) provides a less invasive alternative to the optical imaging methods described earlier. In 1977, Jöbsis was the first to demonstrate the feasibility of transcranial cortical tissue spectroscopy.[47] Unlike open optical imaging that detects light absorption and scattering changes, NIRS makes the assumption that light scattering is relatively constant and therefore relies on light absorption changes within a range of 650 to 950 nm. Most currently used NIRS systems monitor absorption changes associated with changing hemoglobin oxygenation states during cortical stimulation.[28,48]

Two wavelengths of light are used to differentiate changes in Oxy-Hb and Deoxy-Hb. Most imaging systems consist of 20 to 30 source-detector pairs. Since each detector may receive light from multiple sources, light sources are either sequentially switched on and off at high frequency or the incident light from each source is frequency encoded. Each source-detector pair defines a pixel and, through interpolation algorithms, the pixels are smoothed to form a coarse image. To date, most NIRS studies have involved cognitive tasks including different language paradigms. NIRS studies of primary motor, somatosensory, and visual areas have shown that the technique is a feasible alternative to invasive open imaging and other functional imaging techniques. Seizure activity has also been investigated with the NIRS technique in comparison to SPECT/EEG localization and magnetic encephalography (MEG) focus.

5.5 RESULTS OF OPEN HUMAN OPTICAL IMAGING STUDIES

5.5.1 Somatosensory Cortex

The ability to map somatosensory, motor, and language cortex using OI was first demonstrated by Haglund et al. in patients undergoing awake craniotomies for intractable epilepsy.[37] Initially, tongue and palate sensory areas were identified with intraoperative ESM by evoking subjective tingling in those areas. Patients were then instructed to move their tongues from side to side within their closed mouths. During three trials with OI, tongue movements produced the greatest IOS changes within the tongue and palate somatosensory areas as identified by ESM. These IOS changes were similar to those associated with cortical activation after bipolar stimulation, indicating that they reflected somatosensory cortical activation most likely from sensory feedback associated with tongue movements. Motor cortex associated with face movements (as identified by ESM) demonstrated IOS changes in the negative direction during tongue movement. Similar shifts in the IOS of motor cortex were observed during overt speech.[40]

It is tempting to suggest that these negative IOS changes represent decreased neuronal activity in face primary motor cortex during these simple movements. An alternative explanation is that the increased blood flow associated with somatosensory activation caused a shunting of blood flow away from primary motor cortex. We are currently investigating the relationship of blood volume and electrophysiological changes to determine which mechanism underlies this phenomenon.

Others have corroborated the IOS changes observed with somatosensory activation.[38,39] Cannestra et al. elicited somatosensory cortical activation with median nerve transcutaneous stimulation or 110-Hz finger vibration.[39] A close spatial correlation between cortical evoked potentials and IOS changes was noted. Similarly, Shoham and Grinvald elicited somatosensory cortical activation with electrical and tactile peripheral stimulation in 15 patients undergoing brain tumor or AVM resections under general anesthesia.[38] Optical imaging was accompanied by surface evoked potential recording. Due to the presence of optical signal artifacts, they were unable to draw definitive conclusions. However, they were able to obtain reproducible high-resolution somatosensory IOS maps from the hand area in nonhuman primates. The observed IOS changes associated with peripheral tactile stimulation correlated closely with single and multiunit cortical recordings. These findings confirmed the association of positive IOS changes and somatosensory cortical activation.

5.5.2 Language Cortex

Intraoperative ESM under local anesthesia during object naming is a safe, effective way to identify essential language cortex, particularly with the use of modern intravenous propofol anesthesia and local scalp anesthetic block.[49] Stimulation mapping using other, infrequently tested language-related measures such as naming in another language (including American Sign Language), sentence reading, or recent verbal memory have demonstrated dissociation in their cortical representation[50–52] and, under some circumstances, localizing and sparing these other language-related sites are important in avoiding postoperative deficits.[52]

However, mapping many different language functions, particularly when recent memory is included, is quite lengthy. Intraoperative OI may provide greater efficiency and detail during the functional localization of multiple cognitive and language functions.[37,40–42,53]

Haglund et al. performed OI in the inferior frontal language area (Broca's area) and somatosensory cortex of patients undergoing dominant hemisphere temporal lobe resections under local anesthesia.[37] Optical imaging was performed while patients silently viewed blank slides and named objects displayed on slides presented every 2 seconds.

Images obtained during naming showed activation of the premotor cortex, while the sites identified with ESM as demonstrating speech arrest and palate tingling yielded IOS changes in the opposite direction. The area that showed the greatest positive IOS changes during tongue movement was clearly different from the active area in the naming exercise. The premotor cortical areas from which IOS changes occurred during the naming exercise were similar to those identified on PET images obtained during single-word processing studies.[54,55] The IOS changes were greatest in the anatomical area of cortex classically defined as Broca's area (posterior portion of the inferior frontal gyrus) and not as expected in areas where electrical stimulation caused speech arrest.

Further topographical definition of Broca's area was demonstrated by Cannestra et al.[44] Broca's area was defined by ESM in five patients undergoing craniotomy under local anesthesia for the resection of brain tumors and vascular malformations. After identification of Broca's area, OI was performed during object naming (n = 5), word discrimination (n = 4), auditory responsive naming (n = 4), and orofacial movement (n = 3) tasks. Two distinct subregions (anterior and posterior) within Broca's area were identified. Both auditory and visual object naming paradigms were associated with increased IOS changes in both the anterior and posterior Broca's subregions. In contrast, word discrimination produced IOS changes only in the posterior subregion.

The authors concluded that this functional heterogeneity may represent subspecialized cortical networks within Broca's area, with anterior regions subserving semantic functions and posterior regions subserving phonological functions. Similar to the findings of Haglund et al., they noted incomplete agreement between ESM identified language sites and IOS changes because ESM and IOS changes overlapped only in the posterior subregion of the OI defined Broca's area.[37]

Optical imaging of posterior, peri-sylvian essential language sites (i.e., Wernicke's area) demonstrated findings similar to those in Broca's area. Haglund et al. demonstrated that in posterior temporal cortex, IOS changes during object naming originated from the general region where ESM elicited naming errors.[37] Similar to findings in Broca's area, the IOS changes covered a somewhat wider surrounding area compared to essential areas identified during ESM localization. All IOS changes were observed in areas near sites where ESM altered naming. The IOS changes appeared within 2 to 5 seconds of initiating naming and disappeared over a slightly longer time following the termination of naming.

Cannestra et al. demonstrated similar findings among six patients undergoing awake craniotomies for tumor or vascular lesion resection.[41] IOS changes were observed from all ESM-defined peri-sylvian language areas and from adjacent cortex. As in Broca's area, they were able to identify subregions subserving different functions. Object naming (n = 6) activated the central and anterior–inferior Wernicke subregions; whereas word discrimination (n = 5) preferentially activated the central and superior subregions. Auditory-responsive naming preferentially activated the central, anterior–inferior, and superior regions. Additional task-specific activations were observed in the inferior–posterior subregion.

Optical imaging of inferior frontal and posterior peri-sylvian language areas has consistently shown that IOS changes are more diffuse than ESM-identified regions. Cannestra et al. demonstrated that these surrounding regions may represent task-specific subregions.[41] This more diffuse cortical language representation identified by optical imaging may account for the occurrence of deficits following resection of cortex within 1 cm of ESM-identified essential language sites.[56] In one case, the temporal resection was performed very close to an ESM-identified posterior temporal essential language site while testing language, and stopped when naming errors occurred. As often occurs under similar circumstances, the patient's language returned to baseline soon after surgery. Interestingly, the resection extended to the margin of the region of IOS changes, suggesting that OI can provide the reliable localizing information needed to plan safe cortical resections.

5.5.3 Cognitive Function

In more than 20 patients undergoing temporal lobe resections for intractable epilepsy, we studied dominant and nondominant temporal lobe neocortical IOS activation associated with several cognitive tasks.[40] During dominant hemisphere resections, we found that IOS changes associated with short-term memory tasks localized to the posterior–superior temporal gyrus (STG). In these patients, IOS changes associated with object naming overlapped with ESM identified essential language sites.

IOS changes associated specifically with the memory task, however, were immediately anterior to the essential language site. Furthermore, activation with memory input occupied a discrete region that was immediately surrounded by positive IOS changes associated with memory retrieval. In a subset of patients undergoing nondominant hemisphere temporal lobe resections, we performed OI during face matching, complex figure matching, and facial expression interpretation tasks (paradigm described in detail by Ojemann et al).[57] We consistently identified negative IOS changes within the posterior MTG and STG during the tasks.

5.5.4 Epileptiform Activity

Optical imaging can be used intraoperatively to study seizure and interictal activity.[37,58] Prior to imaging, surface EEG is used to roughly localize foci of epileptiform activity. Once localized, the electrocorticogram (ECoG) electrodes are removed from the cortical surface and a glass plate is placed over the site of interest together with an array of recording electrodes about the periphery and a pair of centrally located stimulating electrodes. In addition to imaging spontaneous activity, evoked interictal and seizure activity can be generated through bipolar stimulation at currents above the afterdischarge (AD) potential threshold.

In five patients undergoing surgery for intractable epilepsy, Haglund et al. demonstrated that the IOS intensity, spread, and duration occurring during epileptiform activity evoked from bipolar stimulation correlated with the duration of electrical AD activity.[37] The stimulus was delivered via electrodes separated by 1 cm at an intensity just above the AD potential threshold and the IOS was compared to simultaneous surface EEG recordings. Each stimulation was followed by epileptiform AD activity characterized by varying degrees of intensity and duration.

The spatial spread of the IOS was greatest when associated with long durations of AD activity (12 to 16 seconds) and less when assocaited with short durations of AD activity (<4 seconds). The area of peak IOS intensity during the shorter seizure episode was more limited compared to the much greater spatial extent of IOS changes during the more intense seizure episode. Furthermore, the duration of IOS changes correlated with but lasted longer than the duration of electrical activity. In addition to a greater spatial extent and duration of IOS changes, longer seizure episodes were also associated with a greater magnitude (i.e., greater intensity) of IOS changes. Of interest, but still without a clear mechanism, are the negative IOS changes in the areas surrounding the focus of epileptiform discharges. More detailed studies are needed to determine whether these negative IOS changes represent surround inhibition, shunting of extracellular fluid, shunting of blood volume toward active cortex, or changes in blood oxygenation.

Further analysis, involving comparisons of IOS changes and surface EEG activity during different stages of seizure activity, reveals that the magnitude and direction of IOS changes appear to correlate with changes in electrical activity. IOS changes and surface electrode activity were measured simultaneously at baseline prior to stimulation, after stimulation during the seizure, during postseizure quiescence, and after return to baseline. During baseline activity, the region surrounding the recording electrode demonstrated neutral IOS whereas during the seizure episode this area was clearly activated in the positive direction. During the postseizure period when the electrical activity was quiescent compared to baseline, the area surrounding the recording electrode showed a negative IOS that gradually returned to near baseline. These preliminary observations pointed toward a correlation between the direction of IOS changes and electrical activity where positive IOS changes closely correlate with increases in electrical activity, and negative IOS changes correlate with below-baseline electrical activity.

5.6 UTILITY OF OPTICAL IMAGING

Optical imaging may become a reliable alternative to conventional mapping techniques (e.g., ESM, fMRI, and PET) and may provide a means to better understand the physiologic processes underlying these techniques. However, it requires full operative exposure of the brain. As described in previous sections, fMRI and PET have not yet proven to be reliable alternatives to ESM. The maps generated by OI, on the other hand, demonstrate better colocalization with ESM-generated functional maps compared to those determined by BOLD contrast.[59] Our initial experiences with OI of language and higher cognitive functions indicate that OI will become a valuable means of mapping and precisely pinpointing cortical representations of higher cognitive processes and assessing the temporo-spatial relationships associated with cortical processing during cognitive tasks. To date, mapping these functions with ESM has been difficult at best and is often limited to functional imaging methods often associated with localization errors.

Noninvasive OI (NIRS) is a promising technique, but several limitations must be overcome. For an excellent review of this subject see Obrig and Villringer.[28] In summary, the authors have identified limited spatial resolution, lack of depth resolution, interference artifacts from extracranial oxygenation changes and systemic hemodynamic changes, and lack of adequate statistical data analysis in the majority of published studies. The stereotactic OI method introduced by Giller et al. may ultimately provide a more convenient alternative to identifying deep nuclear structures based on microelectrode recordings. Optical coherence tomography may provide valuable insights into the functions of deep brain nuclei. However, their routine clinical use will have to await further clinical investigation.

Although OI is not yet ready for routine clinical use, it continues to provide insights into normal and pathological cortical function. Furthermore, it has provided insight into the meaning of fMRI BOLD contrast. For example, OI studies show that increases in Deoxy-Hb occur within 2 to 3 seconds after stimulus cessation and may represent the initial negative "dip" seen with decreased BOLD contrast during fMRI. Increases in Oxy-Hb are slower and likely correlate with increased BOLD contrast (decreased Deoxy-Hb). The early IOS changes seen with increased Deoxy-Hb (negative BOLD dip) may be temporally and spatially more localizing than the delayed IOS changes corresponding to the increased Oxy-Hb. Evidence suggests that IOS changes associated with increased blood volume in the vicinity of active neuronal tissue correlate well with stimulus-induced activation compared to IOS changes associated with increased Deoxy-Hb and BOLD contrast.

5.7 CONCLUSION

The clinical utility of OI ultimately depends on the continued development of noninvasive approaches, if at all possible, to avoid the current requirement for open brain exposure. As discussed earlier, noninvasive OI techniques have not yet achieved the specificity and reliability of alternative noninvasive techniques and several technical obstacles remain. Early experiences with intraoperative OI, on the other hand, have demonstrated a combination of spatial and temporal resolution that may be optimal for intraoperative functional mapping and seizure focus localization compared to standard techniques.

REFERENCES

1. Mueller, W.M., Yetkin, F.Z., and Haughton, V.M., Functional magnetic resonance imaging of the somatosensory cortex, *Neurosurg. Clin. N. Amer.*, 8, 373–381, 1997.
2. Binder, J., Functional magnetic resonance imaging language mapping, *Neurosurg. Clin. N. Amer.*, 8, 383–392, 1997.
3. Chapman, P.H., Buchbinder, B.R., Cosgrove, G.R., and Jiang, H.J., Functional magnetic resonance imaging for cortical mapping in pediatric neurosurgery, *Pediatr. Neurosurg.*, 23, 122–126, 1995.
4. FitzGerald, D.B. et al., Location of language in the cortex: a comparison between functional MR and electrocortical stimulation, *Am. J. Neuroradiol.*, 18, 1529–1539, 1997.
5. Jack, C.R. et al., Sensory motor cortex: correlation of presurgical mapping with functional MR imaging and invasive cortical mapping, *Radiology*, 190, 85–92, 1994.
6. Rutten, G.J.M., van Rijen, P.C., van Veelen, C.W.M., and Ramsey, N.F., Language area localization with three-dimensional functional magnetic resonance imaging matches intrasulcal electrostimulation in Broca's area, *Ann. Neurol.*, 46, 405–408, 1999.
7. Schlosser, M.J. et al., Comparative localization of auditory comprehension by using functional magnetic resonance imaging and cortical stimulation, *J. Neurosurg.*, 91, 626–635, 1999.
8. Krings, T. et al., Functional magnetic resonance imaging and transcranial magnetic stimulation: complementary approaches in the evaluation of cortical motor function, *Neurology*, 48, 1406–1416, 1997.
9. Macdonell, R.A.L. et al., Motor cortex localization using functional MRI and transcranial stimulation, *Neurology*, 53, 1462–1467, 1999.

10. Benson, R.R. et al., Language dominance determined by whole brain functional MRI in patients with brain lesions, *Neurology*, 52, 798–809, 1999.
11. Léhericy, S. et al., Functional MR evaluation of temporal and frontal language dominance compared with the Wada test, *Neurology*, 54, 1625–1633, 2000.
12. Yetkin, F.Z. et al., Functional MR activation correlation with intraoperative cortical mapping, *AJNR*, 18, 1311–1315, 1997.
13. Fox, P.T. et al., Mapping human visual cortex with positron emission tomography, *Nature*, 323, 806–809, 1986.
14. Fox, P.T. et al., Non-oxidative glucose consumption during focal physiological neuronal activity, *Science*, 241, 462–464, 1988.
15. Ginsburg, M. et al., Increases in both cerebral glucose utilization and blood flow during execution of a somatosensory task, *Ann. Neurol.*, 23, 152–160, 1988.
16. Hunter, K.E. et al., ^{15}O water positron emission tomography in language localization: a study comparing positron emission tomography visual and computerized region of interest analysis with the Wada test, *Ann. Neurol.*, 45, 662–665, 1999.
17. Gratton, C. et al., Shades of gray matter: noninvasive optical images of human brain responses during visual stimulation, *Psychophysiology*, 32, 505–509, 1995.
18. Hirth, C. et al., Non-invasive functional mapping of the human motor cortex using near-infrared spectroscopy, *Neuroreport*, 7, 1977–1981, 1996.
19. Hochman, D.W., Intrinsic optical changes in neuronal tissue: basic mechanisms, *Neurosurg. Clin. N. Amer.*, 8, 393–412, 1997.
20. Watanabe, E. et al., Noninvasive near infra-red spectroscopic topography in humans, *Neurosci. Lett.*, 205, 41–44, 1996.
21. Yamashita, Y. et al., Noninvasive near-infrared topography of human brain activity using intensity modulation spectroscopy, *Optical Eng.*, 35, 1046–1049, 1996.
22. Johnson, L.J., Hanley, D.F., and Thakor, N.V., Optical light scatter imaging of cellular and sub-cellular morphology changes in stressed rat hippocampal slices, *J. Neurosci. Methods,* 98, 21–31, 2000.
23. Fayuk, D., Aitken, P.G., Somjen, G.G., and Turner, D.A., The relationship between extracellular space and intrinsic optical signals in rat hippocampus *in vitro*: synaptic, spreading depression and osmotic-induced signals, *J. Neurophysiol.,* 87, 1924–1937, 2002.
24. MacVicar, B.A., Feighan, D., Brown, A., and Ransom, B., Intrinsic optical signals in the rat optic nerve: role for K(+) uptake via NKCCl and swelling of astrocytes, *Glia*, 37, 114–123, 2002.
25. Chance, B. and Williams, G.R., The respiratory chain and oxidative phosphorylation, *Adv. Enzymol.,* 17, 65–134, 1956.
26. Heekeren, J.R. et al., Noninvasive assessment of changes in cytochrome C oxidase oxidation in human subjects during visual stimulation, *J. Cereb. Blood Flow Metabol.*, 19, 592–603, 1999.
27. Frostig, R.D., Lieke, E.E., Ts'o, D.Y., and Grinvald, A., Cortical functional architecture and local coupling between neuronal activity and the microcirculation revealed by *in vivo* high resolution optical imaging of intrinsic signals, *PNAS*, 87, 6082–6086, 1990.
28. Obrig, H. and Villringer, A., Beyond the visible: imaging the human brain with light, *J. Cereb. Blood Flow Metabol.*, 23, 1–18, 2003.
29. Lipton, P., Effects of membrane depolarization on light scattering by cerebral cortical slices, *J. Physiol. (Lond.),* 231, 365–383, 1973.
30. MacVicar, B.A. and Hochman, D., Imaging of synaptically evoked intrinsic optical signals in hippocampal slices, *J. Neurosci.,* 11, 1458–1469, 1991.

31. Aitken, P.G., Fayuk, D., Somjen, G.G., and Turner, D.A., Use of intrinsic optical signals to monitor physiological changes in brain tissue slices, *Methods*, 18, 91–103, 1999.
32. Hochman, D.W. et al., Furosemide blockade of epileptiform activity dissociates synchronization from hyperexcitability, *Science*, 270, 99–102, 1995.
33. Blasdel, G.G. and Salama, G., Voltage-sensitive dyes reveal a modular organization in monkey striate cortex, *Nature*, 321, 579–585, 1986.
34. Grinvald, A., Manker, A., and Segal, M., Visualization of the spread of electrical activity in rat hippocampal slices by voltage-sensitive optical probes, *J. Physiol.*, 333, 269–291, 1982.
35. Grinvald, A. et al., Functional architecture of cortex revealed by optical imaging of intrinsic signals, *Nature*, 324, 361–364, 1986.
36. Ts'o, D.Y. et al., Functional organization of primate visual cortex revealed by high resolution optical imaging, *Science*, 249, 417–420, 1990.
37. Haglund, M.M., Ojemann, G.A., and Hochman, D.W., Optical imaging of epileptiform and functional activity from human cortex, *Nature*, 358, 668–671, 1992.
38. Shoham, D. and Grinvald, A., The cortical representation of the hand in Macaque and human area S-1: high resolution optical imaging, *J. Neurosci.,* 21, 6820–6835, 2001.
39. Cannestra, A.F. et al., Temporal spatial differences observed by functional MRI and human intraoperative optical imaging, *Cerebral Cortex,* 11, 773–782, 2001.
40. Hochman, D.W., Ojemann, G.A., and Haglund, M.M., Optical imaging reveals alternating positive and negative changes during cognitive or sensory evoked cortical activity in awake humans, *Soc. Neurosci.*, 20, 5, 1994.
41. Cannestra, A.F. et al., Temporal and topographical characterization of language cortices using intraoperative optical intrinsic signals, *Neuroimage*, 12, 41–54, 2000.
42. Pouratian, N. et al., Optical imaging of bilingual cortical representations, *J. Neurosurg.,* 93, 676–681, 2000.
43. Haglund, M.M., Berger, M.S., and Hochman, D.W., Enhanced optical imaging of human gliomas and tumor margins, *Neurosurgery,* 38, 308–316, 1996.
44. Giller, C.A., Johns, M., and Liu, H., Use of an intracranial near-infrared probe for localization during stereotactic surgery for movement disorders: technical note, *J. Neurosurg.,* 93, 498–505, 2000.
45. Johns, M., Giller, C.A., and Liu, H., Computational and *in vivo* investigation of optical reflectance from human brain to assist neurosurgery, *J. Biomed. Optics*, 3, 437–445, 1998.
46. Fujimoto, J.G. et al., New technology for high-speed and high-resolution optical coherence tomography, *Ann. NY Acad. Sci.*, 838, 95–107, 1998.
47. Jöbsis, F.F., Noninvasive, infrared monitoring of cerebral and myocardial oxygen sufficiency and circulatory parameters, *Science*, 198, 1264–1267, 1977.
48. Williams, I.M., Mortimer, A.J., and McCollum, C.N., Recent developments in cerebral monitoring: near-infrared light spectroscopy, *Eur. J. Vasc. Endovasc. Surg.*, 12, 263–271, 1996.
49. Silbergeld, D.L. et al., Use of propofol (Diprivan) for awake craniotomies: technical note, *Surg. Neurol.,* 4, 271, 1992.
50. Ojemann, G. and Whitaker, H., Language localization and variability, *Brain Lang.*, 6, 239–260, 1978.
51. Ojemann, G.A., Brain organization for language from the perspective of electrical stimulation mapping, *Behav. Brain Sci.*, 6, 189–206, 1983.

52. Ojemann, G.A. and Dodrill, C.G., Intraoperative techniques for reducing language and memory deficits with left temporal lobectomy, *Adv. Epileptol.*, 16, 327–330, 1987.
53. Cannestra, A.F. et al., The characterization of language cortices utilizing intraoperative optical intrinsic signals, *Neuroimage*, 7, 52, 1998.
54. Petersen, S.E. et al., Positron emission tomographic studies of the cortical anatomy of single word processing, *Nature*, 331, 585–589, 1988.
55. Frith, C.D., Friston, K.J., Liddle, P.F., and Frackowiak, R.S., A PET study of word finding, *J. Neuropsychol.,* 29, 1137–1148, 1991.
56. Haglund, M.M. et al., Cortical localization of temporal lobe language sites in patients with gliomas, *Neurosurgery*, 34, 567–576, 1994.
57. Ojemann, J.G., Ojemann, G.A., and Lettich, E., Neuronal activity related to faces and matching in human right nondominant temporal cortex, *Brain,* 115, 1–13, 1992.
59. Schwartz, T.H. and Bonohoeffer, T., *In vivo* optical imaging of epileptic foci and surround inhibition in ferret cerebral cortex, *Nature Medicine*, 7, 1065–1067, 2001.
59. Pouratian, N. et al., Spatial/temporal correlation of BOLD and optical intrinsic signals in humans, *Magn. Reson. Med.*, 47, 766–776, 2002.

6 Pre-ictal Seizure Detection and Demand Treatment Strategies for Epilepsy

Dennis A. Turner, Miguel A.L. Nicolelis, and Kevan Van Landingham

CONTENTS

6.1 TYPES OF EPILEPSY AND INITIAL TREATMENTS

Mechanisms of epilepsy have been explored through a variety of animal models as well as detailed human studies, for more than 70 years.[1–3] Through the animal models, a large number of contributing factors leading to epilepsy have been demonstrated, including conditions that lead to the intermittent, enhanced synchrony leading to partial or generalized seizures. While animal models still have only moderate predictive validity for anticonvulsant therapy development, the mechanisms may potentially apply to the human situation. However, in general, most animal models involve acute seizure development, mirrored in humans as acute convulsions, usually due to systemic or CNS irritants or toxins. For example, a classic convulsion may be seen with an overdose of penicillin or meperidine, and

0-8493-1482-8/05/$0.00+$1.50

convulsions are characterized by a high degree of neuronal electrical synchrony throughout the brain.

In contrast, most human seizure disorders are intermittent (i.e., a few seizures a month), potentially have irregular or variable starting locations, and may involve relatively large "epileptic" zones. The concept of an epileptic focus in humans has come under considerable scrutiny, and a minimum volume of cerebral cortex appears necessary for seizure onset. Once a seizure starts, inherent mechanisms within the brain can either constrain or enhance the spread, often into a generalized tonic–clonic convulsion. Such seizure pathways include the substantia nigra reticulata (SNr), which may be responsible for one part of the generalization. Additionally, the thalamus may enhance synchrony between the two frontal lobes.

Two basic types of spontaneous seizure disorders are recognized: partial and generalized.[1] Partial seizures emanate from specific regions of the brain, for example, partial motor seizures arise from the motor cortex. Partial seizures may be either simple (awareness and memory are maintained) or complex (awareness and memory are lost for a period of time).

Mesial temporal or frontal structures are thought to be involved during a complex partial seizure. A partial seizure may secondarily generalize, resulting in a generalized tonic–clonic seizure, often with versive head or eye movements. The operational definition in epilepsy surgery of the volume of brain important in seizure onset has been defined as how large a brain area must be resected to significantly decrease seizure frequency. For mesial temporal origin seizure disorders, most of the temporal lobe must be resected, including the central gray matter such as amygdala, entorhinal cortex, and hippocampus, for adequate seizure relief.

For neocortical partial seizures, the area may be smaller or difficult to define, depending on the region of cortex involved. However, many partial seizure disorders can be treated by surgical resection if a specific location can be found for the seizures, the area is resectable without too many neurological burdens (speech difficulties, weakness, memory loss) for the patient, and the seizures are not bilateral in origin. Only a small percentage of patients with refractory partial seizure disorders are amenable to resection or are referred at an appropriate age for resection.

The second type of seizure disorder is generalized and is thought to arise from the entire brain simultaneously. These seizures are more common in children and may represent diffuse abnormalities in many regions of the cerebral cortex. For example, some childhood epilepsies may result from subtle (but pervasive and widespread) changes in ion channels regulating inhibition, particularly potassium channels. Many of these channel abnormalities are due to gene mutations, and thus may be hereditary. They may also be developmental in the sense that as the channels change with development, other channels may substitute, so the seizures resolve spontaneously. This type of epilepsy is called an idiopathic generalized seizure disorder. Other types of generalized seizure disorders are designated symptomatic generalized and are due to widespread brain damage; much of the cerebral cortex is diffusely involved, giving rise to multifocal seizures or multiple seizure types. The seizures may worsen with time or respond to various treatments.

Drug treatment is the mainstay of therapy in epilepsy. Many patients take at least one medication. All the available anticonvulsants act primarily to suppress seizures, and none is known to prevent the formation of the epileptic condition, such as epilepsy after a head injury or central nervous system insult. However, most drug treatment interferes with fine cognition, particularly memory and complex thinking skills during growing and learning, thus inhibiting social skills and acquisition of learning skills. Additionally, many patients (potentially as many as 25%) fail to have their seizures controlled by drug treatment and are considered refractory to medical therapy. Even a single occasional seizure prevents activities such as driving a vehicle or pursuing many occupations.

The goal of therapy is to have rare or no seizures, if possible. Therefore, surgical treatment for intractable patients should become a consideration usually when they are children or young adults, when it is beneficial to increase participation in society, and it is clear that medical therapy is not successful. For most epilepsy surgical centers, surgical candidacy is considered after a patient has failed adequate trial with two or more standard medications and had seizures for 2 years or longer. Initial considerations for surgery are detailed localization of seizure onset and laterality and determination of the types of seizures to investigate the feasibility of localizing and removing an epileptogenic zone within the brain.[4]

6.2 SURGICAL TREATMENT

Surgical treatment for seizures began with tumor and obvious lesion resections to remove the irritating sources. However, in many instances the cortex generating the seizures was at a distance from a lesion such as a tumor, so the seizures did not necessarily improve with lesion resection. After the electroencephalogram (EEG) and direct electrocorticogram (ECoG) of the surface of the brain were developed, this technique provided a method to determine localization of brain function, as opposed to structural lesions. By the early 1930s, abnormal areas in the brain could be determined by EEG before the procedure and by ECoG during surgery, and resections of the functionally abnormal areas of the brain could be performed. This technique of preoperative or intraoperative localization of an epileptogenic zone based on an abnormal EEG still constitutes the standard form of epilepsy surgery, including temporal lobectomy procedures and neocortical resections.

A large number of patients present difficult localizations (too diffuse or in a critical area of eloquent cortex, not amenable to resection) or bilateral (multifocal) localization of seizure onset and abnormality. In addition, surgical resectioning of brain areas, even if abnormal, invariably leads to new deficits, however subtle. For many years, numerous attempts have been made to devise alternative surgical procedures that may be less invasive or may achieve benefits in patients not amenable to traditional EEG-guided resections.

These novel treatments fall into two main categories: (1) past treatments, many of which have now been abandoned, and (2) new translational treatments, still in the process of testing and development. Many of these treatments have underlying hypotheses of action not necessarily proven valid in a treatment sense.

Most current medical and surgical treatments for epilepsy are empirical in that they were not hypothesis-based at the time of human application.[1] Drugs are still screened with a basic convulsion model (electroshock therapy in rodents) used to assess ability to prevent death (ED_{50} dose). Many current surgical treatments such as vagus nerve stimulators have no known mechanisms of action.[5–7] The mechanisms of action of most anticonvulsants were studied after their utility in humans was demonstrated, so standard anticonvulsants such as phenytoin, lamotrigine, and carbamazepine, for example, change the properties of the sodium channel involved with action potential generation to favor single action potentials over bursts or groups of potentials. While vagus nerve stimulation appears to have a mild effect on seizure suppression (rather than complete seizure prevention), its mild effect fortunately is balanced by a very low risk profile.

Resective surgery (such as temporal lobectomy) is based on the hypothesis of removing an autonomous, epileptic zone so that abnormal output from the zone cannot influence the remainder of the brain as a result of removing the epileptic influence. Presumably over time more information may be realized about the mechanisms of action of empiric treatments.

6.3 PAST SURGICAL TREATMENTS

One interesting approach stemmed from a basic neuroscience observation: while recording single neurons from nonhuman primate cortex, any neuron could be trained by the animal to respond at a certain rate of neuronal action potential firing.[8] This rate-training capability through biofeedback is now a well-known capability of the brain and it has been extended to EEG biofeedback training. Since neurons in epileptiform regions in the brain tend to have too-high firing rates and fire in abnormal patterns or bursts, considerable effort was made to try to alter the firing rates (and hence suppress the tendency toward seizures) using biofeedback techniques in nonhuman primates with induced seizure disorders.

Although the hypothesis was excellent, the afferent pathways to these abnormal neurons appear to have been altered by the process of seizure disorder induction. Thus, less brain control (and hence less biofeedback control) can be exerted over neurons in epileptic zones. The concept was foiled by the nature of the epileptic process, although much was discovered about afferent denervation in epilepsy from this research. Since then, the concept that an epileptic region is autonomous from normal brain control has developed.[9]

Other observations about epilepsy include the efficacy of a ketogenic diet, usually for childhood epilepsy. A ketogenic diet is characterized by enhanced ketone bodies in the blood stream and decreased glucose. Interestingly, ketone bodies are taken up into the brain via one form of a monocarboxylate transporter (MCT). MCT transporter levels fall rapidly after the neonatal period and weaning because a mother's milk has a high content of ketone bodies and lactate, both requiring MCT-based transport. Thus, in early childhood, uptake of ketone bodies into the brain is lower; the uptake can be upregulated over time on the ketone diet. The mechanisms of the moderate suppression of the ketogenic diet on epilepsy still remain elusive although a switch in central nervous system (CNS) metabolism, possibly to enhanced gamma aminobutyric acid (GABA) levels, may be critical.[10]

Additional conceptual treatments include focal cooling because direct brain cooling at the time of a craniotomy may successfully abort seizures. Other local factors important in suppressing or aborting seizures include enhanced sensory input. Patients may occasionally be able to abort focal motor seizures by increasing sensory input to the affected part of the body, which may suppress the motor cortex through increased inhibition.

One treatment approach that evolved over time is stimulation therapy with electrical current or magnetic flux applied directly to the brain or across the skull. Although cortical stimulation (particularly in regions of hyperexcitable brain) can initiate seizures, cerebellar surface stimulation was suggested initially as a treatment for cerebral palsy and abnormal movement disorders.[11] This technique followed the partial effects of destructive lesions of the dentate nuclei of the cerebellum in cerebral palsy. However, although cerebellar surface stimulation applied to the anterior lobe and placed under the tentorium had little effect on movements, it was noted to have a partial effect on reducing the rate of generalized seizures.

While this effect was empirical initially, numerous stimulators were implanted through the 1970s. Although a randomized trial published later showed minimal clinical benefit, the concept was established that CNS stimulation had potential to improve seizure control.[11] Follow-up nonhuman primate studies showed that the primary effect was on enhancing alertness, and was a direct stimulation effect equivalent to enhancing brainstem reticular system function. Because many seizures occur in a hypnagogic state (toward sleep onset), enhanced alertness may exert a mild anticonvulsant effect. This is an example of a purely empirical treatment (with many advocates), with some insight into potential mechanism of action achieved through basic science studies. It is the complete opposite of a translational approach where ideally the hypothesis is developed first and treatment is second.

Direct cerebellar stimulation waned after demonstration of lack of efficacy (as happens with many empirical clinical treatments), but the concept that nonspecific brainstem stimulation may result in a mild, anticonvulsant effect persisted. Another technique to promote such stimulation is vagus nerve stimulation, which was tested and approved by the U.S. Food and Drug Administration (FDA) based on the earlier concept.[7,12] Although the mechanism is not known (and presumably relates to the same type of nonspecific brainstem effect seen for cerebellar surface stimulation), this empiric treatment has little risk and is well tolerated. However, vagus nerve stimulation is not likely to make an intractable patient seizure-free. It may merely decrease the number of seizures and perhaps their severity. Investigations regarding the relative worth and specific indications for vagus nerve stimulation are ongoing. Despite mild efficacy and lack of a specific hypothesis, the technique has numerous advocates.

6.4 NEW SURGICAL TREATMENTS IN CLINICAL TRIAL OR PRECLINICAL EVALUATION

The vagus nerve stimulator is an example of an open-loop system that is constantly stimulated (at some rate, frequency, and periodicity) without conscious or subconscious feedback from the patient to indicate whether the stimulation is effective.[12] Several other open-loop systems are now anticipated and are in clinical trials, particularly subthalamic nucleus (STN) stimulation and thalamic stimulation.[13–18]

The success of the standard, open-loop deep brain stimulator (DBS) systems from Medtronics for tremor and Parkinson's disease led to many other conceptual uses of the device. STN stimulation may lead to SNr suppression; the SNr is important as a common mediatory for the motor output of a generalized seizure. Thus, constant STN stimulation has been suggested and tested, similar to the type of stimulation used for Parkinson's disease.[13] However, very little is known about the EEG and cortical effects of STN stimulation or its role in mediating seizures in patients that will require study with implanted EEG electrodes. This type of study has been performed with vagal nerve stimulation.[19,20]

Although STN stimulation may result in acute cortical and possible seizure suppression, chronic stimulation may actually result in circuitry changes that are potentially proconvulsant. Thus, intermittent stimulation may be critical, highlighting the sparseness of knowledge about effective stimulation patterns for various uses. Theoretically, an STN system may work best in demand mode, in which some event (e.g., the patient senses an aura or seizure prelude) may trigger the stimulation to prevent a seizure.[21]

Other regions suggested as showing anticonvulsant effects with open-loop stimulation include several areas of the thalamus. The thalamus is known to aid propagation from one hemisphere to another, in other words synchronizing seizures, particularly frontal lobe seizures. Anterior thalamic lesions were suggested to improve generalized (frontal lobe-based) seizures. Following in this vein, DBS of anterior and medial thalamus is now in initial clinical trials to analyze whether any anticonvulsant effects arise within different regions.[16,17,21,22]

Hippocampal stimulation has also been suggested for partial complex seizures of hippocampal origin. These treatment sites use some form of intermittent (but still open-loop or constant) stimulation, theoretically blocking seizure throughput. It is unclear whether thalamic projections to the cortex may also result in some form of long-term suppression of a seizure tendency in the cortex as well.

Other types of treatment in clinical trials include stereotactic radiosurgery for temporal lobe seizures.[23] The dose of the radiation in this case is similar to single-dose treatments for tumors and produces some initial irritation of the brain areas and perhaps mild destruction without necrosis. This treatment is currently in clinical trials at several epilepsy centers; the trial format compares the new treatment to the standard current clinical treatment of temporal lobectomy. Other types of clinical trials also include various approaches to temporal lobectomy, and optimal methods to manage childhood seizures. A number of treatments, particularly for temporal lobe-mediated seizures, are not in small initial clinical trials.

6.5 CLOSED-LOOP OR DEMAND EPILEPSY FEEDBACK SYSTEM

A need still exists to develop alternative methods for controlling chronic epileptic seizures. The concept of an unsupervised method for detecting and treating epileptic activity automatically, in other words a demand system, has been widely proposed.[13,21,24–27] In many ways, such a system is equivalent to a demand for a heart pacemaker or defibrillator. This visionary event-controlled approach is radically different from and inherently more powerful in scope and application than any current treatment for chronic epilepsy.

A seizure feedback system could be an effective treatment scheme for focal seizure disorders, or potentially even for more diffuse generalized seizures, if applied to an output system that controls diffuse propagation. Such a system would function with an afferent limb able to sense the pre-ictal state in a local region of the brain, using either implanted or surface EEG electrodes or multiple single-unit electrodes. Clearly the seizure detection arm would require evaluation of the best area for electrode placement, similar to current invasive schemes. Once the pre-ictal state is reliably detected and processed, then an efferent limb in the region of the seizure can be used to cancel or convert an early pre-ictal rhythm into a less synchronized state.[28–33]

The clinical applicability of this type of system depends on detecting the pre-ictal state before any clinical manifestations occur and converting the state locally into one less likely to lead to seizure occurrence, in other words preventing the seizure from developing. Additionally, this type of system should not introduce any conscious awareness of the efferent treatment effect, such as pain, which would preclude general usefulness. The clinical applicability and current state of development of this type of system will be reviewed.

6.5.1 Demand Epilepsy Treatment System: Implementation

A definitive treatment scheme[21,34,35] involves three main parts (Figure 6.1):

1. A method to detect pre-ictal activity that is representative (specific and sensitive) to the pre-ictal epilepsy state for any patient, usually with implanted electrodes on the surface of the brain or within the brain
2. A technique to process information from these electrodes and generate an output signal with a threshold to indicate a possible seizure state in development
3. An efferent or control method to take the output signal and generate a subconscious effect that would abort or stop the seizure in germination so that it would not occur

The critical aspect of these aims of an ideal system is that the patient should not know about the detection or treatment; if a seizure occurs, the system is a clinical failure. While it is much easier to detect a highly synchronized seizure event, once a seizure has started treatment is too late because the patient has already lost awareness and the clinical implications have already occurred. However, in some seizure states, the goal is not necessarily to prevent all seizures; the goal may be to reduce the number significantly, similar to vagus nerve stimulation.

The need for a demand or feedback control system remains controversial, particularly because several (completely empirical) open-loop stimulation studies in

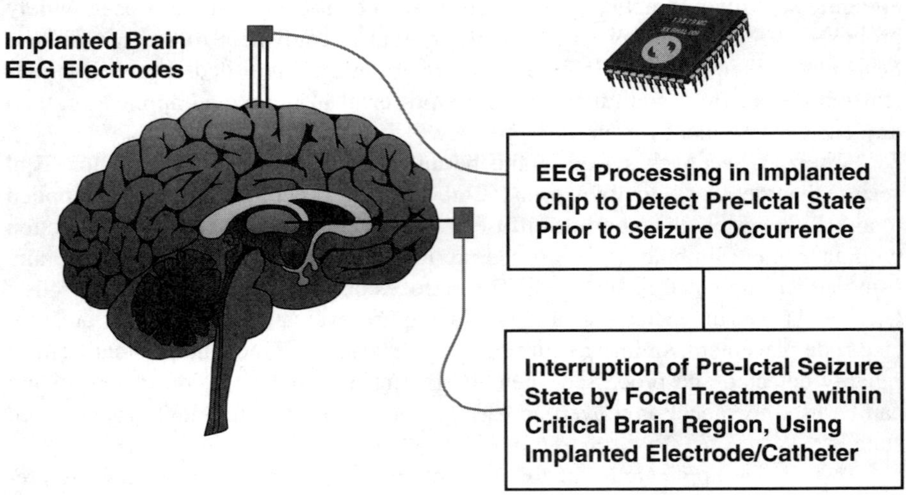

FIGURE 6.1 (See color insert following page 146.) The critical elements of a demand or feedback system for epilepsy include three major components: (1) a sensing arm or group of macro (EEG) electrodes on the brain surface or implanted near the maximally abnormal area that are critical for detecting both the pre-ictal state and the transition into seizure; (2) EEG processing for the detection of the pre-ictal state, usually considered optimal when the seizure is still building and the state can be interrupted in a subconscious fashion; and (3) stimulation or medical treatment to interrupt the seizure. The final component may be most efficient if administered near the origin of the seizure, but could also be a systemic injection, for example.

humans (constant anterior or medial thalamus stimulation, constant or intermittent STN stimulation, and vagus nerve stimulation) are currently in progress. The consensus is that constant stimulation is theoretically much less effective than some form of intermittent stimulation because the networks excited by the stimulation may become refractory and may not provide the necessary inhibition sufficient to prevent seizure development. However, a demand system requires the hypothesis that the development of a seizure may be both detectable and also critically interrupted during the build-up period, before an inter-ictal hyperexcitable occurrence generalizes to include more and more cortex. These aspects will be individually reviewed.

6.5.2 Detection Schemes and Electrodes

EEG and ECoG analysis paradigms have been more or less successful in detecting pre-ictal activity, particularly increased synchronization and decreased level of complexity of the EEG signal (i.e., less chaotic). [24–26,36–40] However, many of these have not been tried in significant animal models or on humans and the effects of noise and extraneous signals have not been adequately included. The location of such electrical recordings depends upon the level at which ictal activity is to be interrupted. For example, if a seizure is already generalized throughout much of the cortex and is bilateral, almost anywhere over the scalp or deep structures may be sufficient because the signal is highly propagated. However, at this late stage, the brain has already been significantly involved with the seizure and will undergo interruption of ongoing function and post-ictal changes.

The more desirable detection would reside in a region close to the hyperexcitable zone, such as hippocampal depth electrodes for mesial temporal lobe complex partial seizures. An event should be detected before the patient becomes aware that a seizure is starting, and the detection method should abort even local seizure propagation. This would allow more subtle methods to be used to control or terminate an event (since less brain is involved), particularly at the subclinical level. These detection algorithms clearly need analysis and assessment.

The critical issues to be resolved for individual patients include electrode placement and how to determine optimal placement, particularly for invasive, implanted electrodes, because all of the envisioned systems would be required to be completely internalized (but with some external remote control possible, similar to current DBS systems). Because epileptogenic zones are often difficult to localize to small regions of the brain, the localization of sufficient electrodes to bracket areas of the brain critical for ictal onset is a continuing challenge.

Detection depends on monitoring sufficient areas of the brain so that all possible pre-ictal states may be recorded, particularly in patients with multiple seizure types. This localization approach is radically different from current epilepsy monitoring, where the point is to localize an epileptogenic zone to a sufficiently small region that a resection may be feasible. These seizure detection electrodes would optimally be near a hyperexcitable region associated with seizure onset (as determined by surface or depth electrodes). For permanent purposes, an implanted depth electrode located near the zone of seizure origin (Figure 6.1) and optimized for EEG detection or stimulation (or both combined in one electrode set) would be ideal.

6.5.3 Processing System

Processing and detection systems are commonly linked to a variety of dynamical approaches where changing ongoing activity slightly may deter the system from proceeding to seizure.[21,25,26,28,31,33,36,38,39,41] The disadvantage of most systems is that the processing time to detect a pre-ictal state is too long for real-time use. In many cases, realistic human data (usually from invasive recordings) has not been used to analyze how successful the various algorithms may be. However, a large variety of approaches, depending on the baseline and ictal activities in particular regions of the brain, may be chosen.

Detection systems may have to cope with a wide variety of pre-ictal and ictal onset patterns, many of which are already well characterized in EEG studies. For example, in many cases temporal lobe seizures involve a desynchronized state before the highly synchronized ictal state appears and many other variations may also exist. This variability in pre-ictal onset, depending on location in the brain and type of seizure, and the likelihood of noise added to any EEG/ECoG system, may contribute to difficult pre-ictal or seizure detection. However, both Litt and Iasemidis suggested that a seizure state may actually be developing for several hours, which may provide sufficient opportunity for perturbing this ictal development at an early stage.[25,37,38] Although may reports have focused on direct ictal detection and subsequent seizure disruption,[28,29,30] some comparative studies are trying different algorithms to assess relative efficacies of seizure detection schemes.[24,26]

Further developments may require multiple detection algorithms employed within a single system. Although it is proprietary, the current Medtronics system in clinical trials uses alternative detection schemes with an external processing box because of the space required for the computer. However, it is anticipated that an internalized system will be critical, although the types of computer processor and algorithm to be used have not yet been confirmed. Clearly, the simpler the algorithm (and the more likely to function in real time), the more likely it will fail or not be applicable to a wide range of pre-ictal states. Thus, the development of small, totally implantable computer systems that can handle these challenges will require sophisticated biomedical engineering support.

6.5.4 Method of Seizure Termination

Many modes of electrical feedback have been applied in both slices and *in vivo* cortex to try to terminate seizures, which usually involve invoking some form of surround inhibition or reversing polarity of an ongoing event. These work far better on a local basis instead of on larger areas of brain.[28–32] Any form of stimulation can also trigger a seizure because the stimulation is usually anticipated to be applied to an epileptogenic zone. Thus, gradients of stimulation current sufficient to disrupt an ongoing ictal build-up and too small to generate a seizure may be required. Local stimulation with a clearly defined regional onset may be most beneficial to smaller regions of brain. Stimulation may also be patterned in such a way as to minimize seizure onset and to maximize seizure disruption, if possible.

There is a large question of the importance of the SNr in the motor output of a generalized seizure, but the effects of stimulation of this region on cortical areas responsible for the seizure or involvement of the cortex in general are unclear. It may be less than helpful to suppress only the motor output of a seizure if the entire cortex has already become involved in a generalized seizure; this will not improve cortical functioning and may only serve to limit damage. This is similar to the role of corpus callosotomy, following which individual cortical areas remain active and undergo ictal events but cannot generalize due to the lack of commissural connections. Cortical functioning (i.e., memory, cognitive functioning) does improve even though safety may not improve due to persistent (and more focal) seizure activity.

Likewise, several areas now suggested for stimulation to prevent seizure are highly nonspecific. Older studies of cerebellar and brainstem stimulation were shown only to heighten awareness via reticular activation. Vagus nerve stimulation may also function in this manner (although it is not well elucidated). Anterior thalamic stimulation may primarily serve to prevent commissural spread in the frontal areas; how it will affect cortical functioning remains unclear. Does a need exist for further empiric studies on alternative nonspecific sites that may lead primarily to mild suppression of diffuse cortical dysfunction? Pursuing a specific therapy that may function on a subclinical level with a feedback or demand loop would seem to be more logical, particularly with the goal of arresting a local positive feedback loop before a significant region of the brain is involved.

Methods to arrest local development of a seizure could also include local parenchymal, intracerebroventricular or system delivery of drugs although few drugs are approved for intrathecal or parenchymal delivery and diffusion is limited.[41,42] The proposed and patented Ludvig system involves local detection and local drug application in a demand sense but, of course, any direct brain delivery of medication will require rigorous study for FDA approval.[41] Systemic delivery of a drug may be effective but would also produce generalized effects. Electrical stimulation of a focal region is attractive, but it would exert a limited field of effectiveness, and if ill-timed could heighten hyperexcitability and thus aggravate seizure activity.

Cardiac demand systems in current clinical use appear to be highly effective at detecting abnormal rhythms and generating sufficient electrical pulses to abort abnormal rhythms and restart more normal beats. It is tempting to suggest that some of this technology could also be applied to pre-ictal detection of seizures. Other modes of diffuse stimulation could include cranial nerve inputs such as trigeminal[30] or the current vagus nerve stimulation, but in intermittent stimulation mode. Magnetic stimulation to some regions may also be inhibitory, as may many of the current sites in thalamus, STN, or hippocampus. Many proprietary systems in development, which are neither publicized nor published, may overcome some of these difficult challenges. However, many of these systems may not become general knowledge unless they are effective or FDA approval is gained.

6.6 REQUIREMENTS FOR CLINICAL APPLICABILITY

One approach is to use current patients who are undergoing video EEG monitoring and add various stimulation sites for short-term assessment of efficacy and cortical effects of a feedback system. This would allow testing of all three aspects of a demand system. Many patients undergoing video EEGs also have implanted depth electrodes, and stimulation of the depth electrodes is currently underused as a possible method for seizure control.

Such research studies would have to be added after sufficient clinical information is garnered to localize seizures. However, pre-ictal detection could be determined *post hoc* with application of known algorithms to optimal EEG or depth electrode signals. The cortical effects and clinical effects of stimulation or medication at different sites could also be determined on a short-term level, with possible targets:

1. Near the seizure initiation zone, such as in the hippocampus
2. The subthalamic nucleus with unilateral or bilateral implanted electrodes
3. The trigeminal nerve if a suitable location could be found for stimulation within or adjacent to the nerve
4. Rapid systemic medication delivery

In addition to short-term studies, a combination of detection, processing, and suppression methods could be tested using a number of types of seizures. It is difficult now to argue for long-term implantation of stimulation electrodes at any of these sites until short-term data on clinical effects and side effects can be obtained. In addition, the stimulation level at which seizures can be suppressed will be critical as will the feasibility of suppression prior to clinical expression of a seizure in a subconscious pre-ictal state.

Clearly, one long-term goal is to design, develop, and clinically test an "intelligent brain-pacemaker" device to detect neural activity preceding clinical manifestations of an epileptic seizure and disrupt this pathological brain state through intermittent electrical stimulation of a brain region or a peripheral cranial nerve.[34,35] Like a modern demand heart pacemaker, a brain pacemaker would operate autonomously to interrupt the development of an epileptic seizure at a critical early time period without intervention from the patient, physician, or other individuals.

6.7 CONCLUSIONS

A number of advances in epilepsy treatment are already in development or are undergoing clinical trials at a time when a large number of new drugs that may decrease the need for new surgical interventions are also available. Additionally, childhood and neonatal seizures and status epilepticus are more aggressively treated early now, as better protocols for status are now in use.

As episodes of febrile seizures and status decline, significantly fewer patients may experience later complex partial seizures because in many instances mesial temporal sclerosis appears to have arisen from early episodes of febrile seizures that terminated in status epilepticus. Thus, better early medical treatment may prevent the development of later intractable epilepsy, and could eventually decrease the population for whom surgical therapy of any type is considered.

Despite a number of new modalities of treatments on the horizon, an ideal system would consist of pre-ictal seizure detection in a critical area of the brain, and then a counteracting influence (local or systemic drug injection, local or diffuse electrical stimulation, etc.) to prevent ictal onset completely. From a clinical perspective, the only effective surgery is one that completely prevents seizure, avoids social stigmas attached to patients with the disorder, and maintains optimal neurological functioning.

REFERENCES

1. Aiken, S.P. and Brown, W.M., Treatment of epilepsy: existing therapies and future developments, *Frontiers Biosci.*, 5, E124–E152, 2000.
2. Jacobs, M.P. et al., Future directions for epilepsy research, *Neurology*, 57, 1536–1542, 2001.
3. Wieser, H.G., Future aspects of epilepsy research, *Acta Neurochir.*, 84S, 1–16, 2002.
4. Maroun, F. et al., Cerebral cortical stimulation and surgery for epilepsy, *Can. J. Neurol. Sci.*, 23, 303–307, 1996.
5. Cohen-Gadol, A.A. et al., Neurostimulation therapy for epilepsy, *Mayo Clin. Proc.*, 78, 238–248, 2003.
6. Karceski, S., Devices in the treatment of epilepsy, *Sem. Neurol.*, 22, 259–268, 2002.

7. McLachlan, R.S., Vagus nerve stimulation for intractable epilepsy, *J. Clin. Neurophysiol.*, 14, 358–368, 1997.
8. Wyler, A.R. and Burchiel, K.J., Operant control of epileptic neurons in chronic foci of monkeys, *Brain Res.*, 212, 309–329, 1981.
9. Lowenstein, D.H., Recent advances related to basic mechanisms of epileptogenesis, *Epilepsy Res.*, 11, 45–60, 1996.
10. Swink, T.D., Vining, E.P., and Freeman, J.M., The ketogenic diet, *Adv. Pedriatr.*, 44, 297–329, 1997.
11. Krauss, G.L. and Fisher, R.S., Cerebellar and thalamic stimulation for epilepsy, *Adv. Neurol.*, 63, 231–245, 1993.
12. DeGiorgio, C.M., Schachter, S.C., and Handforth, A., Prospective long-term study of vagus nerve stimulation for the treatment of refractory seizures, *Epilepsia*, 41, 1195–1200, 2000.
13. Benabid, A.L. et al., Future prospects of brain stimulation, *Neurol. Res.*, 22, 237–246, 2000.
14. Chabardes, S. et al., Deep brain stimulation in epilepsy with particular reference to the subthalamic nucleus, *Epileptic Disorders*, 3S4, S83–S93, 2002.
15. Fisher, R.S. et al., Placebo controlled pilot study of centromedian thalamic stimulation in the treatment of intractable seizures, *Epilepsia,* 33, 841–851, 1992.
16. Velasco, F. et al., Predictors in the treatment of difficult to control seizures by electrical stimulation of the centromedian thalamic nucleus, *Neurosurgery*, 47, 295–305, 2000.
17. Velasco, M. et al., Acute and chronic electrical stimulation of the CM thalamic nucleus, *Arch. Med. Res.,* 31, 304–315, 2000.
18. Vercueil, L. et al., High-frequency stimulation of the subthalamic nucleus suppresses absence seizures in the rat, *Epilepsy Res.,* 31, 39–46, 1998.
19. Olejniczak, P.W., The effect of vagus nerve stimulation on epileptiform activity recorded from hippocampal depth electrodes, *Epilepsia*, 42, 423–429, 2001.
20. Lesser, R.P., How did vagus nerve stimulation become a treatment for epilepsy? *Neurology*, 52, 1117–1118, 1999.
21. Osorio, I. et al., An introduction to contingent (closed-loop) brain electrical stimulation for seizure blockage, to clinical trials and analysis of therapeutic efficacy, *J. Clin. Neurophysiol.,* 18, 533–544, 2001.
22. Mirsky, M.A., Rossell, L.A., Terry, J.B., and Fisher, R.S., Anticonvulsant effect of anterior thalamic stimulation in the rat, *Epilepsy Res.,* 28, 89–100, 1997.
23. Regis, J., Bartolomei, F., Hayashi, M., and Chauvel, P., What role for radiosurgery in mesial temporal epilepsy? *Zentral. Neurochir.,* 63, 101–105, 2002.
24. Iasemidis, L.D., Epileptic seizure prediction and control, *IEEE Trans. Biomed. Engin.*, 50, 549–558, 2003a.
25. Iasemidis, L.D. et al., Adaptive epileptic seizure prediction system, *IEEE Trans. Biomed. Engin.*, 50, 616–627, 2003b.
26. Jerger, K.K. et al., Early seizure detection, *J. Clin. Neurophysiol.,* 18, 259–268, 2001.
27. Tanaka, T. et al., Basic science and epilepsy: experimental epilepsy surgery, *Stereotactic Funct. Neurosurg.,* 77, 239–244, 2001.
28. Bikson, M. et al., Suppression of epileptiform activity by high frequency sinusoidal fields in rat hippocampal slices, *J. Physiol. (Lond.),* 531, 181–191, 2001.
29. Durand, D.M. and Warman, E.N., Desynchronization of epileptiform activity by extracellular current pulses in rat hippocampal slices, *J. Physiol. (Lond.),* 480, 527–537, 1994.

30. Fanselow, E.E., Reid, A.P., and Nicolelis, M.A.L., Reduction of pentylenetetrazole-induced seizure activity in awake rats by seizure-triggered trigeminal nerve stimulation, *J. Neurosci.*, 20, 8160–8168, 2000.
31. Gluckman, B.J., Nguyen, H., Weinstein, S.L., and Schiff, S.J., Adaptive field control of epileptic seizures, *J. Neurosci.,* 21, 590–600, 2001.
32. Lesser, R.P. et al., Brief bursts of pulse stimulation terminate afterdischarges caused by cortical stimulation, *Neurology,* 53, 2073–2081, 1999.
33. Lian, J. et al., Local suppression of epileptiform activity by electrical stimulation in rat hippocampus *in vitro*, *J. Physiol. (Lond.),* 547, 427–434, 2003.
34. Nicolelis, M.A.L., Actions from thoughts, *Nature*, 409, 403–407, 2001.
35. Nicolelis, M.A.L., Brain–machine interfaces to restore motor function and probe neural circuits, *Nature Neurosci. Rev.*, 4, 417–422, 2003.
36. Lehnertz, K. et al., Seizure prediction by nonlinear analysis, *IEEE Engin. Med. Biol.*, Jan./Feb., 57–63, 2003.
37. Litt, B. et al., Epileptic seizures may begin hours in advance of clinical onset: a report of five patients, *Neuron*, 30, 51–64, 2001.
38. Litt, B. and Lehnertz, K., Seizure prediction and the preseizure period, *Curr. Opin. Neurol.,* 15, 173–177, 2002.
39. Navarro, V. et al., Seizure anticipation in human neocortical partial epilepsy, *Brain*, 125, 640–655, 2002.
40. Schiff, S.J. et al., Brain chirps: spectrographic signatures of epileptic seizures, *Clin. Neurophysiol.,* 111, 953–958, 2000.
41. Ludvig, N. and Kovacs, L., Hybrid Neuroprosthesis for the Treatment of Brain Disorders, U.S. Patent 6,497,699, 2002.
42. Stein, A.G. et al., An automated drug delivery system for focal epilepsy, *Epilepsy Res.*, 39, 103–114, 2000.

7 Neuroprosthetics and Clinical Realization of Brain–Machine Interfaces

Dennis A. Turner, Dragan F. Dimitrov, and Miguel A.L. Nicolelis

CONTENTS

0-8493-1482-8/05/$0.00+$1.50

7.1 INTRODUCTION

Neuroprosthetics encompasses a wide variety of interfaces with the nervous system, usually considered in the context of clinical abnormalities or disease. The concept stems from clinical concerns about functional independence and integration of individuals into society and far-reaching visions of direct interactions of the brain and mind and external events.

Conceptually, all devices such as typewriters and cars can be considered brain–machine interfaces (BMIs). The mind controls the machine via buttons, pedals, or wheels. The interface is inherently inefficient. Brain output must be translated into motor movements and then mechanically transmitted to the device. In many disease situations, the brain is preserved but its output mechanisms in the periphery are neither functional nor attached, making interaction with the outside world impossible. Reestablishing a means of interacting with the world by directly connecting to the source — the brain — is the essence of BMI development.

Because all nervous system interaction with the environment normally depends upon both peripheral sensory input and motor output, mind control of action and direct channeling of sensory information into the brain are tantalizing concepts because of the enormous possibilities of control inherent with a more rapid and scalable interface. This visionary approach is rooted in a large number of treatises in the literature, many of which view both positive and negative aspects of "mind control" and particularly suppression of free thought and action.

Current and potential technologies appear rooted in the alleviation of subnormal interactions with the environment in disease conditions, and ethical views of how to apply technology remain highly varied. All aspects of human behavior inherently possess both constructive and destructive sides, including use of extremities for gathering food and participating in combat. An important question is whether technology should be suppressed, solely to prevent ethically inappropriate actions, in spite of potentially significant enhancements to society overall.

This issue is not resolved and should continue to be debated, but the decision as to how to implement technology always rests on individuals who can exert choices. The continuing hope is that ethical considerations can maintain pace with technology advances. For example, ethicist Arthur Caplan argues that enhancing brain function is a natural extension of our human tendency to improve ourselves, in many cases with prosthetics. However, the principles of individual choice without coercion should always be preserved along with freely available access.[1]

In a variety of medical conditions, for example, spinal cord injuries, strokes, or degenerative neurological diseases such as amyotrophic lateral sclerosis (ALS), patients may lose their abilities to use their arms or communicate. They frequently remain alert and maintain cognition, but in many ways they are unable to convert their thoughts into actions. For example, an upper cervical injury patient with quadriplegia needs to activate devices to promote action for activities of daily living such as eating, using a wheelchair, and entering data into a computer. Patients with communication deficits arising from severe left hemisphere infarcts may not be able to signal intents or basic needs to caregivers.

In these situations, cognition and most or all of the cerebral cortex and subcortical structures are intact, but peripheral control has been partially or completely lost. While a large variety of prosthetic aids currently available can enhance function, very few prosthetic devices that can be controlled using existing output channels are available to this group of patients.[2–4] Obviously, the ideal output channel would be a direct bidirectional data stream to and from the patient's brain that would bypass all the inefficiencies associated with today's prosthetic devices. Thus, development of new capabilities for enhanced interaction with the environment and treatment of clinical conditions are high clinical priorities pushing neuroprosthetic developments.

As part of this clinically driven need, a variety of neuroprosthetic devices are available, but in general they are unidirectional and do not take full advantage of brain encoding algorithms for optimal implementation. These devices include cochlear prostheses, deep brain stimulating (DBS) devices for tremor and Parkinson's diseases, and visual (retinal and cortical) and auditory prostheses in development, along with peripheral prostheses for functional electrical stimulation.[5–7]

Technology advancements now in progress, however, may eventually lead to far more complicated brain–machine interactions that could lead to a direct link between a patient's brain and an actuating device, leading to highly efficient and effective ways for certain groups of patients to interact with the world.[8–13] This chapter will initially review current devices and then discuss implementation of conceptual devices for enhanced brain–machine interactions.

7.2 CLINICAL CONDITIONS APPROPRIATE FOR NEUROPROSTHETICS

A large population of individuals must deal with reduced interaction with the environment, for example, because of sensory deficits such as blindness and deafness, motor output limitations such as those caused by ALS, quadriplegia, and severe cerebral infarcts, and communication and speech deficits. All of these factors or conditions prevent full and normal environmental interactions, and in many cases, gainful employment and participation in activities of daily living.

These conditions can be roughly categorized into two groups. The first includes situations in which the supratentorial central nervous system (CNS) is intact but is damaged at either the brainstem or spinal level. The cerebral cortex and cognition are functional (as they are in a quadriplegic or patient with ALS), but the central representation of the periphery is altered due to drastically changed sensory input.

The most severe situation is the "locked-in" patient with a brainstem stroke or damage that has left him or her with normal cortical functioning, but who has virtually no residual interface with the environment except for perhaps eye movements.[4,14,15]

In the second group are patients whose supratentorial CNS suffered damage, as in the case of stroke accompanied by aphasia or hemiparesis. This group includes patients who have impaired communication with the environment and often considerable reorganization of function within the cortex to accommodate the damage.[16,17] Both groups of patients have profound needs for enhanced communication, interaction with the environment, and control of external devices to maintain quality of life, independence, activity of daily living, and output of creative thought.[3,8,11,13,14]

Most current prostheses depend on residual peripheral control, for example, eye movements or residual limb movement, to activate external devices. The devices are highly limited in bandwidth, in terms of ability to transmit effective information between the brain and the environment. For this reason, considerable interest has developed in a direct brain–computer interface that will allow direct brain control of external devices or natural limbs. The potential for this type of interface includes a higher bandwidth and more natural control by using signals generated by the brain to interact with the environment.

7.2.1 Sensory Deficits

Deficits in sensation include both special sensory functions (vision, hearing, vestibular function) and somatic sensation. These deficits can include both inadequate sensation, such as partial or total blindness or the distorted or altered sensation that can occur in various pain syndromes. Clearly, severe deficits arise from blindness and hearing deficits, leading to impetus for development of augmentative devices such as cochlear prostheses. Other approaches to enhancing individual functioning include sign language and Braille communication.[13,15,18]

Some patients also experience distorted or enhanced somatic sensation from a variety of sources (usually referred to as dysesthesias) that commonly result in states perceived as uncomfortable, for example, pain associated with root compression such as radiculopathy and sciatica, neuropathy, benign chronic pain states, and spinal cord injuries. These conditions are highly bothersome to the affected individuals. Even though the conditions are neither life-threatening nor significant in terms of loss of function, patients commonly seek treatments for relief. For example, peripheral nerve, spinal cord and midbrain/thalamic stimulation have been used commonly for more than 30 years for the relief of pain, in part driven by patient suffering and need for treatment.

7.2.2 Motor Deficits

Motor deficits include those arising from CNS sources and those from peripheral lack of control. For example, patients with hemisphere or brainstem strokes may show hemiplegia (inability to move on one side), while patients with spinal cord injuries commonly have upper or lower extremity impairments or both. While lower extremity impairments interfere with walking, the inability can often be overcome by simple use of a wheelchair or other assistive device. Attempts to achieve computer-generated walking through direct muscle stimulation (known as functional electrical stimulation or FES) have shown some ability in aiding muscle movement. Upper extremity and hand function deficits are much more devastating and preclude most tasks; they also have minimal rehabilitation potential and usually require significant assistance even for activities of daily living.

Another type of deficit is caused by ALS, a disease that may also affect the brainstem and upper cervical spinal cord, resulting in intact cognition but impaired speech and hand motion — a severely debilitating combination for interactions with the external world. Cerebral palsy, a severe motor syndrome, affects the basal ganglia. Patients often have intact cognition with almost complete inabilities to express themselves. Peripheral injuries and congenital defects, including lack of upper extremities (iatrogenic or traumatic amputation, for example) may also prevent translation of thoughts into actions. For all these conditions, a residual peripheral output such as a small muscle contraction could be useful for device control, but only in a highly limited format and with minimal information transfer for complex output of thoughts.

7.2.3 Communication Deficits

Communication deficits (in the presence of intact cognition) can vary from direct brain limitations (such as expressive aphasia) to lack of peripheral output (such as a brainstem stroke with a locked-in syndrome) to abnormal peripheral output such as dysarthria. These conditions are all very common. Because most human interactions consist of speech and vocalization, persons with communication deficits may have severe problems defining and stating even their basic daily living needs.

Most current approaches to enhancement of communication problems depend on residual output such as muscle contractions that can then trigger devices to achieve external speech or virtual choice output, but such devices are highly limited in terms of letter and word throughput.[18] Theoretically, the information coded into speech — intentions, thoughts, and opinions — must be neurally coded and could potentially be gathered directly from brain output. However, speech is inherently inefficient. The right word to express thoughts does not always exist. Conceivably, a highly efficient, more direct connection among people could bypass the need for vocalization altogether.

7.2.4 Enhancement of Normal Function

In addition to applications for clinical conditions involving reduced interactions with the external environment, many individuals are interested in augmenting normal functions. Augmentation beyond normal innate human function has been a common thread in the entire history of human development. Tools and devices were designed to improve on normal human sensation and motor function. For example, eyeglasses, laser keratotomy, microscopes, and telescopes all enhance vision beyond normal ability. Hearing can be augmented by speakers, microphones, and other paraphernalia.

Most plastic surgery procedures, joint replacements, and other medical approaches are not always performed only to treat medical conditions; they are intended to improve function beyond what a patient normally experiences. The difference between ordinary augmentations and neuroprosthetics lies in using devices to mimic inherent brain signals for enhanced or direct sensory input into the brain, and decoding of normal brain signals for alternate channeling of motor output function.

Although a variety of methods have been utilized, many current (and projected) neuroprosthetic devices are implanted directly into the brain. Implantation has the advantage of bypassing peripheral inputs and outputs, hence decreasing the time between signal and brain response. For example, a motor output could be channeled directly to a device for enhancing motor control on a microscopic, macroscopic, or larger-than-human level, resulting in considerable scaling of effort, far beyond the capabilities of the ordinary human motor system. Additionally, the time to response could be far less with direct inputs and outputs into the brain by speeding up a reflex loop, assuming the brain can keep pace with such external devices.

Time and physical scaling enhancements have obvious practical importance for extending human control to environments that are hostile to biological tissue or, for example, aiding space exploration by decreasing delay in transmission. As argued in a recent article by noted ethicist Arthur Caplan,[1] such augmentation is a natural extension of the long human interest in tool use and extends our understanding of the universe beyond our meager physical senses and motor capabilities although it potentially requires brain implants to access the nervous system directly.

Implants in other areas of the body, for example, breast implants, are well tolerated by society. The main limitations of a scheme for enhancing brain function are deciphering inherent brain encoding of sensation and motor function and achieving a stable interface between electrodes and the nervous system at a sufficiently small level to be meaningful for brain components, particularly axons and neurons on the micron scale. Excessive stimulation or recording interfaces may lead to unrealistic stimulation of multiple nervous elements, resulting in less-than-specific responses or noise and ranging across too many neural elements for decoding of neural output.

Since enhancement of human performance and nervous system function are commonly employed now, how would such system be perceived and used in a wider arena? Clearly, the ethical issues point to self-determination and use, in other words, coercion to use a device would argue strongly against self-determination and free choice, particularly for implantable devices. Another ethical aspect to consider is universal access to such self-enhancements to prevent unfair advantage. Of course, most current self-enhancement advantages (expensive colleges, SAT preparation courses, etc.) already have limited access, usually based on cost. Whoever applies augmentation technology should bear these ethical principles in mind, particularly for implantable devices, to avoid coercion (as with other types of medical care), maintain individual self-determination, and allow the widest access possible.

7.3 CATEGORIES OF NEUROPROSTHETIC DEVICES

Categories of neuroprosthetic devices are determined by the nature of their interactions with the nervous system. For example, unidirectional sensory stimulation of the nervous system has been used for many years to control pain at the thalamic, midbrain, spinal cord, and peripheral nerve levels; cochlear implants are more recent innovations. Unidirectional control for seizures has also been available for many years. Stimulation of the cerebellum was initially used and vagus nerve stimulation is more recent. Both techniques are used regardless of inter-ictal or ictal state. However, most devices could be improved by expanding the degree of control provided by feedback, which will be discussed in subsequent sections.

7.3.1 Unidirectional Sensory Neuroprosthetic Devices

The development of neuroprosthetic devices has only recently become feasible through advances in many aspects of the required technology. Stimulation of somatic sensory axons at the peripheral nerve, spinal cord, or brain level has been used in a nonspecific fashion to relieve pain for more than 30 years. The decoding of normal sensation has not been attempted, particularly for a complex signal.

Pain stimulation involves the insertion of an abnormal signal (usually perceived as a buzzing feeling, like an electric razor). This abnormal signal, if perceived in the somatopic area of discomfort, can mislead the nervous system into removing the uncomfortable sensation. Other types of sensory neuroprosthetic devices include vagus nerve stimulators for epilepsy. These devices do not rely on conscious perception of a stimulus, but rather subconscious brainstem stimulation, similar to the predecessor device, cerebellar stimulation (see Chapter 6 regarding demand seizure treatments).

An example of a more complex sensory device is a cochlear prosthesis in which microstimulation via platinum/iridium (Pt/Ir) contacts leads to direct activation of the cochlear nucleus, producing "sounds" that can eventually be discriminated by patients after some training. Direct brainstem stimulation of the cochlear nucleus is also being attempted, but the decoding of the input is much more difficult for the patient because natural sensory channels are not directly activated. Other complex unidirectional sensory stimulation devices in development include retinal visual prostheses that stimulate the optical nerve head directly at the back of the retina and direct visual cortical stimulation. These complex sensory stimulation systems clearly require high degrees of specificity of stimulation and considerable training to enable patients to perceive such stimuli.

7.3.2 Unidirectional Motor Neuroprosthetic Devices

Although they were developed initially for pain control, deep brain stimulator (DBS) implants were removed from the U.S. market in the 1980s. However, Medtronics developed a new version of the DBS electrode and received FDA approval for use in movement disorders in 1999.[7] Currently, DBS is approved for tremor control, Parkinson's disease and more recently, dystonia (on a compassionate basis). Because of the common availability of the DBS device and a simple, unidirectional stimulator system (basically, the same types of device control that are available for pain sensory stimulation devices), many additional experimental applications using this device for more than motor control are discussed later.

The current generation of DBS appears to produce motor control through constant stimulation of abnormal motor circuits, similar to the way the common lesions such as thalamotomy (for tremor) and pallidotomy (for control of dyskinesias in Parkinson's disease) worked. The DBS system has now become a common template for considerations of other types of brain implants because of the direct brain electrodes and associated circuitry and telemetry required for external control.

7.3.3 Other Unidirectional Neuroprosthetic Devices

The availability of the DBS system has led to its reconsideration for use in psychosurgery. Previous applications of brain surgery to control disorders of the mind and thought usually involved coercion (i.e., frontal lobotomies performed at mental institutions) and permanent lesions, with high risks of unexpected side effects and death. DBS is usually considered to offer a reversible effect that the patient can turn on or off at will, thus introducing the patient autonomy made possible with medical treatments. The low risk of DBS (approximately 6% total implant risk) also is a considerable improvement over the often brutal forms of frontal lobotomy and psychosurgery previously performed. Thus, DBS may be an excellent alternative for some psychiatric disorders in which medical treatment has failed, particularly in comparison to electroconvulsive therapy (ECT), because of low implant risk and reversibility.[19]

The two most common surgical sites to be considered are the cingulum, in contrast to the lesion-based cingulotomy, and the anterior limb of the internal capsule. The disorders currently being assessed for DBS treatment include obsessive–compulsive disorder (OCD) and severe depression. Both conditions severely interfere with patients' functioning. Depression can lead to suicide risk, which will extend hospitalization. DBS may be a significant improvement over older lesion surgery to treat both conditions because the stimulation can be tailored better and can be simply turned off by the patient if unwanted side effects appear.[19]

Other new applications of currently available DBS systems include constant stimulation for epilepsy, currently in clinical trial as anterior thalamic and substantia nigra reticulata (SNr) stimulation (see Chapter 6 for a detailed discussion). Both types of stimulation appear to affect seizure frequency although the mechanisms are not yet clear. However, both appear to cause tachyphylaxis or loss of stimulation efficacy with constant stimulation, possibly due to plasticity in the circuits stimulated. These early results suggest that a demand stimulation system with intermittent controlled stimulation may be better overall (see Chapter 6).

These devices are all examples of one-way systems. They allow only one-way communication with the CNS without direct capability for feedback.[5,6] Current neuroprostheses utilize indirect channels into the CNS via either microstimulation or macrostimulation of a region or a nerve or motor output detection. The examples discussed illustrate many of the problems and issues of direct interfaces with the nervous system and the need in many instances to provide some form of training to improve the performance of the device.

7.3.4 Bidirectional or Feedback Neuroprosthetic Devices

A feedback system allows a neuroprosthetic device to self-adjust for ongoing circumstances and provides demand control as needed. For example, for seizures, a system would sense a pre-ictal or ictal state and then initiate an anticonvulsant or anti-epileptogenic action (electrical stimulation, drug injection, etc.) rather than provide constant stimulation (see Chapter 6). For motor applications, this would take into account a natural training effect, critical for motor learning, by exerting

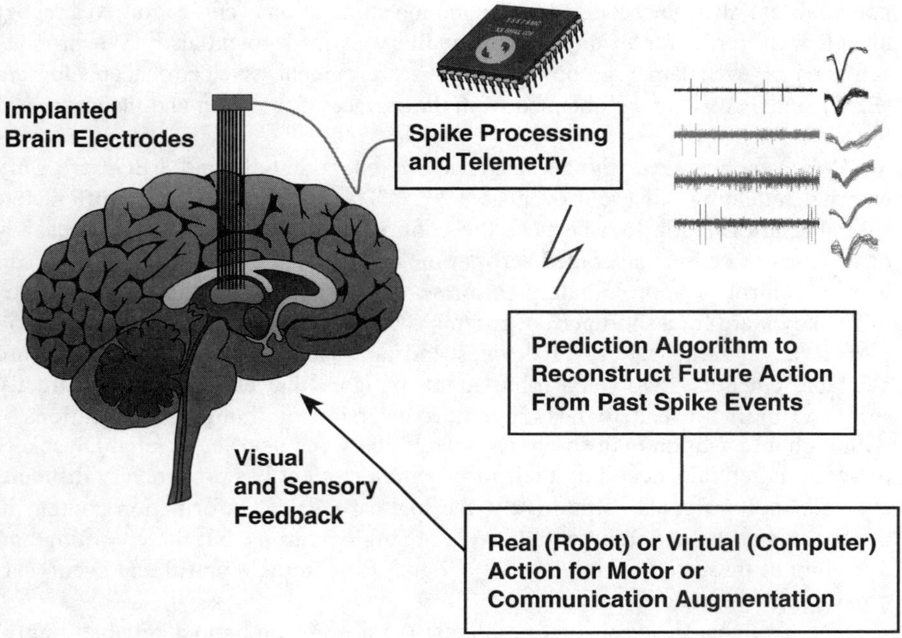

FIGURE 7.1 (See color insert following page 146.) Summary of critical aspects of brain–machine interface. The sensing or afferent arm consists of multiple single neuron electrodes implanted into cortical or subcortical structures. The electrodes detect neuronal single-unit activity (right) that is then sorted for spike occurrence and processed for spike timing information. This more limited data can then be sent via telemetry from an implanted system to a local processing computer where the signals are then converted into a prediction for future action. The third part of the system is the actuator driven by the predictions and honed with visual or sensory feedback to improve functioning on subsequent trials. The actuator may be a real device (robot arm or wheelchair) or a virtual device (computer for speech synthesis or keyboard control).

visual, tactile, or combined control of the device, similar to how motor learning occurs with natural limbs (Figure 7.1). Such feedback is crucial to adjust for different loads, for example, and improve accuracy with motor learning.

7.4 DECIPHERING NERVOUS SYSTEM INFORMATION

Information in the nervous system is primarily routed in action potentials that serve as communication media. Processing of these action potentials occurs at many levels, including presynaptic, postsynaptic and glial–neuronal interactions.[20–25] In some cases, neurons aggregate spatially, leading to common extracellular summation of their individual action potentials if synchronized. These extracellular reflections of hundreds or thousands of neurons occur typically in regions with closely packed neurons arranged in arrays, such as the hippocampus and cerebellum.[26] Evoked potentials are also synchronized by a common stimulation event leading to a recognizable waveform, for example, with auditory evoked potentials.[27] Synchronous activities of even larger groups of neurons are evident as electroencephalogram (EEG) signals that can be obtained from the surface of the brain and the surface of the scalp.

However, averaged signals such as evoked potentials and EEGs are only external reflections of brain events.[4,13,15,26–29] These external signals suffer considerable information loss because the control signal is derived from thousands or millions of neurons averaged across time and space. For example, an EEG can lead to control of approximately six or seven characters per minute on an optimized keyboard for a short period, but this is very limited for most communication purposes.[13] A large variety of devices and approaches to neuroprosthetics are available but none involves a robust control signal that can be derived directly from the brain to lead to fast, reliable conversion of thoughts into actions.[5,6] Although information in the brain is conveyed between neurons in reliable packets or action potentials, decoding their information content has proven very difficult, even for motor signals.[30] Intuitively, the highest level of information content in the brain is at the action potential levels of single neurons, but the recording and decoding of these signals to generate a signal for external control and events has proven highly challenging.[3,8,12,14,25,31,32]

The challenge leads to two problems: (1) a high throughput reliable control signal to directly link the brain with external devices for translation of thought and communication into action,[5,6] and (2) the inherent understanding of what packets of action potentials mean to the brain and how information is transmitted throughout the brain via this common signal, particularly the understanding of concurrent streams of action potentials from multiple neurons as parallel signals between regions. This challenge can be posed from two different angles — the clinical treatment domain of using a control signal (regardless of its meaning if it works) to actuate an external event, and the research domain of interpreting brain coding and networks of neurons involved in coding to explain mechanisms of brain functioning.

7.4.1 Sensory Encoding and Stimulation

Sensory encoding appears to be specific to sensory modality and proximity to the primary receptor. For example, auditory encoding is complex and remains highly controversial, even though many receptors and much of the cochlear nerve are clearly tonotopic. At the peripheral level, many receptors (pressure or temperature receptors) can be measured as having monotonic responses to their input, leading to frequency encoding of the sensory modality. However, at the thalamic level, somatosensory encoding appears to be much more complex due to the processing at intermediate levels. Such processing may also reflect abnormal sensory or pain states, as has been demonstrated in a few patients by thalamic recordings made while they underwent treatment for pain.[20] For many intended motor neuroprosthetic functions, activation of thalamic sensory feedback may be critical to perceive proprioception and sensation for tactile encoding, which may improve motor learning. For example, tactile perception may aid device performance where visual perception fails, for example, objects with different weights and the same appearance.

7.4.2 Motor Encoding and Recording

Electroencephalography has been used to drive devices intended to replace lost motor function.[29] However, the massive summation of electrical activity recorded as an EEG is so general that the output remains unable to generate clinically useful motor movements. Many other techniques exist for studying the output of the brain, although they may not be ideal for use in a BMI designed for use as a human prosthetic. Functional magnetic resonance imaging (fMRI) focuses on blood flow changes that result from metabolic activity areas of the brain. Optical imaging provides information about the activities of neurons by virtue of an intrinsic optical signal generated when neurons are electrically active through changes in tissue swelling (see Chapter 5).

Although these techniques have the advantage that they are noninvasive to the brain, the temporal and spatial resolutions of such techniques preclude their utility in a real-time BMI, where information must be updated at least 10 times per second. For a BMI to demonstrate sufficiently rapid motor output for real-time motion requires at least a 10-Hz response. Also the instrumentation involved with such techniques does not lend itself to something that could be adapted for permanent use by humans. These limitations preclude their effective use for providing a control signal.

7.4.3 Is Information Encoded as an Ensemble?

Measuring the electrical outputs of individual neurons in the brain has been the main technique used by neurophysiologists to study the brain for nearly a century. Since the first implantation of an electrode in the brain by Adrian in 1926, the considerable utility of this technique has been recognized and its use in neurophysiology has blossomed. The benefits of sampling multiple neurons at the same time from a research subject are now more appreciated, and over the past 20 years led to the development of the multielectrode recording technique.[21,23,33,34]

The capability to make such recordings also led to questioning of the older concept of labeled line theory — that an understanding of the functioning of the brain can be traced back to the properties of individual neurons. Rather, the multielectrode technique emphasized the role that populations of neurons play with simultaneous parallel activities. These techniques led to the study of neuronal ensembles and the ways in which multiple neurons participate in generating behavior. Such groups of neurons may be spatially clustered or spread throughout the brain in a functional circuit. The number of neurons involved in naturally encoding even a simple task remains unknown, but preliminary estimates for motor control have suggested more than 500 neurons may form an aggregate that can specify control accurately.[5,6]

7.5 IMPLEMENTATION OF MOTOR NEUROPROSTHETIC DEVICE

Intense ongoing research is focused on understanding the complexities of the mechanisms by which neuronal firing translates into motor activity. It is generally agreed that multielectrodes constitute the most promising technique for acquiring raw data that could be used to drive a useful motor BMI.[21] The resolution of the raw brain signals provided by this technique appears to show the essential attributes of sufficient time resolution (greater than 10 Hz) through interpretation of action potential occurrences. The downside remains that direct implantation of electrodes within the brain is required, with all the inherent risks of any neurosurgical technique including neurological injury, bleeding, and infection (as indicated in Figure 7.1).

Obviously, a noninvasive technique would be ideal, but no suitable candidates for such an externally recorded signal exists. For reasons mentioned above, the multielectrode technique is currently the most suitable for developing a BMI that could be implantable in humans in the near future. Indeed, single neuron versions of such a human BMI using implanted neurotrophic electrodes were implemented and published by Kennedy and Bakay.[14,35] The reason for using a multineuron output requiring an implanted electrode array within the brain instead of an external signal such as an EEG is that information within the brain is specifically relayed through a complex combination of outputs from individual neurons (in the form of action potentials). This goal of real-time multineuron recording has become possible only recently, with the advent of very fast, real-time multiple channel amplification and processing of multineuronal signals to allow updating of the output control stream at up to 100 Hz.

7.5.1 Recording Techniques and Locations

Based on the recent emphasis on studying more neurons in awake and behaving subjects, a multitude of electrode designs have emerged that could potentially be adaptable for use in a BMI. The key factors in electrode design for use in a human brain–machine interface include:

1. Quality and stability of the signal obtained
2. Longevity of signals after the electrode is implanted
3. Number of neurons that can be sampled
4. Sizes of electrodes and biocompatibility of electrode material

Many different approaches to address these issues are underway. The two main types of electrode designs used for research today that could conceivably be adapted for use in a BMI application are microwire arrays and printed circuit silicon microelectrodes. Microwire arrays consist of individual wires made of stainless steel, tungsten, or Pt/Ir, with diameters of 15 to 80 μm. The wires are arranged in a configuration ranging from sixteen to several hundred wires per array. The wires are coated with insulation and the tips are cut bluntly so that the actual recording surface is only the tip of the wire.

Blunt tips have been shown to provide the best long-term recordings, unlike sharp electrodes intended for acute recordings. However, blunt tips do not penetrate the coverings of the brain (pia and arachnoid), in contrast to traditional sharp tungsten microelectrodes. This problem can be overcome for deep (subcortical) recordings by opening the pia surgically and inserting the electrodes through a cannula or guide tube. Surgical opening of the pia can also remove critical blood vessels, irritate the cortex, possibly lead to seizures, and sufficiently damage the underlying cortex that it may be difficult to obtain recordings even from cortical layer V. Thus, how to suitably pass blunt microwires into the cortex remains a challenge. The recording wires are then attached to a connector which, after implantation, is attached to a headstage where amplification, filtering, and processing of the action potential signals begins.

Using the microelectrode wire arrays, each microwire can record up to four single units with an overall yield of approximately one unit per electrode implanted.[34,36] By implanting large arrays across different areas believed to be relevant to generating motor movements, more than 100 neurons have been recorded simultaneously in awake and behaving nonhuman primates. The recordings remained relatively stable with only minimum decay in the number of individual neurons identified over a period of months to years.[36] While the overall number of single cells decays slowly, it appears that the exact same neurons are not recorded each time. This fact becomes important in analyzing and transforming the data stream into a useful output that can adapt itself with a minimum of time and effort on the part of the patient. If different neurons are active daily, does the BMI have to be retrained daily or is the ensemble information sufficiently stable that the overall population retains training information?

Many versions of silicon and biologically inspired electrodes for detecting output signals from single or multiple neurons have been developed but their stability over time and local brain damage have been questioned.[14,21,31,35] Simple microwires have been used successfully for chronic recordings in humans[37] and animals[2,23,34] for many years. The microwires have the advantage of low resistance and ability to detect multiple neurons simultaneously because of their relatively large surface areas.

Pt/Ir microwires show inherently minimal damage at the electrode–brain interface even with constant stimulation and have a long-term history of low toxicity with Medtronics' DBS electrodes.[7] The typical size of these microwires is 30 to 40 μm; they usually have Teflon insulation, and the ends or cut surfaces represent the electrode surfaces.[37]

Whether cortical or subcortical targets would be better for electrode implantation is still unclear. Many years of neurophysiology research generated a large body of knowledge about the motor cortex and related areas, perhaps due to their surface location and availability.[30] This research supports the idea that information about motor control of limbs is coded in the outputs of cortical neurons. Equally compelling data indicate that subcortical locations such as nuclei of the basal ganglia, and particularly the thalamus, contain the information necessary to coordinate motor movements of the extremities.[20,22,24]

Neurosurgeons have decades of experience in passing electrodes to various targets in the thalamus and basal ganglia, including making recordings for the purpose of treating movement disorders. Such procedures have a known track record of minimal morbidity and mortality (only a few percent). In fact, passing electrodes to deep brain targets is now a standard procedure, usually performed for DBS placement.[7] Logistically, it is also easier to place electrodes precisely in a deep brain target compared to the cortex, as a deeply placed electrode must pass more tissue that effectively acts like an anchor. While deep brain targets may be technically less challenging, most research paradigms that could be adapted to a BMI have arisen from work focusing on the cortex, making the cortex a target as likely as the subcortical region for a human BMI.

The ideal location for cortical electrodes is also unclear. Neurons in M1 (Brodman's area 4) are known to be broadly tuned to a variety of different motor movements.[30,34,36] M1 has an expanded layer 5 with large Betz cells making targeting with electrodes easier. Premotor cortical areas involved with motor planning have smaller cells that are more difficult to record from, but are theorized to contain more pertinent information for driving a BMI.[34] Posterior parietal areas involved with associating motor and sensory information are also of interest.

These and other regions have been studied extensively and shown to be important for generating motor commands that produce reaching movements. Wessberg et al. showed that information regarding motor movements is widely distributed across the cortex.[34] However, certain parts of the cortex such as premotor areas apparently contribute a greater amount of information based on their functional specialization.[34]

The emerging picture seems to be that while motor information is widely distributed, certain areas of the cortex would be better for implantation of a BMI in the sense that more information is encoded in these areas or that the BMI may be easier to train. Additionally, most motor information is carried through the thalamus prior to reaching M1, including the output of large motor loops from the basal ganglia (presumably for directing initiation and overall motor plan selection) and the cerebellum (for error correction during the motion), which converge on the motor thalamus. Thus, the thalamus may provide an excellent recording area for a BMI.

Adult human cortical reorganization occurs as a result of deafferentation and motor learning.[16,17,38] The properties of the cortical neurons of a sensory-deprived human such as a spinal cord injury patient or someone suffering from ALS may change and that could impact the ability to extract useful signals from such cortex. It is also known that cortical reorganization can take place based on learning or practice of fine motor tasks of the fingers in terms of expansion of the cortex dedicated to the fingers involved in the motor task.

Rhesus monkeys can intentionally modify the firing rates of single neurons.[39] This ability was confirmed in humans with the implantation of two neurotrophic electrodes in a single ALS patient who was able to modify the firing rate of the neurons on 10 occasions.[14,35] These data lend further support to the idea that an implanted human subject can learn to use a device connected to the central nervous system through training-related modification of cellular properties.

An interesting study suggests that after learning to use tools to reach for items, rhesus monkeys showed changes in visual receptive field firing correlating to incorporation of the tools into their body schemata.[36] Taken together, all these lines of investigation suggest that humans could potentially learn to use new prosthetic limbs connected to their CNSs and incorporate such devices into their body schemata so that the devices would feel like natural extensions of their bodies.

It remains unclear how much effort or training will be required to achieve this. Also, only humans can determine whether the effort to learn to use and manipulate a device would be worth the final function, in other words, determining whether a device is useful for enhancement of their motor or communication functions.[1,13,18,33,35]

7.5.2 Signal Processing and Action Potential Processing

After a stable multineuron signal can be recorded and amplified, a method of combining or interpreting the signals is required (Figure 7.1). For a motor neuroprosthesis, this requires combining or translating the multineuron signals into a robust motor control signal such as three-dimensional arm movement in space that can be replicated on a robot.[2,3,32,34,36] Various approaches to such processing of spike data into a desired target control data set have included linear combination algorithms and recurrent neural networks. However, this final control signal must be translated with sufficient detail that a peripheral device is capable of understanding and acting on the signal.

The presence of visual feedback provides a more rapid training path toward such a virtual task to allow direct feedback on the performance.[12,32] Continued work on action potential processing is needed to resolve critical issues such as spike sorting, spike detection, and whether binning of spikes is desirable. Additionally, knowing whether an adaptive signal processing component is needed to adjust to changing task demands will be critical as opposed to allowing innate brain plasticity to integrate motor and sensory signals into motor learning.

7.5.3 Motion Implementation

Communication with and activation of an external device are required for a motor neuroprosthesis. An appropriate motor control device such as a robot arm with a gripper to be controlled by a brain–computer interface is needed to perform tasks such as eating.[5,6] Many virtual tasks are also of great relevance, for example, an optimized keyboard will allow rapid transmission of characters for communication.[40,41]

Using this paradigm of direct visual feedback, a complex arm movement task can be duplicated using a robot arm and the animal will not have to move its own extremity. The task can be duplicated at a distance, as evidenced by an Internet demonstration of robot arm movement.[34] Convincing preclinical studies indicate the feasibility of a direct brain-to-machine interface using multineuron recording arrays in the cerebral cortices of nonhuman primates. This robust demonstration in nonhuman primates strongly indicates sufficient feasibility to proceed with initial human studies for a similarly designed motor prosthesis. Serious questions remain regarding how many neurons will be required to produce fine motor movements that could replace the functions of fingers. This factor will ultimately decide the utility of such devices and their long-term success.

Several additional aspects of brain–computer interfaces have also been investigated in nonhuman primates. Techniques for electrode array (16 to 128 electrodes per array, each electrode a microwire) implantation into the cortex have been devised; the electrode recordings are stable up to 2 years.[21,36] The microwires provide excellent long-term, multiunit neuronal recordings with typical extracellular triphasic profiles.

Multichannel systems have been devised for recording from a large number of channels (up to 512 currently), using a commercially available system from Plexon (Dallas, TX). These systems include amplification (usual gain of 10,000 larger signal), filtering (300 Hz to 5 kHz), analog-to-digital conversion (12 bits, 40 kHz per channel), spike detection, and spike sorting. The output from the Plexon System focuses on captured waveforms (1 to 2 msec in duration) together with timing; using the captured waveforms allows off-line improved sorting based on waveform detection and clustering algorithms. The timing of these sorted neuron spike waveforms then can be transmitted in real time (usually at 10 Hz updating) to a processing computer that uses past and current neuronal behavior to predict the position of the arm in space. The accuracy of this prediction can then be compared to the real motion in initial studies in which the arm can be directly measured. This comparison shows excellent fidelity of predictions to actual position.

An external actuator including a small robotic arm device (Sensable 1.5 Phantom) or a large robust robotic arm for heavier tasks has been developed. The accuracy of the predictions using a linear algorithm has been excellent, ranging up to 90%, and this entire scheme continues to be updated frequently. Thus, in practice, a nonhuman primate can rapidly learn to control the external device using the brain interface directly, if sufficient visual feedback is provided to properly clue the animal.[5,6,34]

7.5.4 Sensory Feedback

For motor learning, some form of visual or tactile feedback is critical to assess performance. For the nonhuman primate studies, the feedback form can be a video screen with a cursor or a real device that can be visually followed (i.e., a robot arm). For actually gripping and picking up objects, some form of tactile perception related to the object is also needed to enable the user to gauge weight and mass.

Counter-pressure must be placed on the gripper to counteract gravity, depending on the weight of the object, for example a cup containing liquid. Thus, a combination of tactile and visual feedback may lead to more optimal performance. How to introduce this tactile feedback is unclear, but variable pressure on a preserved dermatome (i.e., over the neck or scalp) has been suggested. However, a direct input into a sensory area such as the thalamus may be the critical technique needed, provided that sensory encoding can be deciphered and an appropriate stimulus generated, such that the patient can use this stimulus as representing gripper pressure. The complex integration of the motor output and the visual and tactile inputs into the brain will require considerable plasticity on the part of the brain, clearly requiring significant training for use.

7.5.5 Clinical Applicability and Design Questions

What would a motor neuroprosthetic device look like? Several typical examples are common in the literature, one of which is an electric wheelchair control.[12] A wheelchair could be controlled to move forward or in other directions at a certain rate, mimicking an external joystick with direct brain control, with perhaps audio feedback for collision detection when near an object not otherwise visible.[5]

Another example is a robotic arm for enhancing independent eating to aid a patient with quadriplegia and minimal hand or arm function. The common aspects of such a device include a brain electrode array implanted in one or more areas of the brain with detection and sorting of the action potential data from the electrode array. These data can then be combined in a linear, nonlinear, or other optimized format into a device output stream, for example, a set of coordinates (X, Y, Z) for delivery to the robotic arm to control motion. This type of device has been shown to work well in nonhuman primates for control of an external device such as a robotic arm (Figure 7.1). In addition to motor neuroprostheses, communication aids are also critical for compassionate needs of individuals with disabilities.[13–25,29,35,40] Most communication aids connect to a computer for a virtual screen output or a synthesized speech output, for example.[18,42]

The size and shape of a final, implanted product would likely resemble the current version of Medtronics' DBS electrode that is commonly implanted into brains to control tremors and Parkinson's disease.[7] This DBS electrode currently includes a brain electrode (1 mm in size, 4 contacts), an electrical extension, and a control unit (implanted in the chest wall, similar to a pacemaker).

A likely neuroprosthetic system would include a 1-mm, 64-contact array of microelectrodes with independent microwires implanted into the cortex or a subcortical structure (such as the motor thalamus). An implanted system would then connect to a chip containing preamplifiers and spike sorting, then processed via a chip encoding a motor algorithm and transferred via radio telemetry on a regular wireless computer network frequency to an external device for actuation. Important engineering questions remain regarding the step at which to transmit signals outside the body.

The more processing occurs *in vivo*, the simpler telemetry becomes. However, *in vivo* signal processing has an obvious disadvantage of requiring more complex implantable electronics. An initial system would include visual feedback to identify the accuracy of the intended response and allow correction on subsequent trials. A more sophisticated system could include direct sensory feedback to allow, for example, detection of weight and other properties to enable more sensitive tasks to be performed. This envisioned system would be close to the size, degree of invasiveness, and compassionate use of the DBS system already common in clinical practice. The device is well accepted by patients and presents relatively low risk of untoward brain complications.

Researchers have considerable impetus to work toward a fully implantable system for eventual human application. However, many unknowns remain regarding the needs of such a system for current implementation. For example, the specifications of implanted electrodes — configuration, location, optimal number of microwires — are not known. Other needs include fully implantable amplifiers, filters, and spike sorting devices that can serve as parts of an implantable system. For example, a 96-channel amplifier, filter, and spike sorting circuit could then be routed to a processing chip (possibly a DSP) that would extract a movement code relevant to the task at hand from the spike train data. Finally, the results of this movement code extraction would be broadcast via a standard 802-11b/g wireless computer network to a local computer for control of an external device.

This goal is very different from those associated with other types of neuroprostheses, such as functional neuromuscular stimulation devices intended to recreate artificially a pattern of muscle stimulation for a functional purpose (such as arm use or walking). Rather than attempt to reactivate the body's own musculature that may lack internal nervous system control through disease or injury, the goal of a motor neuroprosthesis is to develop a direct, high-bandwidth signal channel between the brain and the external world. This signal channel could carry intention and communication messages and direct functional motor tasks using an external device (such as a robot arm). While such a signal could eventually be used to reactivate the body's own limbs, this is far from reality, whereas control of an external device is feasible immediately and of significant clinical relevance.

Achieving a fully implantable system is a stepwise goal. The first step is surgical. Electrodes with the appropriate configuration and hardware must be developed and tested intraoperatively until a working prototype is achieved. Much experience in this regard has been gained from experiments with nonhuman primates. Once intraoperative recordings can be made, the next step is to demonstrate, as was done in primates, that long-term stable recordings can be made from human cortex or subcortical structures. Most likely, the first implants will be temporary, connected outside the body via wire connections that can be maintained for weeks in an inpatient setting.[37] Such a setup will answer many important questions such as neuronal yield, stability of signals, and the applicability of previously developed decoding algorithms to these signals. The first generation of actuating devices could also be tested and patient effort and training necessary to use them assessed. Based on such experience, fully implantable designs including telemetry could be conceived.

7.5.6 Incorporation of Device into Body Schemata with Training

In addition to providing information on the technical aspects of device control and adequacy of the signal output from the multineuronal electrode, a considerable intangible component of this project will be further understanding brain function. For example, one concept of brain functioning is that an ensemble of neurons is the critical unit of processing, but the size of such an ensemble is unknown. Another concept to be evaluated is how widespread over the brain the signals representing even a simple motor function may be, both before and after training.[5,6,34]

If a large area of the brain is involved with motor processing, even remotely, then it may be much easier to tap into control signals in many different locations including traditional nonmotor areas such as the frontal lobe and subcortical regions involved with motion. Additionally, motor learning by definition involves incorporation of a device into an extension of the body so that control of the device becomes subconscious.[36] For example, learning to ride a bicycle involves a progressive capability to understand stability, speed and direction, which ultimately becomes automatic. Motor learning by nature is disrupted by conscious thought of the activity. Therefore, a critical aspect to be investigated is whether incorporation of the motor function of a neuroprosthetic device occurs at the subconscious level and becomes a more or less automatic control (similar to a limb).

7.6 IMPLEMENTATION OF A NEUROPROSTHETIC COMMUNICATION DEVICE

In addition to patients with motor dysfunctions, a large group of patients have difficulties with communication.[13,18] In cases of cortical damage, the difficulty could be expressive or receptive aphasia, for example, whereas cerebral palsy involves a global difficulty with motor output and hindrance of voice communication. Currently, such patients often rely on minimal residual motor function for output, such as using a toe for typing or elbow motion to make choices, for example, from a yes/no dichotomy.

An established array of devices using binary choices for speech and activity selection, virtual keyboards, and other computer uses is already available.[40] However, the devices (such as virtual typing on an optimized keyboard) could function far better with improved control signals from the nervous system. Thus, a brain–computer or BMI has been long sought for communication disorders. The first brain implants of single channel electrodes were intended to improve communications in highly disabled patients, those with severe strokes and locked-in syndromes.[14,35]

The goal of the implants in such cases was to offer improved communication output through virtual keyboards and control of computer cursors, and the devices had some success. Thus, a more complex multineuron BMI may offer a more substantial signal throughput to enable more sophisticated communications and integration into society. All the schemes discussed earlier for multineuron implants, signal processing, and connection to external devices apply equally to communication disorders and motor disorders.

7.7 FUTURE DIRECTIONS AND CONCLUSIONS

Development of neuroprosthetic aids follows a clear hypothesis that multineuron outputs can offer improved signals for external device control if decoding of the signals from appropriate brain regions can be accomplished. Thus, neuroprosthetics by and large offers a glimpse into translational neuroscience and neurosurgery, particularly development of devices. Many laboratories have convincingly demonstrated that the goal of an effective motor neuroprosthesis can be accomplished in nonhuman primates, and one group has successfully taken this effort to human application.[43]

Ethically, offering enhanced functional motor or communication independence to a patient deprived of such abilities appears to be a more compassionate goal than relieving tremor, for example, yet brain implants are commonly done for the latter, with excellent patient acceptance.[1] A large number of neuroprosthetic devices are now in common clinical use and many more sensory and motor aids have been conceptualized or are in various stages of development. The technology for increasing sophistication of these implanted devices continues to rapidly grow, anticipating in the near future possibilities for both effective motor control as well as sensory enhancement.

REFERENCES

1. Caplan, A.L., Is better best? A noted ethicist argues in favor of brain enhancement, *Sci. Am.*, 289, 104–105, 2003.
2. Chapin, J.K. et al., Real-time control of a robot arm using simultaneously recorded neurons in the motor cortex, *Nat. Neurosci.,* 2, 664–670, 1999.
3. Donoghue, J.P., Connecting cortex to machines: recent advances in brain machine interfaces, *Nat. Neurosci.,* 5, 1085–1088, 2002.
4. Kubler, A. et al., Brain–computer communication: unlocking the locked in, *Psychol. Bull.,* 127, 358–375, 2001.
5. Nicolelis, M.A.L., Actions from thoughts, *Nature,* 409, 403–407, 2001.
6. Nicolelis, M.A.L., Brain–machine interfaces to restore motor function and probe neural circuits, *Nat. Rev. Neurosci.*, 4, 417–422, 2003.
7. Shrivastava, R.K. and Germano, I.M., Deep brain stimulation for the treatment of Parkinson's disease, *Contemp. Neurosurg.,* 23, 1–10, 2001.
8. Chapin, J.K., Neural prosthetic devices for quadriplegia, *Curr. Opin. Neurol.*, 13, 671–675, 2000.
9. Hoag, H., Neuroengineering: remote control, *Nature*, 423, 796–798, 2003.
10. Joseph, A.B., Design considerations for the brain–machine interface, *Med. Hypotheses,* 17, 191–195, 1985.
11. Loeb, G.E., Neural prosthetic interfaces with the nervous system, *TINS,* 12, 195–201, 1989.
12. Nicolelis, M.A.L. and Chapin, J.K., Controlling robots with the mind, *Sci. Am.,* 287, 46–53, 2002.
13. Wolpaw, J.R. et al., Brain–computer interfaces for communication and control, *Clin. Neurophysiol.*, 113, 767–791, 2002.
14. Kennedy, P.R. and Bakay, R.A., Restoration of neural output from a paralyzed patient by a direct brain connection, *Neuroreport,* 9, 1707–1711, 1998.
15. Kubler, A. et al., The thought translation device: a neurophysiological approach to communication in total motor paralysis, *Exp. Brain. Res.,* 124, 223–232, 1999.
16. Karni, A. et al., Functional MRI evidence for adult motor cortex plasticity during motor skill learning, *Nature,* 377, 155–158, 1995.
17. Lotze, M. et al., Phantom movements and pain: an fMRI study in upper limb amputees, *Brain,* 124, 2268–2277, 2001.
18. Caves, K., Using encoded input strategies to facilitate computer and communication access for individuals with physical disabilities, *Top. Stroke Rehab.,* 7, 12–20, 2000.
19. Nuttin, R. et al., Long-term electrical capsular stimulation in patients with obsessive-compulsive disorders, *Neurosurgery,* 6, 1263–1274, 2003.

20. Hua, S.E. et al., Microelectrode studies of normal organization and plasticity of human somatosensory thalamus, *J. Clin. Neurophysiol.,* 17, 559–574, 2000.
21. Kralik, J.D. et al., Techniques for long-term multineuronal ensemble recordings, *Methods*, 25, 121–150, 2001.
22. Lozano, A.M., Hutchison, W.D., and Dostrovsky, J.O., Microelectrode monitoring of cortical and subcortical structures during stereotactic surgery, *Acta Neurochir.*, S64, 30–34, 1995.
23. Nicolelis, M.A.L. et al., Sensorimotor encoding by synchronous neural ensemble activity at multiple levels of the somatosensory system, *Science,* 268, 1353–1358, 1995.
24. Rodrigues-Oroz, M.C. et al., The subthalamic nucleus in Parkinson's disease: somatotopic organization and physiological characteristics, *Brain,* 124, 1777–1790, 2001.
25. Schmidt, E.M., Single neuron recording from motor cortex as a possible source of signals for control of external devices, *Ann. Biomed. Eng*., 8, 339–349, 1980.
26. Sinkjaer, T. et al., Biopotentials as command and feedback signals in functional electrical stimulation systems, *Med. Eng. Phys.,* 25, 29–40, 2003.
27. Donchin, E., Spencer, K.M., and Wijesinghe, R., The mental prosthesis: assessing a P300-based brain computer interface, *IEEE Trans. Rehab. Eng.,* 8, 174–179, 2000.
28. Makeig, S. et al., A natural basis for efficient brain-actuated control, *IEEE Trans. Rehab. Eng.,* 8, 208–211, 2000.
29. Wolpaw, J.R. et al., An EEG-based brain–computer interface for cursor control, *Electroencephalogr. Clin. Neurophysiol.*, 78, 252–259, 1991.
30. Cheney, P.D. et al., Cortical motor areas and their properties: implications for neuroprosthetics, *Prog. Brain Res*., 128, 135–160, 2000.
31. Isaacs, R.E., Weber, D.J., and Schwartz, A.B., Work toward real-time control of a cortical neural prosthesis, *IEEE Trans. Rehab. Eng.,* 8, 196–198, 2000.
32. Taylor, D.M., Tillery, S.I., and Schwartz, A.B., Direct cortical control of 3D neuroprosthetics devices, *Science,* 296, 1829–1832, 2002.
33. Mussa-Ivaldi, F.A. and Miller, L.E., Brain-machine interfaces: computational demands and clinical needs meet basic neuroscience, *TINS*, 26, 329–334, 2003.
34. Wessberg, J. et al., Real-time prediction of hand trajectory by ensembles of cortical neurons in primates, *Nature,* 408, 361–365, 2000.
35. Kennedy, P.R. et al., Direct control of a computer from the human central nervous system, *IEEE Trans. Rehab. Engin.,* 8, 198–202, 2000.
36. Carmena, J.M. et al., Learning to control a brain-machine interface for reaching and grasping by primates, *Publ. Libr. Sci.*, 1, 1–16, 2003.
37. Ekstrom, A.D. et al., Cellular networks underlying human spatial navigation, *Nature*, 425, 184–188, 2003.
38. Roux, F.E. et al., Virtual movements activate primary sensorimotor areas in amputees: report of three cases, *Neurosurgery,* 49, 736–742, 2001.
39. Wyler, A.R. and Burchiel, K.J., Operant control of epileptic neurons in chronic foci of monkeys, *Brain Res*., 212, 309–329, 1981.
40. Perelmouter, J. and Birbaumer, N., A binary spelling interface with random errors, *IEEE Trans. Rehab. Eng.,* 8, 227–232, 2000.
41. Pfurtscheller, G. et al., Brain oscillations control hand orthosis in a tetraplegic, *Neurosci. Lett.,* 292, 211–214, 2000.
42. Tarr, M.J. and Warren, W.H., Virtual reality in behavioral neuroscience and beyond, *Nat. Neurosci.,* 5, 1089–1092, 2002.
43. Patil, P.G. et al., Ensemble recording of human subcortical signals as a source of motor control signals for a brain-machine interface. *Neurosurgery,* 55, 1–10, 2004.

8 Surgical Treatment of Movement Disorders: DBS, Gene Therapy, and Beyond

Parag G. Patil and Dennis A. Turner

CONTENTS

0-8493-1482-8/05/$0.00+$1.50

8.1 INTRODUCTION

Disorders of movement represent the frontier of understanding of brain function in that the basic mechanisms underlying normal (and abnormal) movement can be ascribed to individual brain structures, but the detailed functions of these structures and their interactions are not well understood.[1–7] The clinical treatment of movement disorders, particularly through neurosurgery, highlights the evolution in understanding nervous system function. In many instances, incompletely proven hypotheses, serendipity, and simple trial-and-error have led to advances in patient treatments prior to a full mechanistic understanding of the disease process or the treatment effect.

For example, the basal ganglia represented completely unknown territory in the 1930s when Russell Meyers began neurosurgical extirpation of the caudate and putamen for various movement disorders. This fascinating history extends to the present day, as radio frequency lesion generation, deep brain stimulation (DBS), and other approaches to disorders of the basal ganglia are proposed and tested in patients with movement disorders.[1,6,8]

Several factors profoundly influence the development and improvement of treatments for movement disorders. First, the neural systems likely to subserve motor control (basal ganglia, globus pallidus, and ventral thalamus) are still only loosely integrated into schemes that can account for normal motor control, although lesions of these structures are clearly associated with pathological disorders of movement.[5] Second, one of the many critical lessons in the treatment of movement disorders is that the functional effects of particular therapeutic interventions may be far different under pathological conditions than they are under normal conditions. Therefore, the impact of many proposed therapies may be somewhat unpredictable.

This chapter provides an overview of the spectrum of movement disorders, discusses the functional connectivity of basic motor-associated circuits in the brain, and reviews current surgical treatments of movement disorders. We hypothesize that the next generation of movement disorder treatments will involve a number of new approaches including more sophisticated sensing and stimulation systems, novel medical delivery systems, new medications, and gene therapy.[1-3,7,8] This chapter introduces several preclinical and clinical investigations along these lines, suggesting that clinical applicability of such therapies may potentially follow within the next few years.

8.2 SPECTRUM OF MOTOR ABNORMALITIES

Traditionally, motor disorders are classified into two main groups of abnormalities: those in which movement is hypokinetic or less than normal, and those in which movement is hyperkinetic or greater than normal. Parkinson's disease is the classic hypokinetic disorder, with bradykinesia and rigidity as hallmarks, although pill-rolling resting tremors are also common.[5] This peculiar mix of decreased capability for motion together with tremor produced the historical term *paralysis agitans* or shaking palsy, still used as a clinical code.

Hyperkinetic movement disorders include tremor, cerebral palsy, chorea, and hemiballismus. Interestingly, in patients with Parkinson's disease who have been on long-term L-dopa therapy, almost all forms of hyperkinetic movement may also be observed. These treatment-related dyskinesias are relatively newly discovered phenomena, noticed only in the past 20 years when patients have remained on L-dopa therapy for longer periods.[5,9,10] *Dyskinesia* includes any type of dystonic posturing or choreoathetotic movement and may be identical to the hyperkinetic features of primary and secondary dystonias or cerebral palsy. This crossover from a predominantly hypokinetic to a predominantly hyperkinetic movement disorder solely due to treatment effects has blurred the traditional distinctions among movement disorders. As understanding of the genetic basis of movement disorders increases, a more proper classification scheme for movement disorders may become available. Particularly for the dystonias, Parkinson's disease, and parkinsonian syndromes, this evolution is already taking place.[7,12]

8.2.1 Parkinsonian Syndromes and Parkinson's Disease

James Parkinson published his observations on shaking palsy in 1817.[5] The chief clinical symptoms of parkinsonian syndromes are tremor, rigidity, bradykinesia, postural instability, autonomic dysfunction, and frequently cognitive impairment. Parkinsonian syndromes can be the result of Parkinson's disease or manifestations of many other nonspecific conditions collectively referred to as secondary parkinsonism. Most patients with Parkinson's disease initially respond to L-dopa therapy; whereas many patients with parkinsonism do not. From this feature arises one initial means of classification for this group of diseases.[5] In both Parkinson's disease and parkinsonism, many axial symptoms such as freezing and nonmotor symptoms such as autonomic dysfunction are resistant to current medical interventions.[1,7,12]

The pathological hallmark of Parkinson's disease is a loss of cells within the pars compacta of the substantia nigra and the presence of Lewy bodies in a fraction of the remaining cells. The clinical features of parkinsonism arise in a wide variety of degenerative disorders including striatonigral degeneration, progressive supranuclear palsy, corticobasilar degeneration, and Shy–Drager syndrome. Classically, parkinsonism has been observed as a postinfectious manifestation of von Economo's encephalitis, a disease that peaked in Europe and North America in the early 1920s. Parkinsonism may also result from toxins such as carbon monoxide, methanol, mercury or MPTP from stroke or from head injury.

Initial therapy for a Parkinson's disease patient can include amantadine, an antiviral agent thought to augment the release of dopamine from striatal neurons; selegiline, a monoamine oxidase inhibitor that slows the intracerebral degradation of dopamine; pergolide, a synthetic ergot derivative that stimulates dopamine receptors; and occasionally vitamin E, an antioxidant.[2,5,10,12] As the disease progresses and the efficacy of these treatments wanes, L-dopa is added to the regimen. L-dopa (or levodopa), a metabolic precursor of dopamine, is the most effective agent for the treatment of Parkinson's disease. It is typically given with the dopamine decarboxylase inhibitor carbidopa to prevent degradation of L-dopa in peripheral tissues. After 8 to 12 years of levodopa–carbidopa (Sinemet) therapy, patients may begin to experience the long-term side effects of these medications, including dyskinesias, and may be considered for DBS.[1,3,8] Stimulating electrodes are placed into the globus pallidus pars interna (GPi) or the subthalamic nucleus (STN). DBS appears to allow a long-term reduction in Sinemet dosage, reducing the severity of medication-induced dyskinesias.[3,13–15]

Responses to therapies such as surgery are very different for Parkinson's disease and the other forms of parkinsonism. As a result, the identification of patients with characteristic histories of Parkinson's disease, including slow progression and L-dopa responsiveness, is very important to the choice of therapy. In addition, not all features of Parkinson's disease, including eye movement abnormalities and dementia, are responsive to surgical intervention.[1,7,15] Therefore, patients with predominantly treatment-resistant symptoms must be excluded from surgery.

Despite incomplete responses to therapy, the wide range of medical and surgical interventions attempted in the treatment of patients with parkinsonism or Parkinson's disease suggests the high level of motivation for treatment present in both the patient and physician populations. Furthermore, the resistance of many symptoms to both medical and surgical therapy, including speech impairments, abnormal postures, gait and balance problems, autonomic dysfunctions, cognitive impairments, and psychiatric disturbances provides goals for the development of new forms of treatment.[7]

8.2.2 Tremor

Tremor is defined as oscillatory movement about a joint. Normal physiologic tremor occurs in a range of 8 to 12 Hz in all muscle groups. Pathological tremor occurs in a range of 4 to 7 Hz and preferentially affects particular muscle groups, such as distal limbs. Pathological tremor may be subclassified into two main categories: action (or postural) tremor and rest tremor.[5] Action tremor is present during voluntary movement and is absent when limbs are at rest. By contrast, rest tremor is present in repose and suppressed during voluntary movement.

The most common form of action tremor is essential tremor. Such tremor often arises in the second decade of life, may worsen with age, and is most pronounced during attempts to maintain a fixed posture. It typically affects the upper extremities and spares the lower extremities. The tremor is typically worsened with emotion, fatigue, or caffeine and is generally improved with alcohol. Pharmacological therapies for essential tremor include the beta-blocker propranolol and the anticonvulsant primidone. Other forms of action tremor may occur with neurological disorders such as multiple sclerosis or meningoencephalitis.

Rest tremor is commonly noted in Parkinson's disease.[5] The coarse 3- to 5-Hz tremor occurs in the distal upper extremities during rest and is absent during sleep. The tremor subsides with action such as lifting a cup, but immediately resumes when the hand is still, such as when a cup is held close to the mouth. In Parkinson's disease, a mild tremor may be the principal manifestation for many years, with few other manifestations of the disorder. The tremor may respond to pharmacological therapy with the phenothiazine derivative ethopropazine (Parsidol) or the anticholinergic trihexyphenidyl (Artane). Other forms of rest tremor may occur with other parkinsonian syndromes or Wilson's disease.

For patients with either action or rest tremor, the condition may be highly disabling. Hence, many patients pursue treatment specifically for the tremor. DBS has been approved for both essential action tremor and parkinsonian rest tremor. The thalamus is the most common target of stimulation.[1,3,11]

8.2.3 Generalized, Focal, and Hemi-Dystonia

Dystonia is a tonic co-contraction of agonist and antagonist muscles in one region of the body, resulting in a transient or persistent extreme of posture.[16] Dystonia may involve a focal group of muscles such as an eyelid (blepharospasm), the head and neck (spasmodic torticollis), a hand (writer's cramp), one side of the entire body (hemi-dystonia), or the entire body (diffuse bilateral dystonia). Manifestations of dystonic conditions may be progressive, initially appearing as mannerisms, and later becoming more persistent.

Many forms of dystonia are idiopathic. However, dystonia also occurs secondary to metabolic disorders such as Wilson's disease, degenerative disorders such as Huntington's disease, drug toxicity such as haldoperidol intoxication, or cerebral hypoxia. No clear pathologic changes are consistently associated with dystonia. A severe form of heritable, generalized dystonia has been associated with mutations of the DYT1 gene. This disorder, termed torsion dystonia of childhood, involves progression from intermittent and focal involuntary movements to persistent contortions of the entire body. In some instances, dystonia may be occupationally related, such as spasms of the hand (writers), spasms of the hand and neck (violinists), and spasms of the lip (trombonists).

Although L-dopa, bromocriptine, benzodiazepines, and other pharmacological interventions may be helpful in some cases, few dystonia patients generally respond to medical management. In many cases of focal dystonia, therapy consists of transient disruption of muscle function with botulinum toxin. In the past, stereotactic lesioning of the ventrolateral thalamus or the pallidum resulted in substantial improvements in axial symptoms for some patients.[11] Recently, pallidal DBS has been applied to the treatment of generalized dystonia.[16] Stimulation of the GPi has been observed to improve dystonia, presumably through effects on pallidal afferents and connections to brainstem nuclei. Interestingly, such pallidal stimulation requires a considerable period before showing treatment effects. Unlike the DBS treatment of tremor or Parkinson's disease, in which symptoms begin to abate within seconds to minutes of DBS lead activation, the symptoms of dystonia may only begin to improve after days to weeks of pallidal stimulation. This slow onset suggests that considerable motor circuitry reorganization is required to achieve observable effects. The mechanisms and motor circuits involved remain unknown.

8.2.4 Chorea and Choreoathetosis

Chorea suggests dance-like rapid, involuntary, short-distance movements that vary from simple to quite elaborate. Athetosis refers to a slow, writhing motion resulting from an inability to maintain a fixed position in space. Chorea and athetosis are observed in Huntington's disease, post-infectious Sydenham's chorea, kernicterus-associated basal ganglia injury, and L-dopa associated dyskinesias.[5,7,9] Patients with choreoathetosis often attempt to incorporate the involuntary motions into voluntary movements, giving them a bizarre, dramatic character. Medical therapies for choreoathetosis are limited. Haldoperidol, a dopamine antagonist, demonstrates some improvement of abnormal movements associated with Huntington's disease. Although many stereotactic surgical lesions have been proposed as treatments, pallidal lesions and DBS have been the only effective treatments for Parkinson's disease-associated symptoms.[1,3,11,15]

8.3 BRAIN CIRCUITS CONCERNED WITH MOVEMENT

Multiple regions of the cerebral cortex, basal ganglia, thalamus, cerebellum, and brainstem are involved in the control of movement. In addition, neuronal circuits within the spinal cord contribute to complex motor control. The roles of these multiple, interacting regions to motor control have been roughly delineated,[5] but the details of the functioning of these regions, particularly of the basal ganglia, remain highly controversial. In general, physiological studies observed the activities of various parts of the brain during the performance of specific, stereotyped two- and three-dimensional movements.

The relationship of regional neuronal activity to the initiation of movement and to the direction and type of motion was then observed across multiple trials. It became apparent that primary motor cortex (M1) activity plays a pivotal role in movement and is highly correlated with subsequent action. However, several other motor areas also contribute to movement including pre-motor cortex (Area 6), posterior parietal cortex (PP), and the supplementary motor area (SMA).[17,18] Two major loops modifying the cortical control of movement include the cortex–basal ganglia–thalamus–cortex loop and the cortex–pons–cerebellum–thalamus–cortex loop reviewed next; brainstem control of axial motion is also examined.

8.3.1 CORTEX–BASAL GANGLIA–THALAMUS–CORTEX LOOP

Among the basal ganglia, the putamen is more involved in motor control than the caudate nucleus, and is tightly linked to the globus pallidus and thalamus. The circuit from cortex to putamen, pallidum, STN, substantia nigra pars reticulata, back to thalamus, and then to the cortex, is clearly involved in motor control. The circuit has an inhibitory effect upon the motor thalamus leading to the theory that the circuit tunes in certain desired actions while suppressing undesired actions. In Parkinson's disease, the depletion of dopamine in the putamen results in altered output from this loop, significantly slowing movement.[5,6]

Treatment with L-dopa leads to normalization of movement velocity by correcting the disordered control effect of this loop upon thalamic and cortical outputs. However, a lesion within the GPi for treatment of Parkinson's disease[11,15] that theoretically should block the output from this loop actually enhances motion. This

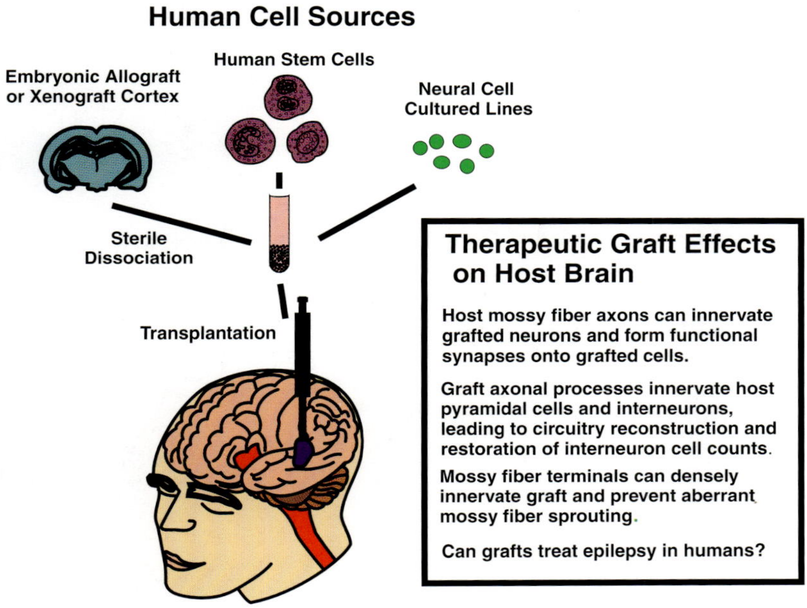

COLOR FIGURE 2.1 Cortical grafting studies. First, cell sources include human or porcine embryonic cortex or hippocampus, various types of progenitor or stem cells, or cultured cell lines, most derived from neuronal tumors. After dissociation and transplantation, the fate of the transplanted cells can be assessed for synaptic integration within the host. In rodent models, therapeutic graft effects on the host include the formation of mossy fiber synapses onto grafted neurons, amelioration of postlesion interneuron loss, and prevention of aberrant mossy fiber sprouting. Whether grafts can ameliorate epilepsy remains to be analyzed in rodents and humans although the framework has been established.

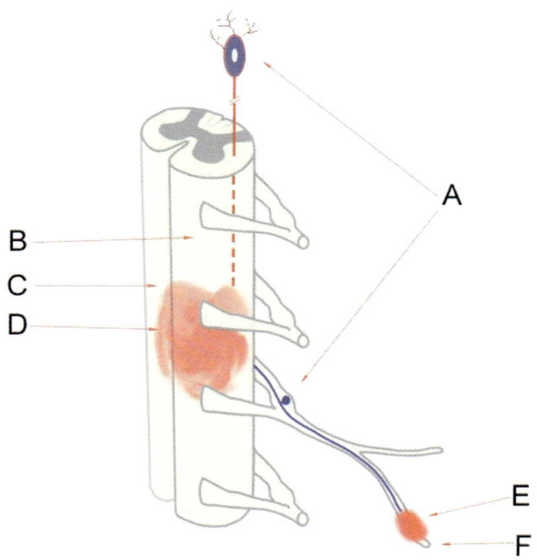

COLOR FIGURE 3.1 Targets for the treatment of spinal cord and peripheral nerve injuries. A: Genetic or small molecule treatments to induce a proregeneration response in the cell body. B: Inhibition of myelin associated growth inhibitors and chondroitin sulfate proteoglycans at the site of injury. C: Neuroprotectants to prevent the spread of injury through secondary mechanisms. D: Transplantation of stem cells, olfactory ensheathing glia, or Schwann cell bridges to span the area of injury and replace lost cell populations; transplantation of macrophages or neurotrophin secreting cells to prevent cell loss and promote regrowth. E: Use of synthetic grafts infused with Schwann cells, extracellular matrix, or neurotrophins to span the area of injury. F: Infusion of neurotrophins or use of electrical stimulation to improve the pace and accuracy of axonal regeneration.

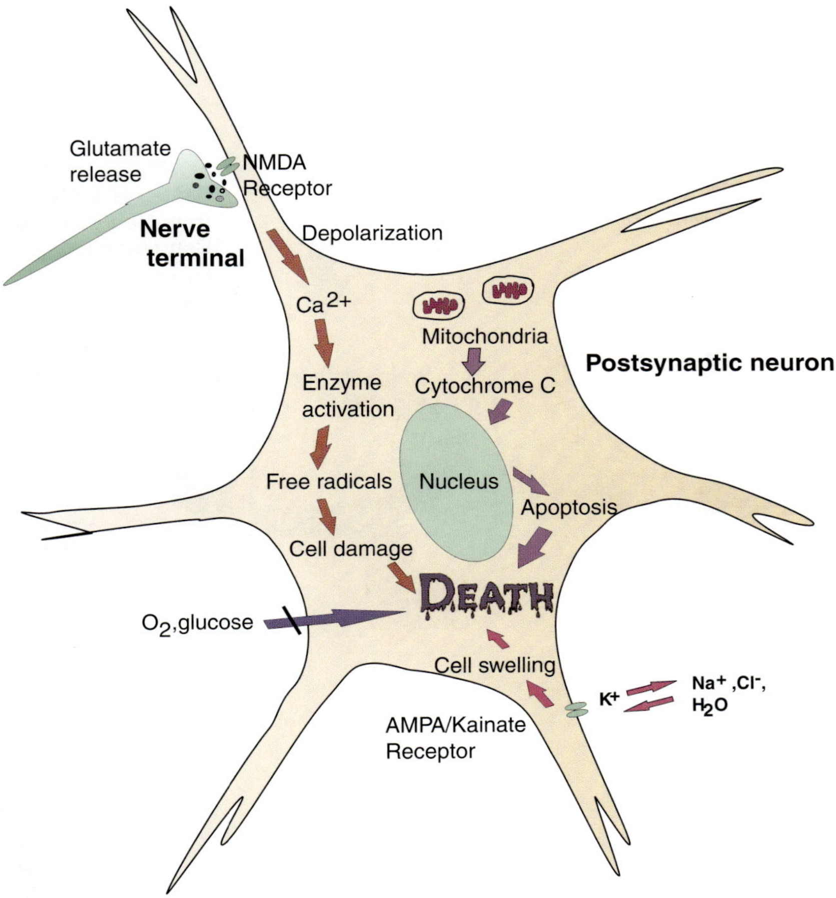

COLOR FIGURE 4.1 Mechanisms of cell death. A typical neuron is represented indicating a variety of perturbed physiological mechanisms leading to cell death. These mechanisms include excess glutamate stimulation and secondary depolarization (excitotoxicity); loss of substrate (oxygen or glucose); free radical formation, particularly following reoxygenation; apoptosis generated by cytochrome C release from mitochondria; and cell swelling induced by water influx.

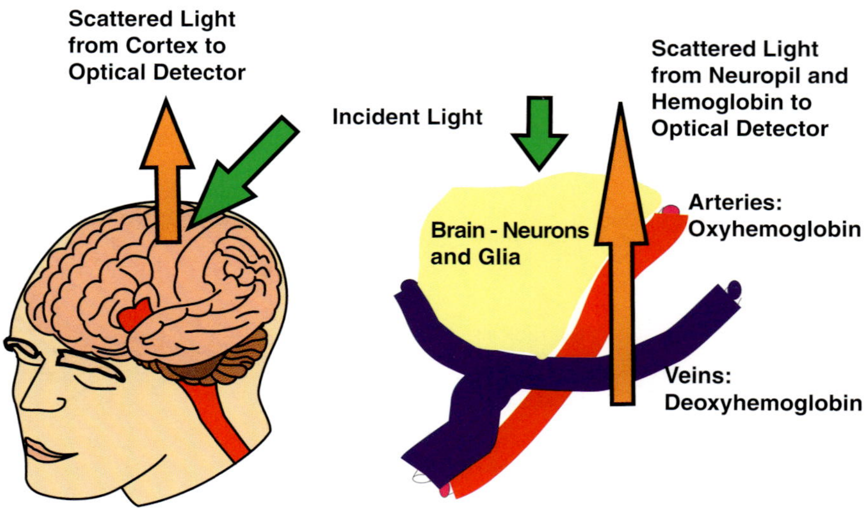

COLOR FIGURE 5.1 Optical imaging at brain and neuropil levels. Two types of optical imaging approaches: at macro (whole brain) level (left) and at neuropil (micro) level (right). In intact brain, incident light is more or less scattered or absorbed by tissue and its contents, with the resulting scattered light detected by an optical system. At the micro level, scattering occurs in distinct elements that include the neuropil (that has its own intrinsic optical signal), and either oxyhemoglobin or deoxyhomoglobin within the blood vessels. Separate absorption and scattering characteristics appear, depending on the relative content of oxy- and deoxy-hemoglobin.

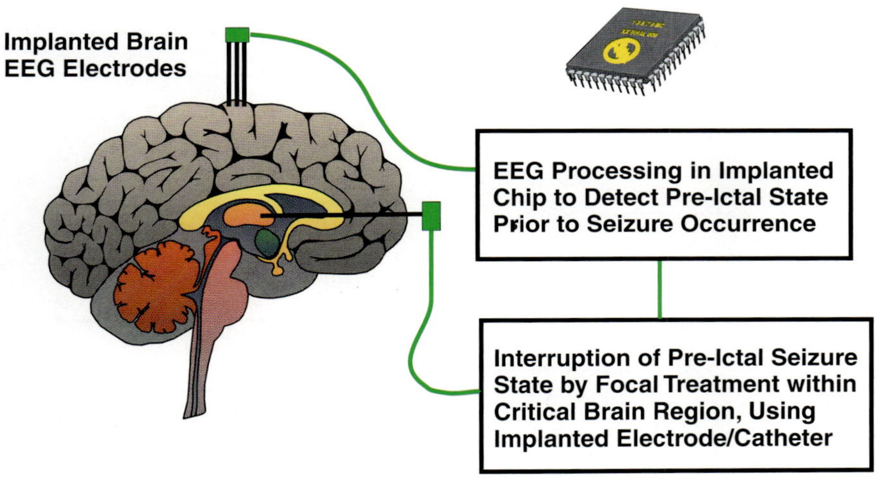

COLOR FIGURE 6.1 The critical elements of a demand or feedback system for epilepsy include three major components: (1) a sensing arm or group of macro (EEG) electrodes on the brain surface or implanted near the maximally abnormal area that are critical for detecting both the pre-ictal state and the transition into seizure; (2) EEG processing for the detection of the pre-ictal state, usually considered optimal when the seizure is still building and the state can be interrupted in a subconscious fashion; and (3) stimulation or medical treatment to interrupt the seizure. The final component may be most efficient if administered near the origin of the seizure, but could also be a systemic injection, for example.

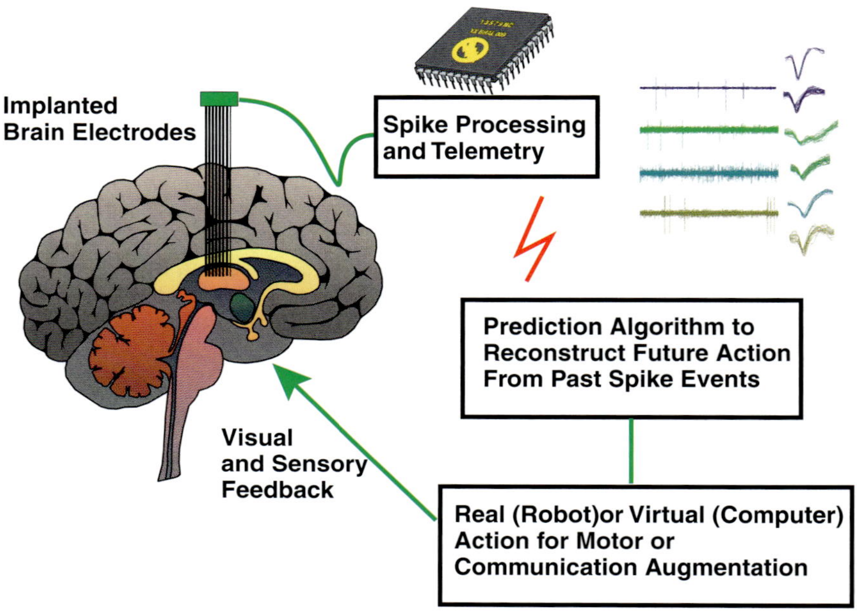

COLOR FIGURE 7.1 Summary of critical aspects of brain–machine interface. The sensing or afferent arm consists of multiple single neuron electrodes implanted into cortical or subcortical structures. The electrodes detect neuronal single unit activity (right) that is then sorted for spike occurrence and processed for spike timing information. This more limited data can then be sent via telemetry from an implanted system to a local processing computer where the signals are then converted into a prediction for future action. The third part of the system is the actuator driven by the predictions and honed with visual or sensory feedback to improve functioning on subsequent trials. The actuator may be a real device (robot arm or wheelchair) or a virtual device (computer for speech synthesis or keyboard control).

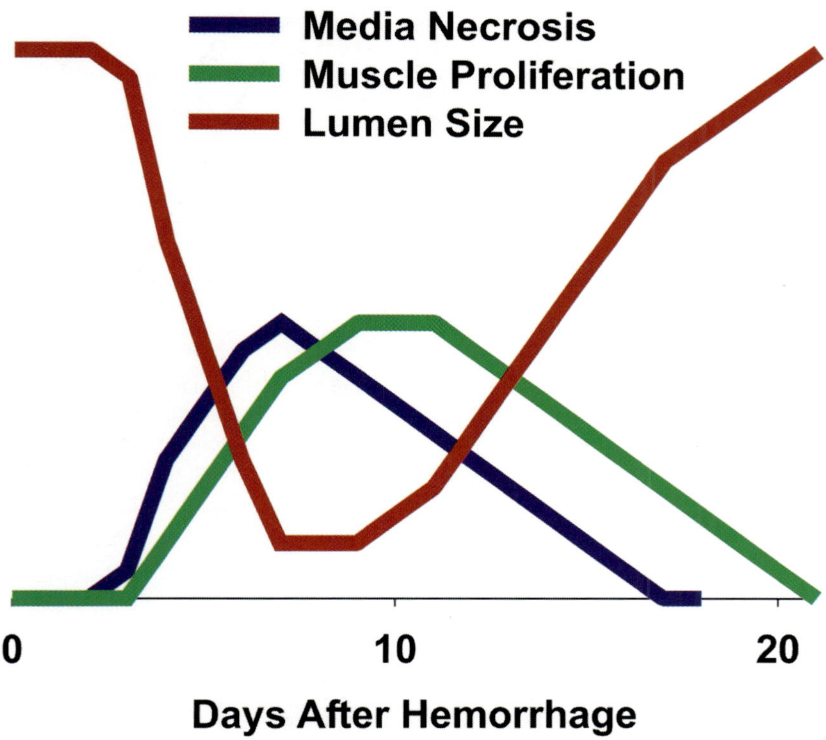

COLOR FIGURE 11.1 Time course of delayed cerebral vasospasm. Two critical processes contribute to delayed cerebral vasospasm. The first is media necrosis that likely begins soon after subarachnoid hemorrhage and peaks at 5 to 7 days. The start of media necrosis acts as a signal to begin smooth muscle cell proliferation for eventual replacement of the smooth muscle cells in the media. However, the additional cells created by new dividing myoblasts and fibroblasts further increase the width of the media. The processes of media necrosis and media cellular proliferation significantly narrow the lumen, beginning early, but peaking at the 5- to 10-day range.

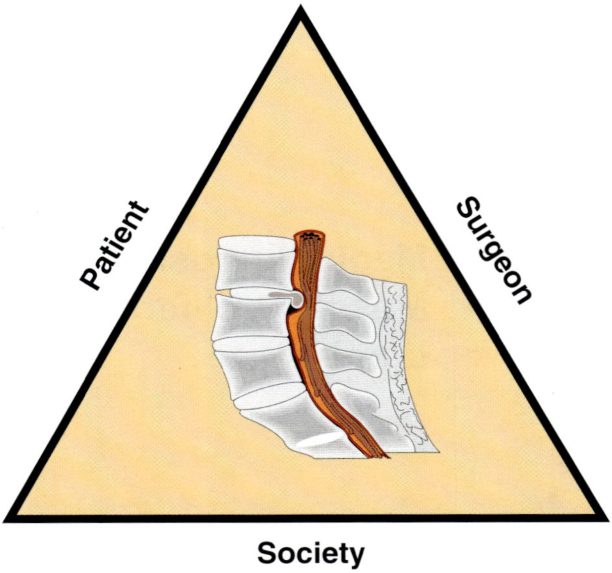

COLOR FIGURE 14.1 Multiple perspectives on lumbar disc herniation. The center component is one of the most common spine problems: lumbar disc herniation. There are three different perspectives on the disease and surgical treatment, depending upon the vantage point. The patient, the surgeon, and society all have different views of definition of outcome and whether a surgical procedure is worth undertaking.

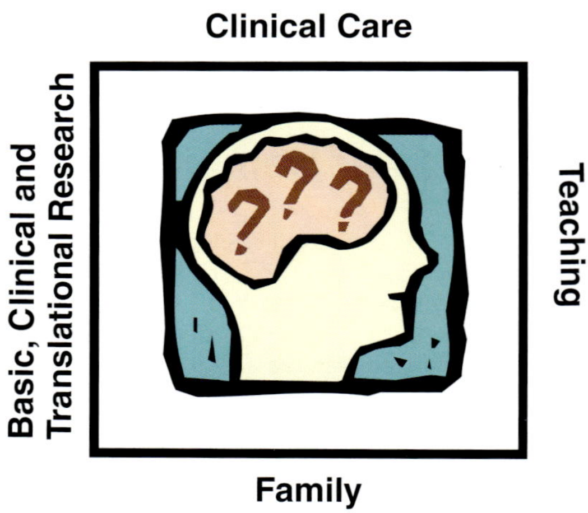

COLOR FIGURE 16.1 Dilemmas of academic neurosurgery. The center figure represents a typical neurosurgeon trying to decide on a career in academia, or after embarking on such a career, deciding how to pay appropriate and adequate attention to multiple critical aspects of life simultaneously. He or she has no easy choices.

indicates that a reorganization of normal motor control circuits must occur in Parkinson's disease so that motor output is subserved via alternative parallel pathways.

It is hypothesized that a decrease in dopaminergic input to the striatum in Parkinson's disease results in reduced direct inhibition of the GPi. In addition, the lack of GPi inhibition of the STN leads to overexcitation of the GPi, particularly because cortical excitatory input to the STN is preserved.[6,15] With less inhibition from the putamen directly upon the GPi and increased excitation of the GPi from the STN, inhibitory output from the GPi to the thalamus is markedly increased, resulting in a suppression of movement output from the thalamus. This model of basal ganglia function suggests that GPi lesions may improve parkinsonian symptoms and thalamic lesions should not. However, thalamic lesions help reduce parkinsonian tremors, suggesting that this model may be incomplete.

8.3.2 Cortex–Pons–Cerebellum–Thalamus–Cortex Loop

Cortical efferents from multiple regions project upon ipsilateral pontine nuclei. These nuclei then project into the cerebellum. Cerebellar outputs project to the lateral and posterolateral thalamic nuclei that, in turn, project upon the primary motor cortex. This loop is thought to be important in motor control, particularly during motion. In functional MRI studies comparing real and imagined motions, the cortex and basal ganglia are active in both situations; whereas the cerebellum is only active during real motion.[19]

The inputs to the cerebellum from the periphery are proprioceptive fibers, activated during motion. Many physiological studies suggest that the cerebellum stores motor learning for sequential actions and serves to compare the stored plan for intended movement with the proprioceptive evidence of actual movement. If an error or deviation from the desired action occurs, the cerebellum is proposed to help to restore the intended path by modulating the activity of the motor thalamus.

The cerebellum particularly coordinates multijoint movements. Hence, cerebellar dysfunction is associated with ataxic movement, decomposition of movement into single-joint components, and reduced correction of movement errors. No pharmacological treatments for cerebellar disorders currently exist. The neurotransmitters involved (glutamate and gamma aminobuteric acid or GABA) are highly nonspecific and serve the entire CNS. Furthermore, little improvement of function follows cerebellar injury, unlike neocortical injury. Cerebellar lesions therefore often result in permanent ataxia and gait abnormalities.

8.3.3 Brainstem Control of Movement

Both the basal ganglia and cerebellar loops impact motor output through motor thalamic projections to the cortex. By contrast, brainstem nuclei have much more direct effects. The motor cortex (particularly M1) has major direct efferents that project to multiple brainstem and spinal cord nuclei. These brainstem nuclei are particularly important for axial motor control. The red nucleus gives rise to the rubrospinal pathway, the reticular nuclei of the pons and midbrain give rise to the reticulospinal pathway, and the lateral vestibular nucleus gives rise to the vestibulospinal pathway. The pedunculopontine nucleus lies in a region whose stimulation elicits walking movements. Brainstem lesions result in unwanted flexor and extensor reflex posturing. Such posturing is believed to result from unbalanced brainstem nuclei inputs to the spinal cord, without sculpting and control by the cortex. Lesions of the cortex, basal ganglia, or thalamus result in maintained extremity movement and reduce volitional movement. It appears, therefore, that the brainstem is critical to the maintenance of unconsciously maintained antigravity tone.

Due to complex interactions with the brainstem, abnormalities of axial movement such as dystonia, are more resistant to treatment.[16] Thus, one of the current frontiers of understanding motion is defining the relationship between the cerebral cortex and the brainstem nuclei and explaining how the contributions of these two regions combine and influence spinal cord activities.[1,7,15] Among the cortex, the globus pallidus, and the thalamus, the globus pallidus is thought to have a greater influence upon motor control. GPi therefore becomes the primary site to treat axial abnormalities associated with dystonia. However, considerable further research is required to assess whether direct interventions in brainstem areas might prove more effective for the control of axial movement.

8.4 CURRENT SURGICAL TREATMENTS

Surgical treatments of movement disorders have varied widely over time, offering a fascinating history of hypothesis-driven surgical therapy and the evolution of effective therapeutic targets.[1,6,15] Early surgical treatments of movement disorders consisted of ablative procedures of the known motor system, ranging from ventral rhizotomy to precentral corticectomy. For example, beginning in 1932, Bucy performed subpial resections of the precentral cortex for the treatment of choreoathetosis and tremor. In 1939, Meyer performed a transventricular ablation of the caudate head and body to treat a patient with parkinsonian tremor. Later, Cooper, in attempting to perform a mesencephalic pedunculotomy for parkinsonian tremor, inadvertently tore the anterior choroidal artery. Although the procedure was halted, the patient awoke from anesthesia free of tremor. This led to the discovery that ablation of the medial globus pallidus could relieve parkinsonian tremor.[5,11,15]

Along with extirpation of the ansa lenticularis, the abolition of abnormal movements through lesions of the basal ganglia represented a major advance because patients were spared the hemiparesis that accompanied corticectomy, mesencephalic pedunculotomy, lateral cordotomy, and ventral rhizotomy.

8.4.1 Lesions: Thalamotomy and Pallidotomy

With the advent of stereotactic localization techniques in the late 1940s, lesions could be made in the basal ganglia without the risks of open craniotomy. Lesions were produced through freezing with liquid nitrogen cryoprobes or thorough heating with microwave radio frequency probes. Until the early 1950s, the globus pallidus was the stereotactic target of choice for the treatment of parkinsonian tremor. In 1954, Hassler and Riechert reported dramatic improvement of parkinsonian tremor following placement of a lesion in the ventrolateral thalamus.[5,11] Over subsequent years, the thalamus replaced the globus pallidus as the stereotactic target of choice for Parkinson's disease. Until the early 1990s, the primary surgery performed for any type of movement disorder was thalamotomy, the placement of a lesion in the motor thalamus.[11] However, Leksell continued to place lesions in the ventral–posterior pallidum for Parkinson's disease. Eventually these patients were studied as a group, sparking a resurgence of pallidal stereotactic surgery in the 1990s. Laitenen then recognized that the posterior aspects of the pallidum are more important in Parkinson's disease than the anterior aspects that are more involved in cognitive and frontal lobe function.[11,22]

The exact target coordinates of a stereotactic lesion depend on treatment purpose (tremor or rigidity) and surgeon preference.[1,11] Because the radiological landmarks used in stereotactic surgery do not bear a constant relationship to the target nuclei, most surgeons employ physiological monitoring to locate targets. Although some lesion placement is guided by changes in tissue impedance or the effect of transiently cooling tissue, most surgeons monitor involuntary movements, paresthesias, and tremor suppression resulting from transient electrical stimulation.

Outcome studies demonstrate excellent results of lesion surgery in the relief of tremor.[8,11] In one study, 72% of patients were nearly free of tremor. However, one quarter experienced transient or minor complications including worsening of speech (1.3%), transient contralateral hypotonia (7%), subjective finger or mouth numbness (12%), transient confusion (12%), transient neglect or ataxia of hand (5%), and transient foot dystonia (3%). Radiofrequency lesions carry a risk of hemorrhage, particularly in patients with preexisting hypertension where damage to the vessels of the basal ganglia and thalamus may exist prior to surgery. Leksell reported that stereotactically placed lesions in the posteroventral pallidum produced good long-term mitigation of tremor, bradykinesia, and rigidity in 19 of 20 parkinsonian patients (95%) followed for 1 to 5 years.[20] In 1992, Laitinen reported a series of 38 patients who had undergone the Leksell posteroventral pallidotomy, monitored postoperatively for 2 to 71 months. At follow-up, 34 (89%) were improved and 92% noted relief of hypokinesia.[20] Interestingly, patients experienced relief of bilateral symptoms from unilateral lesions. Adverse effects included central homonymous visual field deficits in six patients and transient facial weakness and dysphasia in one patient. Based upon these data, posteroventral (GPi) pallidotomy became the procedure of choice for Parkinson's disease, particularly because it improved L-dopa-induced dyskinesias.[11]

Intraoperative high-frequency stimulation during lesion surgery resulted in transient suppression of tremor. This inspired the development of chronic DBS for tremor and Parkinson's disease.[1–3,11,15] Enthusiasm for DBS as a treatment of movement disorders increased after the late 1990s, primarily due to perceived lower risks of placement and the possibility of reversibility, compared to the permanent lesions used in thalamotomy and pallidotomy. Despite the general trend away from lesion surgery, however, it should be noted that a randomized trial of pallidotomy versus best medical therapy was stopped early due to the higher than expected efficacy of pallidotomy in relieving Parkinson's symptoms.[21] Thus, in spite of the waning enthusiasm for pallidotomy procedures, particularly among patients, the lesions appear to provide excellent long-term relief of many Parkinson's symptoms, and in many instances may represent a good alternative to DBS.[11,21] In addition, considerable interest exists for performing lesions instead of placing stimulating electrodes in the STN.[22] A potential disadvantage of STN lesions is the hemiballismus known to arise following strokes in the region of the STN. However, this may prove to be a more theoretical concern.

In a study of subthalamotomy, only one in 21 patients experienced unmanageable dyskinesias after surgery and proceeded to DBS placement.[22] Advantages of lesions over DBS include considerable reductions in surgical costs, the permanent effect of the lesion, the lack of required postoperative care, and higher patient throughput. However, side effects also tend to be permanent, and many believe that DBS therapy is likely to have fewer permanent risks. Of course, this advantage may be balanced by more problems with stimulator programming, infections, late electrical dysfunction, and the need for surgical battery replacement.

8.4.2 Deep Brain Stimulation: Thalamic, Pallidal, and Subthalamic

DBS for the treatment of disabling tremor and Parkinson's disease rose to prominence in the late 1990s.[1,3,13–15,23,24] Initially, Benabid attempted to suppress disabling tremor with chronic stimulation of ventral intermediate nucleus (VIM) in 26 patients suffering from Parkinson's disease. Twenty-three patients with thalamic stimulators (67%) experienced total suppression of tremor when assessed an average of 13 months following electrode placement.

The first commercial DBS system was FDA-approved for placement into VIM for tremor in 1999 and for placement into GPi or STN for Parkinson's disease in 2002. Currently, practice patterns have shifted considerably with most Parkinson's patients receiving unilateral or bilateral STN DBS stimulation,[14] while tremor patients typically receive VIM stimulation. In 2003, DBS was approved for placement into GPi for dystonia.[16] The mechanism by which DBS achieves its functional effect remains a topic of active research.[15]

8.4.3 Neural Tissue Grafts

Basic mechanisms underlying the integration of embryonic tissue into the adult brain have been studied intensively for more than 30 years, particularly with a view to ameliorating parkinsonism in experimental animal models (see Chapter 2 for discussion of neural grafting for other indications).[25–29] However, few procedures were performed in human patients with Parkinson's disease until the mid-1980s. Enthusiasm for tissue grafting into the human brain rose rapidly in 1987, following a dramatic report from Mexico that adrenal medulla autografts into the caudate could improve motor performance in patients with Parkinson's disease.[30]

Although the procedure did not follow known principles on tissue preservation and little was known about the chances of survival and integration of the grafts in the brain, there was a rush to replicate the findings. The attempts were unsuccessful, confirming that the transplant conditions were nonphysiological and supporting the established literature mechanisms on transplant survival in the brain.[31]

Several studies of embryonic grafting were done in the United States. Results of the first long-term studies were published recently[29,32,33] and showed modest effects on parkinsonian symptoms. Several patients in each study exhibited new, unexpected side effects, particularly dyskinesias.[9] Several patients required further surgery to control these otherwise untreatable side effects. The recent Swedish experience corroborated both the findings of modest symptom improvements and occurrence of side effects.[26,34] It also led to considering how to alter grafting conditions and donor cells to improve the clinical outcome, but a clear dose–response relationship comparing cell survival with clinical outcome has not yet been established.[28,34–36]

In addition to clinical outcome questions, many scientific and ethical issues surround the placement of embryonic human neural tissue grafts into the striatum for Parkinson's disease.[37] It is difficult to characterize donor tissue sources. Because of mixing of individual cadaveric specimens, the grafts exhibit immunological diversity, potentially low-cell recovery rates, low graft-cell survival, and lack of cellular migration. Furthermore, acquisition of embryonic tissue is difficult and ethically complex.[37] No method of standardization of the dose delivered (numbers of surviving cells and their eventual location) exists. In addition, funding for such experimental surgery has been challenging because of the absence of a corporate sponsor. Furthermore, the clinical trial format usually requires a double-blind, placebo-controlled approach.[38]

Because of the shortage of human embryonic allograft tissue, xenograft (particularly porcine) tissue has been suggested as an alternative.[28,39,40] However, a trial of porcine embryonic cell therapy by Diacrin/Genzyme resulted in cancellation due to high cost and lack of efficacy.[39] Finally, the appearance of side effects with embryonic transplants curtailed much of the enthusiasm for further trials.[9] Whether this pessimism will extend to potential neural stem cell transplantation strategies remains to be seen because the technologies remain under development. Whether the current pessimistic outlook for development of neural grafts as a treatment for clinical disorders will extend also to stem cells remains a significant question.

8.4.4 GDNF: Ventricular/Putaminal Infusions and Gene Therapy

Glial-derived neurotrophic factor (GDNF) has been studied extensively as a treatment for Parkinson's disease due to its specific enhancement and support of dopaminergic neurons.[41] GDNF was studied as an intraventricular infusion in nonhuman primates with MPTP-induced parkinsonism. In these model animals, striatal dopaminergic neurons demonstrated considerable regrowth, suggesting a role for GDNF in restorative therapy. The results of initial human trials for intraventricular GDNF therapy were disappointing[42] due to intolerable side effects at doses below the therapeutic threshold. Side effects included intractable nausea and vomiting resulting in significant weight loss and diffuse paresthesias, likely due to GDNF stimulation effects upon sensory ganglia. No improvements in parkinsonian symptoms were noted.

Despite these initial disappointing results, investigators have adopted new approaches for delivery of GDNF to the brain and continue to express optimism that GDNF may provide benefit if delivered to appropriate regions.[43,44] Results of direct putaminal GDNF infusion have been recently reported.[45] In this study of five patients, no serious clinical side effects were noted and improvements occurred in both motor symptoms and activities of daily living. In addition, significant increases in dopamine storage in the putamen were observed by positron emission tomography. Both the direct infusion and gene therapy approaches for GDNF delivery to the brain have re-energized the field since considerable dopaminergic fiber regrowth may be noted following adequate GDNF therapy.[41,43,44]

Direct putaminal infusion of GDNF versus placebo is currently under study in a randomized, double-blinded, multicenter study sponsored by Amgen and Medtronics. Should this study confirm the preliminary results, further pivotal studies may follow. Future studies may considerably further our understanding of long-term drug delivery within the brain and lead to improvements in drug delivery systems. Planning software based upon MRI studies of the brain that consider the relative diffusion of water and therapeutic molecules is currently under development. Such planning programs may eventually allow determination of the precise volume of distribution of a treatment molecule from a point source, taking into account tissue heterogeneity, the structural properties of the treatment molecule, and the rate of administration.

8.5 EVOLVING SURGICAL TREATMENTS

A number of new approaches are now being considered for initial human clinical trials, often following promising results from preliminary animal studies. As with many surgical interventions, the level of evidence needed to transition from animal to human feasibility trials varies considerably, depending on sponsorship and regulation. Preliminary human studies tend to be more common when considerable commercial interests are available to initiate and fund research efforts. By contrast, investigator-initiated studies tend to follow a slower pace. The ethics of experimental surgical interventions remains an issue of considerable interest and concern, particularly with regard to the amount of preclinical data required, the nature of the preclinical animal models, and the amount of time allowed to pass before suggesting human trials.[1,3,8,37]

8.5.1 Advances in DBS: New Targets and Stimulation Paradigms

Many manifestations of Parkinson's disease are not routinely improved by current DBS or lesion-generating surgery.[7,15] The manifestations include axial and gait abnormalities, cognitive decline, and autonomic disturbances. Because the motor symptoms of the disease can be extremely disabling, searches for new targets and stimulation paradigms for DBS are ongoing. Novel stimulation targets in the brainstem may provide potential improvements in axial symptoms. However, few studies at present suggest appropriate targets in humans. In addition, potentially important but poorly localized brainstem nuclei such as the pedunculopontine nucleus, may be substantially more difficult to target than large prominent nuclei such as the red nucleus. Furthermore, the lower brainstem may be a difficult region in which to target and position stimulating electrodes safely. The search for additional targets of DBS may prompt further investigation of the role of brainstem nuclei in axial motor control.[1,15]

In addition to finding new targets for current DBS technologies, many potential improvements to the DBS device are under consideration. The number of channels and the degree of control over stimulation paradigms could be considerably increased. A large number of ongoing human studies are attempting to improve stimulation methods, for example, by using patterning. Implantation of the DBS might be made easier and more accurate with an advanced frameless stereotactic system that can decrease the time required to sample different targets. The hardware also may be improved. Advanced Bionics markets a cochlear stimulator that can be both flat and skull-mounted and intends to convert the stimulator to a new form of DBS device. Smaller devices would be easier to implant near a burr hole, for example, obviating the current need to tunnel electrode wires long distances to stimulator units in the chest or abdomen. Finally, control of motor abnormalities may become more efficient through the development of feedback-control systems that sense abnormal motions and provide corrective response stimulations (see Chapter 6 and Chapter 7). Such feedback-control systems may work particularly well for tremor control, as opposed to the current, invariant stimulation pattern. Of course, such changes would necessarily increase the complexity of the implanted DBS circuitry.

8.5.2 Viral Therapy: Subthalamic Glutamate to GAD Conversion

A popular model of Parkinson's disease suggests that reduced dopaminergic regulation of the striatum leads to STN overactivity. One novel approach under study is to introduce genes into the STN that will induce the production of the inhibitory neurotransmitter, GABA.[46] The genes under study are GAD-65 and GAD-67 and they are introduced by viral vector to the STN. This approach demonstrates considerable promise in a rodent model of Parkinson's disease.[47] STN was effectively transformed from an excitatory to an inhibitory phenotype following GAD transfection. In addition, GAD transfection appeared to provide some neuroprotective inhibition of 6-OHDA-induced parkinsonian asymmetry. Of course, many residual questions remain regarding the mechanisms of phenotypic effects. For example, the overall impact of changing the phenotype of the STN on overall basal ganglia function is not clear. In addition, the relevance of the rodent model to human disease with respect to neuroprotection is also uncertain.

Investigators pursuing this viral approach have argued that the system is sufficiently developed in the animal model to begin human testing.[46] Difficulties with viral approaches in the past have included a lack of persistent transfection over several months and toxicity due to the viral vectors. The relative absence of a strong immune response to the viral vector in rodents may not translate to humans. As in all viral and gene therapy trials, numerous theoretical safety concerns arise. Hence, this highly innovative, promising approach not only may have applicability in the clinical setting, but also may have considerable (and unforeseen) consequences as human feasibility studies proceed.

8.5.3 Stem Cell Approaches

The degeneration of a specific population of neurons in Parkinson's disease makes it an attractive target for stem cell therapeutic approaches.[48] Many varieties of self-renewing stem cells have been described. Embryonic stem (ES) cells are pluripotent cells derived from a preimplantation blastocysts; they give rise to all cells in an organism. Multipotent stem cells, such as neural stem cells, are derived from individual organs. The adult human brain contains stem cells capable of forming new neurons and glia. Cells obtained from adults tend to have more limited capacities for development, and are often restricted to lineages for a particular region such as the hippocampus or spinal cord. Neurospheres or balls of cells that contain certain percentages of a clonal population have been derived from most brain regions of embryonic or adult individuals, and can be propagated almost indefinitely in culture. However, differentiation of the cells from neurospheres can be challenging, particularly to obtain neurons. In addition, although appropriate neurospheres have been obtained from many brain regions, cells of dopamine lineages, appropriate for Parkinson's cell transplants, have been much more difficult to find and culture.[48,49]

Stem cell approaches have shown some promise in the treatment of Parkinson's disease, but are accompanied by considerable technical, political, and commercial difficulties. ES cells have been shown to differentiate into functional dopaminergic neurons after transplantation in a rat model of Parkinson's disease.[28] However, transplanted stem cells may have high teratogenic potential. A patient with Parkinson's disease died of ventricular obstruction and brainstem compression following transplantation with embryonic mesencephalic dopamine neurons. An autopsy demonstrated teratomas throughout the ventricular system.[9]

Further impediments to progress in stem cell approaches arise from a lack of availability of stem cells for study, highly variable definitions, the different and often proprietary methods to produce stem cells, the political climate against human ES cell research, and the difficulty of producing differentiated cells from undifferentiated precursors. In addition, many of the proprietary stem cell lines propagated in culture are accompanied by specific and severe legal restrictions upon their use, further inhibiting developments in neural grafting. Stem cell neural grafts therefore remain promising future tools, but a decade or more may be required before a clinically effective treatment regimen becomes available.[48,49]

8.5.4 Novel Drug Delivery Methods

Medications to treat movement disorders are often limited by oral dosing schedules and systemic fluctuations that can lead to considerable motor variability.[7,10,12] Although new medications are in development, most act upon dopaminergic signaling. In addition to dopamine agonists and inhibitors of dopamine degradation, new classes of drugs may include dopamine uptake inhibitors, neuroprotective medications, and opioid or nicotinic receptor modulators. Other approaches may include direct intracerebral infusion of drugs that are not absorbed orally or are unable to cross the blood–brain barrier, similar to the system implemented for GDNF delivery into the brain.[45]

An advantage of local drug delivery into the brain is that the regional concentration may be maintained at a high level, reducing nonspecific remote actions or systemic side effects. The development of an effective intracerebral infusion system and an accurate pharmacological modeling program to guide device placement could yield substantial therapeutic benefits. In contrast to other schemes, drug infusions could be easily halted if side effects developed. However, the FDA has not approved any drugs for direct, intracerebral infusion, although several (e.g., morphine, baclofen) have been approved for intrathecal infusion into CSF. Pharmaceutical firms will have to demonstrate significant benefits to obtain such approval, particularly compared to traditional oral medications, because of the high degree of invasiveness. Such systems will require direct catheter placement into the brain, usually performed stereotactically, as well as permanent implantation of one or more programmable pumps.

8.5.5 Neuroprosthetic Approaches

No effective treatments to reverse the abnormality or improve the motor output scheme currently exist for many movement disorders. In cerebral palsy, for example, extensive damage to the basal ganglia and motor system defies medical and surgical correction even with more sophisticated DBS and other treatment modalities. However, in many instances, the cortex remains normally functional. In such scenarios, a neuroprosthetic approach that obtains motor signals directly from the cortex or from subcortical structures and bypasses damaged regions of the brain may be highly effective (see Chapter 7 for further examples and discussion).

A neuroprosthetic may be able to drive external actuators to perform desired tasks that a patient is unable to perform alone (see Chapter 7). The signals obtained from the cortex might be direct neuronal recordings or local field potential recordings that may require a large number of neurons to produce a signal with sufficient information bandwidth for device control. Such approaches currently work for the control of robotic arms, for example, in nonhuman primates.

8.6 CONCLUSIONS

The surgical treatment of movement disorders and Parkinson's syndrome and disease in general has developed in concert with clinical and basic science knowledge about the roles of various motor structures.[1,3,6,15] In many cases, treatments have been performed first, driving further insight into the structures and their functions, particularly with precentral corticectomy, mesencephalic pedunculotomy and pyramidotomy, and later with basal ganglia and thalamic lesions.[11,21] As the use of lesions has waned, neural tissue transplants have demonstrated the possibility of true restorative surgery, to be further developed along with various types of growth factor enhancements and stem cell transplants.[50]

DBS is currently the most frequently performed type of movement disorder surgery, and further development may include additional targets and improved designs, particularly with intermittent demand systems rather than constant stimulation. Further surgical treatments are in development to more radically prevent cell loss in the early stages of Parkinson's disease, for example, or to switch phenotypes to alter function. It is likely that the fascinating history of surgical treatments and availability driving basic research developments in motor systems will continue for some time, with neurosurgeons potentially leading many advances, due to patient demands for improved treatments.

REFERENCES

1. Betchen, S.A. and Kaplitt, M., Future and current surgical therapies in Parkinson's disease, *Curr. Opin. Neurol.*, 16, 487–493, 2003.
2. Dawson, T.M. and Dawson, V.L., Neuroprotective and neurorestorative strategies for Parkinson's disease, *Nature Neurosci.*, 5S, 1058–1061, 2002.
3. Eskandar, E.N., Cosgrove, C.R., and Shinobu, L.A., Surgical treatment of Parkinson's disease, *JAMA*, 286, 3056–3059, 2001.
4. Isacson, O., Models of repair mechanisms for future treatment modalities of Parkinson's disease, *Brain Res. Bull.*, 6, 839–846, 2002.
5. Lang, A.E. and Lozano, A.M., Parkinson's disease, *NEJM*, 339, 1044–1053, 1998.
6. Lozano, A.M. and Lang, A.E., Parkinson's disease, *NEJM*, 339, 1130–1143, 1998.
7. Rascol, O. et al., Limitations of current Parkinson's disease therapy, *Ann. Neurol.*, 53 (Suppl. 3), S3–S15, 2003.
8. Stowe, R.L. et al., Surgery for Parkinson's disease: lack of reliable clinical trial evidence, *J. Neurol. Neurosurg. Psych.*, 74, 519–521, 2003.
9. Hagell, P. et al., Dyskinesias following neural transplantation in Parkinson's disease, *Nature Neurosci.*, 5, 627–628, 2002.
10. Nomoto, M., Clinical pharmacology and neuroprotection in Parkinson's disease, *Parkinsonism Rel. Dis.*, 9, S55–S58, 2003.
11. Hariz, M.I., From functional neurosurgery to "interventional" neurology: survey of publications on thalamotomy, pallidotomy and deep brain stimulation for Parkinson's disease from 1966 to 2001, *Movement Dis.*, 18, 845–852, 2003.
12. Siderowf, A. and Stern, M., Update on Parkinson disease, *Ann. Intern. Med.*, 138, 651–658, 2003.
13. Krack, P. et al., Five-year follow-up of bilateral simulation of the subthalamic nucleus in advanced Parkinson's disease, NEJM, 349, 1925–1934, 2003.
14. Houeto, J.L. et al., Subthalamic stimulation in Parkinson's disease, *Arch. Neurol.*, 60, 690–694, 2003.
15. Lozano, A.M. et al., Deep brain stimulation for Parkinson's disease: disrupting the disruption, *Lancet Neurol.*, 1, 225–231, 2002.
16. Coubes, P. et al., Treatment of DYTI dystonia by stimulation of the globus pallidus, *Lancet*, 355, 2220–2221, 2000.
17. Wessberg, J. et al., Real–time prediction of hand trajectory by ensembles of cortical neurons in primates, *Nature*, 408, 361–365, 2000.
18. Carmena, J.M. et al., Learning to control a brain–machine interface for reaching and grasping by primates, *Publ. Libr. Sci.*, 1, 1–16, 2003.

19. Mattay, V.S. and Weinberger, D.R., Organization of the human motor system as studied by fMRI, *Eur. J. Radiol.*, 30, 105–114, 1999.
20. Laitenen, L.V. et al., Leksell's posteroventral pallidotomy in the treatment of Parkinson's disease, *J. Neurosurg.*, 76, 53–61, 1992.
21. Vitek, J.L. et al., Randomized trial of pallidotomy versus medical therapy for Parkinson's disease, *Ann. Neurol.*, 53, 558–569, 2003.
22. Patel, N.K. et al., Unilateral subthalamotomy in the treatment of Parkinson's disease, *Brain*, 126, 1136–1145, 2003.
23. Abosch, A. et al., Movement-related neurons of the subthalamic nucleus in patients with Parkinson's disease, *J. Neurosurg.*, 97, 1167–1172, 2002.
24. Rodrigues-Oroz, M.C. et al., The subthalamic nucleus in Parkinson's disease: somatotopic organization and physiological characteristics, *Brain* 124, 1777–1790, 2001.
25. Bjorklund. A., and Lindvall, O., Cell replacement therapies for central nervous system disorders, *Nature Neurosci.,* 3, 537–544, 2000.
26. Bjorklund, A. et al., Neural transplantation for the treatment of Parkinson's disease, *Lancet Neurol.*, 2, 437–445, 2003.
27. Isacson, O., The production and use of cells as therapeutic agents in neurodegenerative diseases, *Lancet Neurol.,* 2, 417–424, 2003.
28. Isacson, O. et al., Cell implantation therapies for Parkinson's disease using neural stem, transgenic or xenogeneic donor cells, *Parkinsonism Rel. Dis.*, 7, 205–212, 2001.
29. Kordower, J.H. et al., Fetal nigral grafts survive and mediate clinical benefit in a patient with Parkinson's disease, *Movement Dis.,* 13, 383–393, 1998.
30. Madrazo, I. et al., Development of human neural transplantation, *Neurosurgery,* 29, 165–177, 1991.
31. Goetz, C.G., Stebbins, G.T., III, and Klawans, H.L., United Parkinson Foundation Neurotransplantation Registry on adrenal medullary transplants: presurgical, and 1- and 2-year follow-up, *Neurology,* 41, 1719-1722, 1991.
32. Freed, C.R. et al., Transplantation of embryonic dopamine neurons for severe Parkinson's disease, *NEJM*, 344, 710–719, 2001.
33. Olanow, C.W. et al., A double-blind controlled trial of bilateral fetal nigral transplantation in Parkinson's disease, *Ann. Neurol.*, 54, 403–414, 2003.
34. Hagell, P. and Brundin, P., Cell survival and clinical outcome following intrastriatal transplantation in Parkinson's disease, *J. Neuropathol. Exp. Neurol.*, 60, 741–752, 2001.
35. Brundin, P. et al., Improving the survival of grafted dopaminergic neurons: a review of current approaches, *Cell Transplant.,* 9, 179–195, 2000.
36. Barker, R.A. and Dunnett, S.B., Functional integration of neural grafts in Parkinson's disease, *Nature Neurosci.,* 2, 1047–1048, 1999.
37. Turner, D.A. and Kearney, W., Scientific and ethical concerns in neural fetal tissue transplantation, *Neurosurgery,* 33, 1031–1037, 1993.
38. Redmond, D.E. and Freeman, T., The American Society for Neural Transplantation and Repair considerations and guidelines for studies of human subjects, *Cell Transpl.*, 10, 661–664, 2001.
39. Schumacher, J.M. et al., Transplantation of embryonic porcine mesencephalic tissue in patients with PD, *Neurology,* 54, 1042–1050, 2000.
40. Brevig, T., Holgersson, J., and Widner, H., Xenotransplantation for CNS repair: immunological barriers and strategies to overcome them, *Trends Neurosci.,* 23, 337–344, 2000.
41. Brundin, P., GDNF treatment in Parkinson's disease: time for controlled clinical trials? *Brain*, 125, 2149–2151, 2002.

42. Nutt, J.G. et al., Randomized, double-blind trial of glial cell line-derived neurotrophic factor (GDNF) in PD, *Neurology*, 60, 69–73, 2003.
43. Kordower, J.H. et al., Neurodegeneration prevented by lentiviral vector delivery of GDNF in primate models of Parkinson's disease, *Science*, 290, 767–773, 2000.
44. Kordower, J.H., *In vivo* gene delivery of glial cell line-derived neurotrophic factor for Parkinsons's disease, *Ann. Neurol.,* 53 (Suppl. 3), S120–S132, 2003.
45. Gill, S.S. et al., Direct brain infusion of glial cell line-derived neurotrophic factor in Parkinson disease, *Nature Med.*, 9, 589–595, 2003.
46. During, M.J., Kaplitt, M.G., Stern, M.B., and Eidelberg, D., Subthalamic GAD gene transfer in Parkinson disease patients who are candidates for deep brain stimulation, *Human Gene Therap.*, 12, 1589–1591, 2001.
47. Luo, J. et al., Subthalamic GAD gene therapy in a Parkinson's disease rat model, *Science*, 298, 425–429, 2002.
48. Gerlach, M. et al., Current state of stem cell research for the treatment of Parkinson's disease, *J. Neurol.*, 249 (Suppl. 3), 33–35, 2002.
49. Storch, A. and Schwarz, J., Neural stem cells and Parkinson's disease, *J. Neurol.*, 249 (Suppl. 3), 30–32, 2002.
50. Hodge, C.J. and Boakye, M., Biological plasticity: the future of science in neurosurgery, *Neurosurgery*, 48, 2–16, 2001.

9 Novel Therapeutic Approaches for High-Grade Gliomas

Kent C. New, David Corey Adamson, Lee Selznick, and John Sampson

CONTENTS

0-8493-1482-8/05/$0.00+$1.50

9.1 INTRODUCTION

The incidence of primary brain tumors is increasing. Approximately 18,000 new cases were expected in the U.S. in 2003.[1] High-grade gliomas (HGGs) including glioblastoma multiforme (GBM) and anaplastic astrocytoma (AA) are the most common primary tumors of the central nervous system (CNS). HGGs remain refractory to treatment and have dismal prognoses. The median survival for AA patients is approximately 2 years, the median is 9 months for GBM patients.[2] Neurosurgeons remain intimately involved in the care of patients with HGG and with research into new treatments for this deadly disease. This stems in part from the fact that surgical resection continues to be an important treatment for HGG.[3]

The hypothesis of this chapter is that improved understanding of the biology of gliomas and the discovery of novel cancer treatment modalities will lead to therapies that will significantly improve the prognosis for patients with HGG. The avenues of ongoing research into novel mechanisms of cancer therapy that may eventually lead to new treatments for HGGs are vast. It is, of course, impossible to predict which of many ongoing areas of research will lead to successful therapies for brain tumors and coverage of all potential areas of future treatment is beyond the scope of this chapter.

We have chosen to limit our discussion to three brain tumor therapies that are the most promising experimental treatments involving local delivery of agents to the brain. The treatment modalities to be discussed include oncolytic viruses, gene therapy, and convection-enhanced delivery (CED) of targeted toxins and other agents. Many neurosurgeons have been and continue to be involved in the development of these novel therapies for brain tumors, and the fact that they involve local delivery of agents to the brain makes them of interest to all neurosurgeons who treat brain tumors.

Many promising areas of cancer research that will not be covered in this chapter may, of course, lead to new treatments for brain tumors. These include development of new chemotherapeutic treatments,[4–6] molecular therapies,[7,8] immunologic therapies,[9–11] and therapies targeted at blood–brain barrier disruption.[12,13] In addition, developments in intraoperative imaging to guide surgical resections will not be covered.[14–16] Please see cited references for further reading on these topics.

9.2 ONCOLYTIC VIRUSES FOR BRAIN TUMOR THERAPY

9.2.1 Introduction

The revolution in molecular biology that culminated in completion of the Human Genome Project spurred an explosion of interest in various forms of gene therapy for brain tumors over the past decade. In fact, brain tumors were some of the first tumors tested in experimental gene therapy models.[17,18] Unfortunately, this early enthusiasm has not yet led to the development of any tangible treatment modalities (see Section 9.3 on gene therapy). However, work with the viruses used as gene therapy vectors demonstrated that they may be powerful oncolytic tools in and of themselves.

Initial viral vector gene therapy strategies utilized viruses rendered replication-defective to reduce their neurovirulence. These studies led to the hypothesis that replication-competent viruses might serve as treatment agents for tumors, not by delivering specific genes into tumors (as is the case with gene therapy), but by directly infecting and lysing the tumor cells. The neurovirulence of the virus, however, must be reduced by creating a neuro-attenuated mutant that can replicate in tumor cells but not in normal brain. Viral replication within targeted brain tumor cells results in production of viral progeny and lyses the infected cells in the process. The progeny may then infect neighboring cells to extend the effects of the virus beyond the initially infected cells. Although the use of viruses as oncolytic agents was explored many years ago, the hypothesis that mutated viruses might selectively replicate in tumor cells was validated by Martuza et al in 1991.[19] Their initial study led to the development of a new field of research based on the idea of using replication-competent viruses as cancer therapies.

9.2.2 Herpes Simplex Virus Type-1 (HSV-1) Vectors

To better appreciate where oncolytic viral therapy may go in the next decade, it is perhaps instructional to examine just how far science has taken us over the past decade in this exciting area of brain tumor therapy. As one of the first oncolytic viruses used in the treatment of HGG, HSV-1 has remained one of the most widely studied because:

1. It has affinity for numerous cell types including a natural neurotropism.
2. It is naturally cytolytic during its replication and virion production life cycle.
3. It contains nonessential viral genes that can be replaced with large transgenes (up to 30 kb).
4. The HSV genome contains several known genetic determinants of neurovirulence encoded by nonessential genes that may be deleted or replaced.
5. The viral genome remains as an episome in the target cell, eliminating the possibility of insertional mutagenesis.[20]
6. It is susceptible to several antiviral medications such as acyclovir and gancyclovir.

The susceptibility of HSV to medications provides a safeguard against uncontrollable infection during attempted therapeutic uses of the virus.

Oncolytic HSV-1 viral therapy has gone through several modifications since the initial viral vectors were engineered in the early 1990s.[20] The first generation versions involved mutations of viral enzymes involved in nucleotide metabolism, namely thymidine kinase (TK) and ribonucleotide reductase.[19,21,22] These viral enzymes possess cellular homologues that are upregulated in actively dividing tumor cells but not in nondividing cells. *Dls*ptk was one such HSV-1 virus with a mutated TK gene.[19] This virus was efficacious against mice inoculated with human HGGs, but its neurovirulence at high titers and resistance to antiherpetic drugs hampered enthusiasm for clinical use.

The second generation HSV-1 vectors were mutated in selected genes to maximize safety as well as attempt to provide improved specificity and efficacy. The investigations found that R3616, an HSV-1 mutant in the *γ34.5* gene, could act as an oncolytic virus with greatly attenuated neurovirulence as well as maintained susceptibility to antiviral medications.[23] *γ34.5* is present in two copies on the HSV genome and encodes a major determinant of neurovirulence.[24] The discovery that R3616, a virus with a deletion in both *γ34.5* genes that makes it highly neuro-attenuated, could still replicate in and kill brain tumors increased excitement about this potential therapeutic modality.

The next advance in oncolytic viruses came with the development of a multimutated HSV-1 virus, G207.[25] G207 is an oncolytic viral vector derived from R3616 by insertion of the *E. coli lacZ* gene in the *UL39* gene. The *UL39* gene encodes the large subunit of ribonucleotide reductase, a key enzyme in nucleotide metabolism and viral DNA synthesis in nondividing cells but not in dividing cells.[26,27] Mutation of the *UL39* gene results in a virus that is both neuro-attenuated and hypersensitive to antiviral medications.[28,29] Combining the *γ34.5* and *UL39* gene mutations in one virus resulted in G207, an oncolytic HSV-1 mutant with multiple desirable properties:

1. An infinitely small chance of reversion to wild type, particularly as the *γ34.5* mutation is a gene deletion.
2. Highly attenuated neurovirulence, in fact, no neurotoxic dose of the virus could be attained in a highly HSV-sensitive primate model.[30]
3. Hypersensitivity to antiviral medications in case an HSV infection occurred during cancer therapy.[22]
4. A method of tracking viral spread through chemical detection of the inserted *lacZ* gene, a common marker enzyme.[25]

The combination of these properties of G207 led to its becoming the first replication-competent HSV-1 mutant used in human clinical trials for brain tumors.[31] Shortly thereafter, a *γ34.5* mutant vector similar to R3616 was also approved for clinical trials in glioma patients (see next).[32]

The original studies with G207 utilized human brain tumor implants into nude mice in order to avoid immunologic rejection of the foreign tumors. These models had the disadvantage that the effects of the immune systems of the host animals on the tumors could not be assessed. Further experiments investigated inoculation of G207 into immunocompetent mice and yielded very interesting results. The injection of G207 induced a systemic immunity to the tumor that resulted in regression of distant tumors and resistance to rechallenge with tumor cells in inoculated animals.[33] This introduced the novel concept of a viral vector acting as a tumor vaccine therapy, eliciting a CD8+ T cell-mediated immunity in this particular case.

Further advances in replication-competent HSV-1 mutants for use in tumor therapy are ongoing. HSV-1 mutants have been engineered to utilize gene promoters or enhancer sequences to specifically express an essential viral gene in tumor cells but not in normal brain.[34,35] This approach relies on the tumor specificity of the gene promoter to create a virus that can replicate only in tumor cells. Such HSV-1 and adenovirus vectors have been created using tissue- or tumor-specific promoters such as α-fetoprotein, kallikrein, L-plastin, midkine, prostate-specific antigen, tyrosinase, and calponin; however, more work must be done to identify specific glioma promoter/enhancer sequences before this concept can be applied to brain tumors.[36–38]

To date, two oncolytic HSV-1 vectors have been approved for use in clinical trials for HGGs G207 and 1716. Data from a Phase I trial using G207 showed encouraging results in patients with recurrent HGGs refractory to chemotherapy and radiation.[31] No toxicity or serious adverse events attributable to the virus were noted in 21 patients injected with G207 in escalating doses. Eight of 21 patients showed reduced tumor volumes and two patients survived more than 4 years. G207 is now in a Phase Ib trial and is being inoculated into the tumor bed, followed by tumor resection 2 days later and subsequent inoculation in the resection cavity. The clinical trial using the *γ34.5* mutant 1716 also showed encouraging results in a small number of patients with recurrent HGGs.[32] These trials require future expansion and, importantly, must include patients with newly diagnosed HGGs.

9.2.3 Adenovirus (Ad) Vectors

Adenovirus (Ad) vectors have gained popularity as gene therapy tools for several fundamental reasons:

1. They are unaffected by the complement system.
2. They remain episomal and therefore lack the risk of insertional mutagenesis.
3. Large viral titers can be generated more easily than with HSV.
4. They demonstrate broad infectivity of dividing and nondividing cells.
5. They generate high levels of transgene expression.
6. They are less neurotoxic than HSV vectors.[39–41]

The smaller genome of the adenovirus confers the advantage that recombinant vectors are more easily generated. However, one disadvantage is that the size of the potential gene insert is less than the size of the HSV vector. A significant drawback to the use of Ad vectors for gene therapy is that they induce potent inflammatory and specific immune responses that may damage infected tissues and limit repeated use.[42,43] In fact, such a response may have been responsible for the death of a patient in a gene therapy clinical trial using an Ad vector.[44]

While Ad vectors have been used more commonly than HSV vectors in gene therapy applications, their use in oncolytic viral therapy has been more limited despite the fact that Ad was tested as an oncolytic agent shortly after its isolation in 1953 because of its ability to grow in epithelial cells.[45] These early studies showed initial tumor regression that was only transient and the idea of using Ad as an oncolytic agent was abandoned until recently.

As with HSV, two basic strategies have been employed to attempt to develop Ad mutants that can selectively replicate in and lyse tumor cells. The first strategy was to delete viral genes unnecessary for replication in tumor cells. The first example of such a mutant was the E1b 55-kDa deleted Ad dl1520 (also known as Onyx-015) designed to replicate better in p53 negative cells — a common mutation in tumors.[46] Wild type Ad requires E1b binding to p53 to permit viral replication, and therefore E1b-deleted Ad is theoretically only able to replicate in p53-deficient cells.

Unfortunately, further studies found a lack of correlation between p53 expression in the host cell and replication of dl1520.[47,48] These studies were extended to p53 negative gliomas, with similar disappointing results.[49] However, dl1520 does replicate well in tumor cells, and has a low level of toxicity that allowed its approval for clinical trials in head and neck, colorectal, lung, and other cancers.[50] In a similar strategy, an Ad mutant with deletion of the E1a gene, a retinoblastoma tumor suppressor binding site, was developed and studied for its ability to lyse glioblastoma cells.[51] Further work must be done to determine whether this Ad mutant has sufficiently low toxicity to be appropriate for testing in clinical trials of brain tumor patients.

As mentioned above and as with HSV vectors, a second strategy has been tested for development of oncolytic Ad mutants. It employs tumor-specific promoter or enhancer gene sequences to drive expression of an essential viral gene product in tumor cells but not normal tissues. Typically this work has used putative tumor-specific promoter sequences such as α-fetoprotein, prostate specific antigen, and others to drive expression of E1a, an Ad gene product essential for replication of the virus.[52,53] Unfortunately, very low levels of E1a are adequate for replication of Ad, and thus tumor-specific promoter vectors are not as tumor-specific as was hoped. Neither HSV-1 nor Ad vectors designed with this strategy have reached clinical trials to date, and further work is needed to create a truly tumor-specific virus by this method.

9.2.4 Other Oncolytic Viruses

Although most work in the area of oncolytic viral therapy has focused on the use of HSV-1 or Ad vectors, several other viruses have also been investigated for oncolytic potential. They include vesicular stomatitis virus,[54] reovirus,[55] and poliovirus.[56] Interestingly, the highly neurotoxic poliovirus may be among the most promising oncolytic viruses for glioma therapy. The neurotoxicity of poliovirus can be greatly attenuated by replacement of the internal ribosome entry site element with that of a human rhinovirus.[57] The safety of the attenuated virus has been extended to studies in nonhuman primates.[58] Meanwhile, increased expression of the cellular receptor for poliovirus on glioma cells may make the tumor particularly susceptible to poliovirus infection. Gromeier et al. demonstrated that a neuro-attenuated poliovirus can replicate in and lyse human gliomas *in vitro* and *in vivo*.[56] Further work is ongoing to begin clinical trials with attenuated poliovirus in human glioma patients.

9.2.5 Future Potential of Oncolytic Viral Therapy

There are numerous ways that oncolytic viral therapy may be improved upon over the next decade in a quest for the ideal viral vector and hopefully a cure for malignant brain tumors. One potential strategy for improved targeting of tumor cells may be through the use of cell surface molecular markers unique to tumors that may be exploited to create virus mutants able only to infect tumor cells. Modification of viral knob, fiber, or coat proteins can alter viral tropism and enhance tumor transduction in Ad vectors.[59,60] Unfortunately, HSV-1 utilizes multiple cell surface glycoproteins for viral–cell interactions, and it will be more difficult to restrict viral infection only to tumor cells.

Another potential strategy to increase the effectiveness of oncolytic viruses is to combine the use of replication-competent vectors with gene therapy strategies. In this scenario, an oncolytic virus has a gene inserted in its genome that directly or indirectly enhances tumor cell killing. An example of this strategy proved efficacious when the antiviral drug sensitizing gene TK was inserted in the E1b mutant Ad virus dl1520.[61] The virus is then used to infect a tumor and the lytic effect of the viral therapy is enhanced by administration of gancyclovir.

Another example of the potential of combining gene therapy with oncolytic viral therapy is through addition of immunomodulatory gene products in the oncolytic virus. For example, HSV-1 vectors have been engineered to express immunostimulatory cytokines such as interleukin-2, interleukin-12, or soluble B7.1-Ig.[62–64] Other vectors have been designed to inhibit viral-induced down-regulation of major histocompatibility complex (MHC) class I in an effort to increase immune-mediated tumor cell killing.

Suboptimal viral spread within tumors has been a challenge because of many physical and antiviral immune barriers. Extracellular matrix proteins, tight gap junctions, fibrosis, necrosis, neutralizing antibodies, and cell-mediated immunity are all significant impediments to the spread of even replication-competent viruses within a tumor. Viruses that spread via cell-to-cell transmission may prove less affected by neutralizing antibody and other extracellular factors.

Another factor effecting virus spread is, of course, method of delivery. To date, oncolytic viruses have been primarily delivered locally via injection into the tumor or tumor resection cavity. Other routes of delivery, such as intravenous, intra-arterial, lymphatic, intraperitoneal, and local vascular perfusion, have proven successful in animal models with oncolytic HSV-1 vectors and may proceed to clinical trials in the near future.[65–68]

For systemic administration to be an effective delivery method for oncolytic viruses, additional obstacles such as viral inactivation from instability, absorption, homing to nonspecific cells, clearance by the reticuloendothelial system, innate immunity, preexisting immunity with antibodies, and complement-mediated inactivation must be addressed. It is possible that armored vectors able to avoid clearance by the reticuloendothelial and immune systems could be tested systemically.

Viral vectors coated with polyethylene glycol to avoid interaction with macrophages are already under investigation.[69] Other nonviral vectors that avoid neutralizing antibodies and allow repeated administration are in development. Undoubtedly, the next decade will witness the design of several "trojan horse" viral vectors with cell carriers, chemical coatings, and other ways of bypassing the immune system.[68,70]

The past decade has witnessed the birth of a novel therapy for treating malignant brain tumors with oncolytic viral therapy, and hopefully, the next decade will demonstrate its full therapeutic potential. Clinical trials have demonstrated that replication-competent viruses can be administered safely in humans. The next decade will see the continued refinement and clinical testing of oncolytic viruses for use in brain tumor therapy.

9.3 GENE THERAPY FOR BRAIN TUMORS

9.3.1 INTRODUCTION

Gene therapy is broadly defined as the transfer of genetic material into a patient's cells for therapeutic purposes. It is an elegant conceptual approach for the treatment of many diseases that are largely due to genetic aberrations, including brain tumors. The resistance of gliomas to current treatment modalities has stimulated interest in new therapeutic approaches, and raises the prospect of gene therapy as a novel component of multimodal therapy for these extremely aggressive tumors.

The scientific progress of gene therapy and its ultimate translation into clinical benefit depend upon four key steps. First, genes encoding products that specifically destroy or inhibit the growth of tumor cells must be discovered. Second, vectors that deliver one or more genes effectively to tumors must be developed. Third, methods of reliably delivering vectors to target cells with minimal toxicity must be designed. Finally, preclinical data must be translated into well-designed clinical trials in order to test the safety and efficacy of this emerging technology. Recent developments and ongoing research related to these four steps will be discussed next.

9.3.2 GENE TARGETING: TUMOR SUPPRESSION, APOPTOSIS, AND DRUG SUSCEPTIBILITY

Recent advances in molecular biology techniques and the completion of the Human Genome Project provided a wealth of potential targets for gene therapy. One method to conceptualize the strategies being developed is to group potential therapeutic genes into those that induce one of three effects on targeted cells, tumor growth suppression, cellular suicide (apoptosis), or drug susceptibility. There are several mechanisms by which gene therapy may induce suppression of tumor growth. One paradigm aims to directly suppress genetic alterations responsible for the molecular progression of normal cells to HGG cells. Other indirect mechanisms for inhibiting tumor growth via gene therapy, for example, by inducing host immunity to the tumor or inhibiting angiogenesis will not be discussed here.

The specific genetic mutations that occur in brain tumors correlate with tumor type and may be utilized for further subclassifications of tumors. For example, primary malignant astrocytomas develop via molecular pathways distinct from secondary HGGs that develop from low-grade lesions. The most frequent genetic alteration in primary malignant astrocytomas is in the gene for the epidermal growth factor receptor (EGFR) located on chromosome 7.[71,72] The mutated form of EGFR (most often a truncated product known as EGFR-Viii) has a high level of tyrosine kinase activity in the absence of receptor ligand. This amplified signal overrides the normal negative regulation of a tumor suppressor, resulting in uncensored cellular growth. Over-expression of EGFR in glioma cell lines has been correlated with tumor invasiveness and inhibition of this receptor *in vivo* can eliminate this malignant property.[73]

Secondary malignant astrocytomas are most often associated with a missense mutation or allelic loss of chromosome arm 17p.[72,74] At this site is the gene TP53 that encodes the p53 protein; p53 normally functions as a key transcription factor in the regulation of cellular growth and arrest. If a genetic aberration occurs, p53 arrests the damaged cell in the G1 phase of the cell cycle to allow repair mechanisms to commence. If the cell cannot be repaired, p53 induces programmed cell death. Loss of p53 occurs early in secondary malignant astrocytomas and is largely responsible for the accumulation of other genetic errors that ultimately lead to the malignant phenotype. The replacement of wild-type p53 in glioma cell lines induces massive apoptosis *in vitro*, increases sensitivity to ionizing radiation, and inhibits tumor growth *in vivo*.[75,76]

Although there is some genetic overlap with astrocytomas, the malignant transformation of oligodendrogliomas most often involves distinct molecular pathways. Vascular endothelial growth factor (VEGF) is an endothelial cell-specific mitogen that promotes angiogenesis in many solid tumors and over-expression of this gene is the most common aberration in high grade oligodendrogliomas. VEGF antisense gene therapy in glioma cell lines results in fewer blood vessels, more necrosis, and inhibited tumor growth *in vivo*.[77,78] Expression of VEGF and its receptor correlate strongly with oligodendroglioma tumor grade and patient survival.[72] Likewise, allelic loss of 1p and 19q is nearly 100% predictive for drug sensitivity and survival in patients with oligodendrogliomas.[72] Ongoing research on the genes that promote proliferation, angiogenesis, and invasion of HGGs will continue to provide new targets for tumor suppression in the future.

A second gene therapy strategy involves targeting genes that are directly involved in the induction of apoptosis. Apoptosis, also called programmed cell death or cellular suicide, involves the activation of an intrinsic proteolytic cascade that terminates with the activation of cell death effectors. As discussed above, secondary HGGs often contain mutations in TP53 that prevent the normal activation of programmed cell death. This offers one potential target of gene therapy for gliomas. Many other signals and downstream effectors of apoptosis may also be delivered to tumor cells via gene therapy to oppose the unchecked growth of malignant cells.

Fas ligand and its receptor form a well-studied upstream signal in the apoptotic cascade. Fas is a transmembrane protein of the nerve growth factor/tumor necrosis factor receptor family, binding of the Fas ligand triggers apoptosis. Fas-associated protein with death domain (FADD) binds to the intracellular domain of Fas and is the immediate downstream signal in the cascade. Most malignant astrocytomas express high levels of Fas in contrast to their nonmalignant counterparts.[79] Gene therapy with both Fas ligand and FADD effectively inhibit *in vitro* and *in vivo* survival of HGGs via induction of apoptosis.[80,81]

Ultimately, the apoptotic signal induces activation of caspases, a family of cysteine proteases, and the downstream effectors of apoptosis. Caspase 8 is a well-characterized effector activated by the Fas/Fas ligand signal. Transfer of this gene preferentially induces apoptosis in glioma cell lines when compared to endothelial cells, fibroblasts, and nonmalignant neuronal cells.[82]

It is logical that the transfer of genes that induce tumor suppression or apoptosis may also confer susceptibility to chemotherapy and radiation, which has been demonstrated.[75,76] Similarly, gene therapeutic techniques may involve the transfer of exogenous genes that sensitize tumor cells to a specific drug or prodrug. The classic example of this approach is delivery of the HSV-1 thymidine kinase (HSV-TK) gene and subsequent therapy with gancyclovir. Unlike the human TK protein, HSV-TK is able to monophosphorylate gancyclovir, a nucleoside analog that is then incorporated into DNA during synthesis and halts cell division. This strategy was first demonstrated to be effective in 1986 by Moolten and has since been reliably reproduced in several model systems.[83] However, clinical trials utilizing this strategy have been limited by poor and variable gene expression with only anecdotal benefit observed thus far.

Numerous preclinical studies have confirmed that each of these gene therapy strategies shows promise in tumor models *in vitro* and *in vivo*. By transferring genes capable of inducing tumor suppression, apoptosis, or drug susceptibility, researchers have been able to decrease tumor size, aggressiveness, and resistance to chemotherapy or radiation.[76] The limiting factor in the application of these strategies is the development of vectors and gene promoters that will allow sufficient and specific expression of therapeutic genes in targeted cells. The recent development and rapid improvement in techniques such as serial analysis of gene expression (SAGE) and microarray gene expression analysis now allow the simultaneous determination of differential expression of thousands of potential targets.[84] Therefore, the identification of novel gene targets is a productive area of research and it is not considered the rate-limiting step for successful gene therapy.

9.3.3 Vectors: Viral and Nonviral Systems

Once a target gene is identified, a vehicle for transport to the targeted cell population must then be selected. Vectors for use in gene therapy can be viral or nonviral. Any vector for gene therapy of HGGs would ideally:

1. Be stable and relatively easy to produce
2. Be capable of carrying transgenes large enough for desired applications
3. Be able to transfect target cell efficiently
4. Express a gene of interest in sufficient amounts and for a sufficient time
5. Exert minimal inflammatory effect
6. Demonstrate minimal toxicity to surrounding tissue

None of the vectors employed to date satisfies all these criteria, in fact, most are lacking in several areas. Thus optimizing a vehicle for gene delivery is probably the rate-limiting step for the success of gene therapy as a therapeutic modality.

9.3.4 Retroviruses

Several viruses have been engineered to be nonpathogenic replication-deficient gene therapy vectors. Most clinical trials to date have used retroviruses due to several attributes of these vectors. Retroviruses are RNA viruses that integrate their genetic information into the genomes of replicating cells. This offers two theoretical advantages when treating brain tumors. First, therapeutic genes should be expressed for the life of a cell. Second, gene expression should be highly specific because tumor cells divide and surrounding cells do not.

However, since only a small portion (approximately 10 to 15%) of glioma cells replicate at any particular time, making transfection with retroviruses highly inefficient.[85] Interestingly though, tumor cells can be affected even without expressing the delivered gene in a phenomena called the "bystander effect."[86] The effect is likely mediated when tumor cells surrounding a transfected cell pick up either the TK gene product or the activated metabolite of the antiviral medication.

Insertion of the retrovirus genome into the host genome carries a distinct disadvantage as well. Whenever a retroviral vector is administered, there is a risk of insertional mutagenesis from random-site insertion that may induce or potentiate neoplastic transformation in a transfected cell. This risk was thought to be very low in humans until a retrovirus gene therapy trial in France was halted when one patient developed leukemia as a result of insertional mutagenesis.[87]

Other important features of retroviruses for gene therapy applications are: (1) retrovirus genomes are small and can only accommodate 9 kb of exogenous information, thus limiting the repertoire of genes they can carry, and (2) the partial immune-privileged nature of the CNS should reduce any immune response to the murine packaging cells necessary for high-titer production of engineered retroviruses.[88]

9.3.5 Adenoviruses

While retroviruses have been the most commonly studied viruses for use in gene therapy applications overall, adenoviruses are the most commonly used DNA viral vectors. Adenoviruses have double-stranded linear DNA genomes and can be engineered to be replication-defective vectors via deletion of early genes that encode transcriptional regulators.[89] Adenoviruses offer some specific advantages over retroviral vectors. The most significant is that they can be produced with relative ease at high titers in cell-free preparations, while retrovirus gene therapy *in vivo* requires injection of a viral packaging cell line. In addition, their DNA genome makes adenovirus vectors significantly more stable.

As with all vectors, adenoviruses have several properties that confer advantages and disadvantages for use in gene therapy applications. First, the adenovirus genome remains episomal in an infected cell, which means that it does not integrate into the host DNA. While this eliminates the risk of insertional mutagenesis, it appears to result in a shorter duration of transgene expression. Second, adenovirus vectors have the ability to infect both dividing and nondividing cells. This increases the potential targets of adenoviral gene therapy, but makes them less tumor-specific than retroviral vectors. Third, the adenovirus genome is large, and thus can accommodate the insertion of much more genetic material than retroviruses. However, the larger genome is somewhat more difficult to manipulate. Fourth, adenoviral target cell tropism is controlled by interaction between components of the viral capsid with its receptor. Thus manipulation of the viral capsid may be used to target the vectors to desired cells.[59,60] However, expression of the adenovirus receptor is absent on many tumor cells, making them resistant to adenovirus infection.[83]

Another important feature of adenoviruses cited in the section on oncolytic viruses is the large immune response generated to the virus. While elimination of adenovirus gene transcription significantly reduces the adaptive immune response, a rapid innate response to the viral particle or capsid remains.[90,91] This innate immune response is potent enough to result in a substantial loss of adenoviral vectors within 24 hours after injection *in vivo*. Given the various advantages and disadvantages of each vector system, we have no way to know whether adenoviruses are better suited for gene therapy of gliomas without a direct comparison to retroviruses.

Fortunately, adenoviral vectors and retroviral vectors were compared in a head-to-head trial for gene therapy of brain tumors via delivery of HSV-TK followed by administration of gancyclovir.[92] This Phase I trial with seven patients in each group revealed a significant survival advantage in those treated with adenovirus vectors compared to retrovirus vectors. Patients treated with adenoviruses also had significant increases in side effects, with seizures occurring in two patients, fevers in two, and an increased anti-adenovirus antibody titer in four. The side effects observed in adenovirus-treated patients in this trial underscore the need for improvements in these vectors before more widespread use can be considered.

9.3.6 Herpes Simplex Virus

HSV is an enveloped, double-stranded DNA virus that has long been considered a possible agent for delivering genes into mammalian cells. The virus infects virtually all cell types and vertebrate species when tested *in vitro*, making it potentially valuable as a vector for use in any organ system or animal model. The HSV genome is very large, 152 kb, and as many as 50 kb are available for deletion and replacement with desired transgenes.

Two primary methods have been developed for using HSV as a vector for gene transfer into mammalian cells. The first method (recombinant vectors) consists of HSV particles in which the gene of interest is inserted into a portion of the viral genome.[93] Early studies showed the potential of recombinant HSV vectors to efficiently deliver transgenes into cells of the CNS using marker genes.[94,95] Because of the large viral genome, these vectors were classically constructed by homologous recombination, an arduous process that limited their use. However, this trend has changed recently and recombinant HSV vectors may still develop into important tools for gene therapy in neurons.[96]

Most work with HSV in gene therapy applications has utilized a second system (defective HSV vectors). Defective HSV vectors are generated by creation of a DNA plasmid containing the desired transgene along with an HSV origin of DNA replication and packaging sequence. This plasmid is transfected into cells along with a replicating "helper" HSV that then packages the desired transgene into nonreplicating "defective" particles.[97,98] Several studies demonstrated the potential of defective HSV vectors for gene therapy in the CNS.[99–102] Most studies using defective HSV vectors for gliomas utilized the TK/gancyclovir mechanism of tumor cell lysis.[103,104] However the vectors have also been used to deliver other therapeutic genes.[105,106]

9.3.7 Other Vectors

The shortcomings of retroviral, Ad, and HSV vectors have fueled searches for other gene therapy vectors including viral and nonviral constructs. Other viruses studied include adeno-associated viruses,[107,108] lentiviruses,[109] and foamy viruses.[110] Nonviral contructs seek to transfer desired transgenes without the associated toxicity of viruses. Encapsulation of plasmid DNA into liposomes is a promising nonviral mechanism for gene therapy that has been applied to brain and other tumors.[111–113]

Plasmid–liposome complexes have many distinct advantages compared to viral vectors. They (1) transfer genes of essentially unlimited size, (2) cannot recombine to form an infectious agent, (3) protect DNA from the extracellular environment, and (4) evoke weaker inflammatory responses because they lack proteins.[114] However, transfection via plasmids is highly inefficient and is not selective. Gene expression has been found to be relatively transient.

Liposomes can also be used as vectors for delivery of antisense DNA. This is a unique gene therapy approach that uses therapeutic strands of DNA to bind a complementary sequence within a target gene to block synthesis at the transcription level. Antisense oligonucleotides have shown therapeutic benefit (and minimal toxicity) against brain tumors.[77,115,116]

9.3.8 Delivery, Neurogenetic Surgery

Delivery is a potential obstacle to the success of gene therapy for HGGs due to the presence of the blood–brain barrier (BBB). Although several systemic approaches are aimed at bypassing or disrupting the BBB,[78] local delivery is a logical and efficient approach for these locally aggressive tumors. If a tumor is accessible to surgery, the surgical cavity can be lined with the vector or vector-producing cells. Alternatively, repeated delivery of a vector may be performed through placement of an Ommaya or Rickham reservoir at the time of surgery. MRI-guided stereotactic injection is a reasonable alternative for surgically inaccessible tumors. Using these methods, gene therapy may be delivered via traditional surgical approaches, prompting some to refer to this mode of delivery as neurogenetic surgery.[117] A more detailed discussion of delivery methods is presented next in the section on convection-enhanced delivery (CED).

9.4 TRANSLATION INTO CLINICAL TRIALS: HUMANS ARE NOT LARGE MICE

The success of *in vitro* and *in vivo* preclinical studies led to cautious optimism in regard to the potential clinical utility of gene therapy for HGGs. It has been only 10 years since the first human trial of gene therapy for HGGs.[17] Since then, multiple Phase I and Phase II clinical trials have demonstrated the safety and feasibility of this approach in humans. Unfortunately, the dictum that humans are not large mice holds true for the translation of *in vitro* and *in vivo* experimental successes. Despite tumor regression and improved survival in animal models, significant clinical benefit in humans has yet to be achieved.

Only two gene therapy strategies have been tested in clinical trials for HGG, the p53 tumor suppression approach and the HSV-TK/gancyclovir susceptibility approach. A single Phase I trial of p53 gene therapy utilized an adenoviral vector.[118] Delivery was via a two-stage approach, first with stereotactic injection after tumor biopsy followed by tumor resection and direct injection of vector into the remaining tumor bed. This study demonstrated low toxicity and successful transfection of functional p53 gene to tumor cells. Ad vector injection successfully resulted in apoptosis of p53 transfected cells. However, extension of transfection only reached 5 mm from the injection sites. Further studies are needed to improve distribution of this agent prior to Phase II studies designed to determine therapeutic efficacy.

The majority of clinical trials of gene therapy for any tumor utilize transduction of HSV-TK followed by systemic administration of gancyclovir. Phase I and Phase II trials have demonstrated the safety of this approach in brain tumors via both stereotactic and open surgical delivery.[17,119-121] These trials have utilized retroviral or adenoviral vectors. As mentioned, in a Phase I trial that directly compared retrovirus and adenovirus delivery, the adenovirus group showed statistically significant improved survival along with dose-limiting inflammatory side effects.[92]

The bystander effect demonstrated in preclinical experiments may also be significant in human clinical trials. A small Phase I trial demonstrated MRI evidence of tumor regression after transfection with HSV-TK and subsequent treatment with gancyclovir, despite transfection of only a small cluster of cells and variable expression of the HSV-TK gene.[122] However, in the only Phase III trial of this approach in brain tumor patients, no significant differences in time to progression or overall survival were observed in gene therapy-treated patients.[123] Major limitations to therapeutic efficacy include poor distribution of the viral vector and poor penetration of gancyclovir across the BBB.

Despite limited clinical success, gene therapy is now a tenable goal and will very possibly become a standard part of multimodal therapy for patients with HGGs in the future. The past decade has provided major conceptual and technological advances in brain tumor biology and molecular genetics; the years to come will provide many new breakthroughs. This vigorous area of basic science research has already been translated into clinical trials (http://clinicaltrials.gov/ct/gui) and hopefully the best combination of transgene, vector, and delivery method to benefit brain tumor patients will be discovered soon.

9.5 CONVECTION-ENHANCED DELIVERY OF TARGETED TOXINS AND OTHER AGENTS

9.5.1 INTRODUCTION

Delivery of therapeutic agents to HGGs is a difficult task that has perplexed neurosurgeons and brain tumor researchers for several decades. The effectiveness of some chemotherapeutic agents against gliomas *in vitro* has been recognized for many years, but the BBB minimizes the amount of drug that penetrates tumors when administered systemically, even with highly lipophilic nitrosureas.[124]

Toxicity limits how high a systemic dose can be given and prevents satisfactory levels of agents from reaching tumors in the brain. This circumstance led to many attempts to treat brain tumors with intratumoral or local injections of methotrexate or nitrosureas in the 1960s and 1970s, all with minimal beneficial responses.[125–129] Despite the lack of therapeutic benefit, these early investigations were encouraging because they found that intratumoral injections of chemotherapeutic agents resulted in lower systemic toxicity.[129,130]

The revolution in molecular biology techniques and other scientific advances are leading to a dramatic increase in discoveries of potential therapeutic agents for the treatment of cancer. These agents include traditional chemotherapies, molecular therapies, targeted toxins, viruses, liposomal–DNA complexes, viral packaging cells, stem cells, and others.[131–133] Although few of the new therapeutic modalities have achieved mainstream use in cancer therapy as yet, it is likely that some will do so soon. To allow brain tumor patients to benefit from these exciting new developments, a method to deliver therapeutic agents to the brain in a safe and effective manner must be developed. It is possible that this stumbling block to progress in the treatment of HGGs will be overcome by promising developments in CED.

9.5.2 CONVECTION-ENHANCED DELIVERY

Traditional means of delivering agents to the brain have involved direct injection into the parenchyma or cerebrospinal fluid. These injections rely on diffusion of the delivered agent to reach brain tissue away from the injected site. Unfortunately, multiple studies demonstrated that diffusion of agents in the brain is extremely limited, particularly with high molecular weight or polar molecules.[134–136]

Attempts have been made to overcome this limitation with use of multiple intraparenchymal catheters.[137] One study involving cisplatin infusion via 68 catheters still did not produce a significant impact on the patient's prognosis. This suggests that far too many catheters would be required to treat gliomas in this fashion. A more feasible approach is to use fewer catheters and increase the volume of diffusion through each catheter using CED.

CED uses sustained intracerebral infusion to induce a convective interstitial fluid current that has the potential to homogeneously distribute even large molecules great distances within the brain by displacing interstitial fluid.[138] In animal models, CED achieved high homogeneous concentrations of various macromolecular therapeutic agents throughout large regions of the brain that were several orders of magnitude greater than those obtainable by systemic delivery.[139] The potential benefit of CED in the treatment of brain tumors in animal models has been demonstrated in several studies.[139–141]

A significant limitation to interpreting data from CED experiments comes from the fact that human brains are much larger than those of the animal models routinely used. Although a few studies have been conducted using CED in humans,[142] no data are available on the actual distribution of agents delivered in the human brain via this method. Data recently submitted for publication demonstrate distribution of at least 10% of the injected concentration of a macromolecule within a nearly spherical radius over 4 cm from the catheter tip throughout the gray and white matter surrounding a tumor resection cavity (D. Bigner, personal communication, 2004).

In addition to this encouraging data on distribution of agents in the human brain using CED, two clinical trials demonstrated the efficacy of CED in treating human brain tumor patients. In a clinical trial by Laske et al., 9 of 15 malignant brain tumor patients had greater than 50% reductions in tumor volume after receiving therapeutic agents via CED.[142] Although local toxicity was seen at the highest dose administered, no systemic toxicity was observed, suggesting CED is an effective way to deliver therapeutic toxins to the human brain. In a trial by Rand et al., 7 of 9 patients treated with CED had increased tumor necrosis as evidenced by reduced gadolinium enhancement on MRI following therapy.[143] One patient survived more than 18 months after therapy.

Although these results are encouraging, several limiting factors remain as obstacles to the use of CED in the treatment of HGG patients. First, although a distribution of agent 4 cm from the catheter tip is encouraging, the technique still requires infusion via multiple catheters and careful optimization and planning to deliver therapeutic agent to the region surrounding a tumor or its resection cavity. Second, tumors clearly alter the fluid dynamics in the brain and the effect of this alteration on CED is poorly understood. Despite these limitations, further studies aimed at optimizing catheter design and infusion parameters should identify modifications capable of effectively addressing these issues now that the potential utility of this approach has been established in humans.

9.5.3 Targeted Toxins

Although CED could be used to deliver any of a number of therapeutic agents to treat brain tumors, the majority of work to date has utilized targeted toxins. A targeted toxin is attached to a receptor ligand; an immunotoxin consists of a toxin attached to an antibody that recognizes a receptor. In both cases, receptors selected for targeting are over-expressed on tumor cells (for simplicity, this chapter will use the term "targeted toxin" in reference to both moieties). Targeted toxins allow targeted delivery of potent toxins to tumors with relative sparing of normal tissue.[133] The specificity of these agents is enhanced and systemic toxicity reduced by delivery to an anatomically isolated compartment, such as the intracranial or intrathecal space.[144]

Bacterial and plant toxins are potent cytotoxic agents that have been exploited in targeted toxin therapy. Such toxins have at least two important advantages over most chemotherapeutic agents: (1) they are far more potent, while most chemotherapies require $>10^4$ molecules to kill a single tumor cell, many toxins require only one,[133] and (2) they are active against hypoxic and nondividing cells, making them potentially effective against tumors that are resistant to chemotherapy and radiation.[145]

The powerful potential of targeted toxins derives from a combination of the high potency and toxicity of the toxin with the highly selective binding of a receptor ligand or antibody. Critical to the success of targeted toxin therapy is the identification of a receptor that is ubiquitously highly expressed on the tumor but not on surrounding tissue. This has been accomplished in tumors outside the CNS. Clinical trials using targeted toxin therapy have targeted interleukin-2 receptors in hematologic malignancies[146] and interleukin-13 receptors in squamous cell carcinomas.[147] Other trials have used tumor-specific antibodies to target ovarian, breast, and colon cancers.[148,149]

In order for targeted toxin therapy to be effective against HGGs, a receptor that is commonly over-expressed on the tumors must be identified and targeted. It has been known for several years that HGGs frequently over-express EGFR.[150] Over-expression is often associated with amplification of the EGFR gene. A simultaneous examination of GBM samples for EGFR gene amplification, mRNA, and protein found approximately one-third had gene amplification, all had mRNA, and 85% had detectable EGFR protein[151] (McLendon et al., personal communication, 2004). By contrast, EGFR was found in only very low levels in surrounding brain — a circumstance that lends it to targeted toxin treatment with minimal unwanted toxicity.[152]

EGFR has two natural ligands, epidermal growth factor and transforming growth factor alpha (TGF-α). A targeted toxin for the EGFR was designated TP-38. It is a recombinant chimeric protein composed of TGF-α and a genetically engineered form of the pseudomonas exotoxin PE-38. Encouraging results of a Phase I clinical trial examining treatment of patients with recurrent HGGs using CED of TP-38 have recently been submitted for publication.[153]

Other receptors over-expressed on HGGs have been identified. Targeted toxins for interleukin-4 and interleukin-13 receptors showed therapeutic efficacy against HGGs.[154,155] Further work using sophisticated molecular biology techniques will undoubtedly identify other potential receptors for toxin targeting and enhance the potential of this novel therapy for HGG patients.

9.6 CONCLUSION

The relatively recent revolution in molecular biology techniques has in fact led to many significant discoveries of underlying mechanisms of the development of HGGs, only a few of which were covered here. Even more importantly, a variety of scientific advances led to the development and rapid translation to clinical trials of many novel forms of cancer therapy, broadly increasing the landscape of potential therapies far beyond the traditional modes of surgery, chemotherapy, and radiation.

Although we have not yet discovered the combination of novel therapy and better understanding of underlying tumor mechanisms that will lead to an efficacious new treatment of HGGs, many promising new therapies are on the horizon. In this environment of rapid new discovery, it remains of utmost importance that neurosurgeons are involved in and informed of the development of these exciting new therapies that may soon allow us to better serve our sickest patients.

REFERENCES

1. Central Brain Tumor Registry of the U.S., *Primary Brain Tumors in the United States, Statistical Report, 1995-1999* and *Statistical Report, 2002-2003*, Chicago.
2. Laws, E.R., Parney, I.F., Huang, W. et al., Survival following surgery and prognostic factors for recently diagnosed malignant glioma: data from the Glioma Outcomes Project, *J. Neurosurg.*, 2003, 99(3), 467–473.
3. Lacroix, M., Abi-Said, D., Fourney, D.R. et al., A multivariate analysis of 416 patients with glioblastoma multiforme: prognosis, extent of resection, and survival [comment], *J. Neurosurg.*, 2001, 95(2), 190–198.
4. Gururangan, S. and Friedman, H.S., Innovations in design and delivery of chemotherapy for brain tumors, *Neuroimaging Clin. N. Amer.*, 2002, 12(4), 583–597.
5. Dunkel, I.J. and Finlay, J.L., High-dose chemotherapy with autologous stem cell rescue for brain tumors, *Crit. Rev. Oncol. Hematol.*, 2002, 41(2), 197–204.
6. Ciordia, R., Supko, J., Gatineau, M. et al., Cytotoxic chemotherapy, advances in delivery, pharmacology, and testing, *Curr. Oncol. Rep.*, 2000, 2(5), 445–453.
7. Fathallah-Shaykh, H., New molecular strategies to cure brain tumors, *Arch. Neurol.*, 1999, 56(4), 449–453.
8. Karpati, G., Li, H., and Nalbantoglu, J., Molecular therapy for glioblastoma, *Curr. Opin. Molecular Therap.*, 1999, 1(5), 545–552.
9. Soling, A. and Rainov, N.G., Dendritic cell therapy of primary brain tumors, *Molecular Med.*, 2001, 7(10), 659–667.
10. Fecci, P.E. and Sampson, J.H., Clinical immunotherapy for brain tumors, *Neuroimaging Clin. N. Amer.*, 2002, 12(4), 641–664.
11. Virasch, N. and Kruse, C.A., Strategies using the immune system for therapy of brain tumors, *Hematol. Oncol. Clin. N. Amer.*, 2001, 15(6), 1053–1071.
12. Rapoport, S.I., Advances in osmotic opening of the blood–brain barrier to enhance CNS chemotherapy, *Expert Opin. Invest. Drugs*, 2001, 10(10), 1809–1818.
13. van Vulpen, M., Kal, H.B., Taphoorn, M.J. et al., Changes in blood–brain barrier permeability induced by radiotherapy: implications for timing of chemotherapy (review), *Oncol. Rep., 2002*, 9(4), 683–688.
14. Alexander, E., III, Optimizing brain tumor resection: midfield interventional MR imaging, *Neuroimaging Clin. N. Amer.,* 2001, 11(4), 659–672.
15. Metzger, A.K. and Lewin, J.S., Optimizing brain tumor resection: low field interventional MR imaging, *Neuroimaging Clin. N. Amer.,* 2001, 11(4), 651–657.
16. Tummala, R.P., Chu, R.M., Liu, H. et al., Optimizing brain tumor resection: high field interventional MR imaging, *Neuroimaging Clin. N. Amer.,* 2001, 11(4), 673–683.
17. Oldfield, E.H., Ram, Z., Culver, K.W. et al., Gene therapy for the treatment of brain tumors using intra-tumoral transduction with the thymidine kinase gene and intravenous gancyclovir, *Human Gene Therap.,* 1993, 4(1), 39–69.

18. Takamiya, Y., Short, M.P., Ezzeddine, Z.D. et al., Gene therapy of malignant brain tumors: a rat glioma line bearing the herpes simplex virus type 1-thymidine kinase gene and wild type retrovirus kills other tumor cells, *J. Neurosci. Res.,* 1992, 33(3), 493–503.
19. Martuza, R.L., Malick, A., Markert, J.M. et al., Experimental therapy of human glioma by means of a genetically engineered virus mutant, *Science*, 1991, 252(5007), 854–856.
20. Varghese, S. and Rabkin, S.D., Oncolytic herpes simplex virus vectors for cancer virotherapy, *Canc. Gene Ther.*, 2002, 9, 967–978.
21. Kaplitt, M., Tjuvajev, J., Leib, D.A. et al., Mutant herpes simplex virus-induced regression of tumors growing in immunocompetent rats, *J. Neurol. Oncol.*, 1994, 19, 137–147.
22. Mineta, T., Rabkin, S.D., and Martuza, R.L. Treatment of malignant gliomas using ganciclovir-hypersensitive, ribonucleotide reductase-deficient herpes simplex viral mutant, *Cancer Res.,* 1994, 54(15), 3963–3966.
23. Chambers, R., Gillespie, G.Y., Soroceanu, L. et al., Comparison of genetically engineered herpes simplex viruses for the treatment of brain tumors in a SCID mouse model of human malignant glioma, *PNAS*, 1995, 92(5), 1411–1415.
24. Chou, J., Kern, E.R., Whitley, R.J. et al., Mapping of herpes simplex virus-1 neurovirulence to gamma 134.5: a gene nonessential for growth in culture, *Science*, 1990, 250(4985), 1262–1266.
25. Mineta, T., Rabkin, S.D., Yazaki, T. et al., Attenuated multi-mutated herpes simplex virus-1 for the treatment of malignant gliomas, *Nature Med.,* 1995, 1(9), 938–943.
26. Goldstein, D.J. and Weller, S.K., Herpes simplex virus type 1-induced ribonucleotide reductase activity is dispensable for virus growth and DNA synthesis, isolation and characterization of an ICP6 lacZ insertion mutant, *J. Virol.,* 1988, 62(1), 196–205.
27. Goldstein, D.J. and Weller, S.K. Factor(s) present in herpes simplex virus type 1-infected cells can compensate for the loss of the large subunit of the viral ribonucleotide reductase: characterization of an ICP6 deletion mutant, *Virology,* 1988, 166(1), 41–51.
28. Jacobson, J.G., Leib, D.A., Goldstein, D.J. et al., A herpes simplex virus ribonucleotide reductase deletion mutant is defective for productive acute and reactivatable latent infections of mice and for replication in mouse cells, *Virology,* 1989, 173(1), 276–283.
29. Coen, D.M., Goldstein, D.J., and Weller, S.K. Herpes simplex virus ribonucleotide reductase mutants are hypersensitive to acyclovir, *Antimicrob. Agents Chemother.*, 1989, 33(8), 1396–1399. Erratum in *Antimicrob. Agents Chemother.*, 1989 33(10),1827.
30. Hunter, W.D., Martuza, R.L., Feigenbaum, F. et al., Attenuated, replication-competent herpes simplex virus type 1 mutant G207: safety evaluation of intracerebral injection in non-human primates, *J. Virol.*, 1999, 73, 6319–6326.
31. Markert, J.M., Medlock, M.D., Rabkin, S.D. et al., Conditionally replicating herpes simplex virus mutant G207 for the treatment of malignant glioma: results of a phase I trial [comment], *Gene Ther.,* 2000, 7(10), 867–874.
32. Rampling, R., Cruickshank, G., Papanastassiou, V. et al., Toxicity evaluation of replication-competent herpes simplex virus (ICP 34.5 null mutant 1716) in patients with recurrent malignant glioma [comment], *Gene Ther.,* 2000, 7(10), 859–866.
33. Toda, M., Rabkin, S.D., Kojima, H. et al., Herpes simplex virus as an *in situ* cancer vaccine for the induction of specific anti-tumor immunity, *Hum. Gene Ther.* 1999, 10, 385–393.

34. Miyatake, S., Iyer, A., Martuza, R.L. et al., Transcriptional targeting of herpes simplex virus for cell-specific replication, *J. Virol.*, 1997, 71, 5124–5132.
35. Chung, R.Y., Saeki, Y., and Chiocca, E.A., B-myb promoter retargeting of herpes simplex virus gamma 34.5 gene-mediated virulence toward tumor and cycling cells, *J. Virol.*, 1999, 73, 7556–7564.
36. Yamamura, H., Hadhio, M., Noguchi, M. et al., Identification of transcriptional regulatory sequences of human calponin promoter and their use in targeting a conditionally replicating herpes vector to malignant human soft tissue and bone tumors, *Cancer Res.*, 2001, 61, 3969–3977.
37. Alemany, R., Gomez-Manzano, C., Balague, C. et al., Gene therapy for gliomas: molecular targets, adenoviral vectors, and oncolytic adenoviruses, *Exp. Cell Res.*, 1999, 252(1), 1–12.
38. Rodriguez, R., Schuur, E.R., Lim, H.Y. et al., Prostate attenuated replication competent adenovirus (ARCA)CN706: a selective cytotoxic for prostate-specific antigen-positive prostate cancer cells, *Cancer Res.*, 1997, 57, 2559–2563.
39. Alemany, R., Balague, C., and Curiel, D.T., Replicative adenoviruses for cancer therapy, *Nature Biotech.*, 2000, 18, 723–727.
40. Chen, S.H., Shine, H.D., Goodman, J.C. et al., Gene therapy for brain tumors: regression of experimental gliomas by adenovirus-mediated gene transfer *in vivo*, *PNAS,*1994, 91(8), 3054–3057.
41. Adachi, Y., Tamiya, T., Ichikawa, T. et al., Experimental gene therapy for brain tumors using adenovirus-mediated transfer of cytosine deaminase gene and uracil phosphoribosyltransferase gene with 5-fluorocytosine, *Hum. Gene Ther.,* 2000, 11(1), 77–89.
42. Dewey, R.A., Morrissey, G., Cowsill, C.M. et al., Chronic brain inflammation and persistent herpes simplex virus 1 thymidine kinase expression in survivors of syngeneic glioma treated by adenovirus-mediated gene therapy: implications for clinical trials, *Nature Med.,*1999, 5(11), 1256–1263.
43. Yang, Y., Li, Q., Ertl, H.C. et al., Cellular and humoral immune responses to viral antigens create barriers to lung-directed gene therapy with recombinant adenoviruses, *J. Virol.*, 1995, 69, 2004–2015.
44. Hollon, T., Researchers and regulators reflect on first gene therapy death, *Nature Med.,* 2000, 6(1), 6.
45. Smith, R.R., Huebner, R.J., Rowe, W.P. et al., Studies on the use of viruses in the treatment of carcinoma of the cervix, *Cancer*, 1956, 9, 1211–1218.
46. Bischoff, J.R., Kirn, D.H., Williams, A. et al., An adenovirus mutant that replicates selectively in p53-deficient human tumor cells [comment], *Science*, 1996, 274(5286), 373–376.
47. Rothmann, T., Hengstermann, A., Whitaker, N.J. et al., Replication of ONYX-015, a potential anticancer adenovirus, is independent of p53 status in tumor cells, *J. Virol.*, 1998, 72(12), 9470–9478.
48. Dix, B.R., Edwards, S.J., and Braithwaite, A.W., Does the antitumor adenovirus ONYX-015/dl1520 selectively target cells defective in the p53 pathway? *J. Virol.*, 2001, 75(12), 5443–5447.
49. Geoerger, B., Grill, J., Opolon, P. et al., Oncolytic activity of the E1B-55 kDa-deleted adenovirus ONYX-015 is independent of cellular p53 status in human malignant glioma xenografts, *Cancer Res.,* 2002, 62(3), 764–772.
50. Ganly, I., Kirn, D., Eckhardt, G. et al., A phase I study of Onyx-015, an E1B attenuated adenovirus administered intratumorally to patients with recurrent head and neck cancer, *Clin. Cancer Res.,* 2000, 6(3), 798–806. Erratum to correct author's name in *Clin. Cancer Res.*, 2000 6(5), 2120.

51. Fueyo, J., Gomez-Manzano, C., Alemany, R. et al., A mutant oncolytic adenovirus targeting the Rb pathway produces anti-glioma effect *in vivo*, *Oncogene,* 2000, 19(1), 2–21. Erratum in *Oncogene*, 2000, 19(43), 5038.
52. Yu, D.C., Sakamoto, G.T., and Henderson, D.R., Identification of the transcriptional regulatory sequences of human kallikrein 2 and their use in the construction of calydon virus 764: an attenuated replication competent adenovirus for prostate cancer therapy, *Cancer Res.,* 1999, 59(7), 1498–1504.
53. Hallenbeck, P.L., Chang, Y.N., Hay, C. et al., A novel tumor-specific replication-restricted adenoviral vector for gene therapy of hepatocellular carcinoma, *Hum. Gene Ther.,* 1999, 10(10), 1721–1733.
54. Stojdl, D.F., Lichty, B., Knowles, S. et al., Exploiting tumor-specific defects in the interferon pathway with a previously unknown oncolytic virus, *Nature Med.*, 2000, 6, 821–825.
55. Coffey, M.C., Strong, J.E., Forsyth, P.A. et al., Reovirus therapy of tumors with activated Ras pathway, *Science,* 1998, 282, 1332–1334.
56. Gromeier, M., Lachmann, S., Rosenfeld, M.R. et al., Intergeneric poliovirus recombinants for the treatment of malignant glioma [comment], *PNAS,* 2000, 97(12), 6803–6808.
57. Gromeier, M., Alexander, L., and Wimmer, E., Internal ribosomal entry site substitution eliminates neurovirulence in intergeneric poliovirus recombinants, *PNAS,* 1996, 93(6), 2370–2375.
58. Gromeier, M., Bossert, B., Arita, M. et al., Dual stem loops within the poliovirus internal ribosomal entry site control neurovirulence, *J. Virol.* 1999, 73(2), 958–964.
59. Wickham, T.J., Targeting adenovirus, *Gene Ther.,* 2000, 7, 110–114.
60. Yoshida, Y., Sadata, A., Zhang, W. et al., Generation of fiber-mutant recombinant adenoviruses for gene therapy of malignant glioma, *Hum. Gene Ther.,* 1998, 9(17), 2503–2515.
61. Freytag, S.O., Rogulski, K.R., Paielli, D.L. et al., A novel three-pronged approach to kill cancer cells selectively: concomitant viral, double suicide gene, and radiotherapy [comment], *Hum. Gene Ther.,* 1998, 9(9), 1323–1333.
62. Toda, M., Martuza, R.L., Kojima, H. et al., *In situ* cancer vaccination: an IL-12 defective vector/replication-competent herpes simplex virus combination induces local and systemic antitumor activity, *J. Immunol.*, 1998, 160(9), 4457–4464.
63. Todo, T., Martuza, R.L., Dallman, M.J. et al., *In situ* expression of soluble B7-1 in the context of oncolytic herpes simplex virus induces potent antitumor immunity, *Cancer Res.,* 2001, 61(1), 153–161.
64. Carew, J.F., Kooby, D.A., Halterman, M.W. et al., A novel approach to cancer therapy using an oncolytic herpes virus to package amplicons containing cytokine genes, *Molecular Ther.,* 2001, 4(3), 250–256.
65. Walker, J.R., McGeagh, K.G., Sundaresan, P. et al., Local and systemic therapy of human prostate adenocarcinoma with the conditionally replicating herpes simplex virus vector G207, *Hum. Gene Ther.,* 1999, 10(13), 2237–2243.
66. Wong, R.J., Joe, J.K., Kim, S.H. et al., Oncolytic herpes virus effectively treats murine squamous cell carcinoma and spreads by natural lymphatics to treat sites of lymphatic metastases, *Hum. Gene Ther.,* 2002, 13(10), 1213–1223.
67. Coukos, G., Makrigiannakis, A., Montas, S. et al., Multi-attenuated herpes simplex virus-1 mutant G207 exerts cytotoxicity against epithelial ovarian cancer but not normal mesothelium and is suitable for intraperitoneal oncolytic therapy, *Cancer Gene Ther.,* 2000, 7(2), 275–283.

68. Ikeda, K., Ichikawa, T., Wakimoto, H. et al., Oncolytic virus therapy of multiple tumors in the brain requires suppression of innate and elicited antiviral responses, *Nature Med.,* 1999, 5(8), 881–887.
69. Chillon, M., Lee, J.H., Fasbender, A. et al., Adenovirus complexed with polyethylene glycol and cationic lipid is shielded from neutralizing antibodies *in vitro, Gene Ther.,* 1998, 5(7), 995–1002.
70. Coukos, G., Makrigiannakis, A., Kang, E.H. et al., Use of carrier cells to deliver a replication-selective herpes simplex virus-1 mutant for the intraperitoneal therapy of epithelial ovarian cancer, *Clin. Cancer Res.,* 1999, 5(6), 1523–1537.
71. Ekstrand, A.J., Sugawa, N., James, C.D. et al., Amplified and rearranged epidermal growth factor receptor genes in human glioblastomas reveal deletions of sequences encoding portions of the N- and/or C-terminal tails, *PNAS,* 1992, 89(10), 4309–4913.
72. Shapiro, J.R., Genetics of nervous system tumors. *Hematol. Oncol. Clin. N. Amer.,* 2001, 15(6), 961–977.
73. Lal, A., Glazer, C.A., Martinson, H.M. et al., Mutant epidermal growth factor receptor up-regulates molecular effectors of tumor invasion, *Cancer Res.,* 2002, 62(12), 3335–3339.
74. Lang, F.F., Miller, D.C., Koslow, M. et al., Pathways leading to glioblastoma multiforme: a molecular analysis of genetic alterations in 65 astrocytic tumors, *J. Neurosurg.,* 1994, 81(3), 427–436.
75. Lang, F.F., Yung, W.K., Sawaya, R. et al., Adenovirus-mediated p53 gene therapy for human gliomas, *Neurosurgery,* 1999, 45(5), 1093–1094.
76. Broaddus, W.C., Liu, Y., Steele, L.L. et al., Enhanced radiosensitivity of malignant glioma cells after adenoviral p53 transduction, *J. Neurosurg.,* 1999, 91(6), 997–1004.
77. Saleh, M., Stacker, S.A., and Wilks, A.F., Inhibition of growth of C6 glioma cells *in vivo* by expression of antisense vascular endothelial growth factor sequence, *Cancer Res.,* 1996, 56(2), 393–401.
78. Zlokovic, B.V. and Apuzzo, M.L., Cellular and molecular neurosurgery: pathways from concept to reality: part II: vector systems and delivery methodologies for gene therapy of the central nervous system, *Neurosurgery,* 1997, 40(4), 805–813.
79. Weller, M., Malipiero, U., Rensing-Ehl, A. et al., Fas/APO-1 gene transfer for human malignant glioma, *Cancer Res.,* 1995, 55(13), 2936–2944.
80. Kondo, S., Ishizaka, Y., Okada, T. et al., FADD gene therapy for malignant gliomas *in vitro* and *in vivo* [comment], *Hum. Gene Ther.,* 1998, 9(11), 1599–1608.
81. Shinoura, N., Yoshida, Y., Asai, A. et al., Adenovirus-mediated transfer of p53 and Fas ligand drastically enhances apoptosis in gliomas, *Cancer Gene Ther.,* 2000, 7(5), 732–738.
82. Shinoura, N., Koike, H., Furitu, T. et al., Adenovirus-mediated transfer of caspase-8 augments cell death in gliomas: implication for gene therapy, *Hum. Gene Ther.,* 2000, 11(8), 1123–1137.
83. Fecci, P.E., Gromeier, M., and Sampson, J.H., Viruses in the treatment of brain tumors, *Neuroimaging Clin. N. Amer.,* 2002, 12(4), 553–570.
84. Polyak, K. and Riggins, G.J., Gene discovery using the serial analysis of gene expression technique, implications for cancer research, *J. Clin. Oncol.,* 2001, 19(11), 2948–2958.
85. Eck, S.L., Alavi, J.B., Alavi, A. et al., Treatment of advanced CNS malignancies with the recombinant adenovirus H5.010RSVTK: a phase I trial, *Hum. Gene Ther.,* 1996, 7(12), 1465–1482.
86. van Dillen, I.J., Mulder, N.H., Vaalburg, W. et al., Influence of the bystander effect on HSV-tk/GCV gene therapy: a review, *Curr. Gene Ther.,* 2002, 2(3), 307–322.

87. Hacein-Bey-Abina, S., von Kalle, C., Schmidt, M. et al., A serious adverse event after successful gene therapy for X-linked severe combined immunodeficiency [comment], *New Eng. J. Med.*, 2003, 348(3), 255–256.
88. Culver, K.W., Ram, Z., Wallbridge, S. et al., *In vivo* gene transfer with retroviral vector–producer cells for treatment of experimental brain tumors [comment], *Science*, 1992, 256(5063), 1550–1552.
89. St George, J.A., Gene therapy progress and prospects: adenoviral vectors, *Gene Ther.,* 2003, 10(14), 1135–1141.
90. Liu, Q. and Muruve, D.A., Molecular basis of the inflammatory response to adenovirus vectors, *Gene Ther.*, 2003, 10(11), 935–940.
91. Lowenstein, P.R. and Castro, M.G., Inflammation and adaptive immune responses to adenoviral vectors injected into the brain: peculiarities, mechanisms, and consequences, *Gene Ther.*, 2003, 10(11), 946–954.
92. Sandmair, A.M., Loimas, S., Puranen, P. et al., Thymidine kinase gene therapy for human malignant glioma using replication-deficient retroviruses or adenoviruses, *Hum. Gene Ther.,* 2000, 11(16), 2197–2205.
93. Shih, M.F., Arsenakis, M., Tiollais, P. et al., Expression of hepatitis B virus S gene by herpes simplex virus type 1 vectors carrying alpha- and beta-regulated gene chimeras, *PNAS*, 1984, 81(18), 5867–5870.
94. Chiocca, E.A., Choi, B.B., Cai, W.Z. et al., Transfer and expression of the lacZ gene in rat brain neurons mediated by herpes simplex virus mutants, *New Biol*., 1990, 2(8), 739–746.
95. Dobson, A.T., Margolis, T.P., Sedarati, F. et al., A latent, nonpathogenic HSV-1-derived vector stably expresses beta-galactosidase in mouse neurons, *Neuron,* 1990, 5(3), 353–360.
96. Glorioso, J.C. and Fink, D.J., Use of HSV vectors to modify the nervous system, *Curr. Opin. Drug Discovery Dev.,* 2002, 5(2), 289–295.
97. Spaete, R.R. and Frenkel, N., The herpes simplex virus amplicon: a new eucaryotic defective virus cloning–amplifying vector, *Cell*, 1982, 30(1), 295–304.
98. Geller, A.I. and Breakefield, X.O., A defective HSV-1 vector expresses *Escherichia coli* beta-galactosidase in cultured peripheral neurons, *Science*, 1988, 241(4873), 1667–1669.
99. New, K.C., Martuza, R.L., and Rabkin, S.D. Defective herpes simplex virus vectors for the study of promoter and gene function in the CNS, *Gene Ther.*, 1994, 1 (Suppl. 1), S79.
100. New, K.C., Gale, K., Martuza, R.L. et al., Novel synthesis and release of GABA in cerebellar granule cell cultures after infection with defective herpes simplex virus vectors expressing glutamic acid decarboxylase, *Mol. Brain Res.,*1998, 61(1–2), 121–135.
101. During, M.J., Naegele, J.R., O'Malley, K.L. et al., Long-term behavioral recovery in parkinsonian rats by an HSV vector expressing tyrosine hydroxylase [comment], *Science,* 1994, 266(5189), 1399–1403.
102. Aghi, M. and Chiocca, E.A., Genetically engineered herpes simplex viral vectors in the treatment of brain tumors: a review, *Cancer Invest.,* 2003, 21(2), 278–292.
103. Burton, E.A., Fink, D.J., and Glorioso, J.C., Gene delivery using herpes simplex virus vectors, *DNA Cell Biol.,* 2002, 21(12), 915–936.
104. Miyatake, S., Martuza, R.L., and Rabkin, S.D., Defective herpes simplex virus vectors expressing thymidine kinase for the treatment of malignant glioma, *Cancer Gene Ther.,* 1997, 4(4), 222–228.

105. Hoshi, M., Harada, A., Kawase, T. et al., Antitumoral effects of defective herpes simplex virus-mediated transfer of tissue inhibitor of metalloproteinases-2 gene in malignant glioma U87 *in vitro*: consequences for anti-cancer gene therapy, *Cancer Gene Ther.,* 2000, 7(5), 799–805.
106. Kanno, H., Hattori, S., Sato, H. et al., Experimental gene therapy against subcutaneously implanted glioma with a herpes simplex virus-defective vector expressing interferon-gamma, *Cancer Gene Ther.,* 1999, 6(2), 147–154.
107. Yoshida, J., Mizuno, M., Nakahara, N. et al., Antitumor effect of an adeno-associated virus vector containing the human interferon-beta gene on experimental intracranial human glioma, *Japn. J. Cancer Res.,* 2002, 93(2), 223–228.
108. Mizuno, M., Yoshida, J., Colosi, P. et al., Adeno-associated virus vector containing the herpes simplex virus thymidine kinase gene causes complete regression of intracerebrally implanted human gliomas in mice in conjunction with ganciclovir administration, *Japn. J. Cancer Res.,* 1998, 89(1), 76-80.
109. Naldini, L., Blomer, U., Gallay, P. et al., *In vivo* gene delivery and stable transduction of nondividing cells by a lentiviral vector [comment], *Science*, 1996, 272(5259), 263–267.
110. Russell, D.W. and Miller, A.D., Foamy virus vectors, *J. Virol.*, 1996, 70, 217-222.
111. Okada, H., Okamoto, S., and Yoshida, J. Gene therapy for brain tumors: cytokine gene therapy using DNA/liposome (series 3), *No Shinkei Geka*, 1994, 22(11), 999–1004.
112. Aoki, K., Yoshida, T., Matsumoto, N. et al., Gene therapy for peritoneal dissemination of pancreatic cancer by liposome-mediated transfer of herpes simplex virus thymidine kinase gene, *Hum. Gene Ther.,* 1997, 8(9), 1105–1113.
113. Yagi, K., Ohishi, N., Hamada, A. et al., Basic study on gene therapy of human malignant glioma by use of the cationic multilamellar liposome-entrapped human interferon beta gene, *Hum. Gene Ther.,* 1999, 10(12), 1975–1982.
114. Crystal, R.G., Transfer of genes to humans: early lessons and obstacles to success, *Science*, 1995, 270(5235), 404–410.
115. Zhang, Y., Jeong Lee, H., Boado, R.J. et al., Receptor-mediated delivery of an antisense gene to human brain cancer cells, *J. Gene Med.,* 2002, 4(2), 183–194.
116. Yung, W.K., New approaches in brain tumor therapy using gene transfer and antisense oligonucleotides, *Curr. Opin. Oncol.,*1994, 6(3), 235–239.
117. Carter, B.S., Zervas, N.T., and Chiocca, E.A., Neurogenetic surgery: current limitations and the promise of gene- and virus-based therapies, *Clin. Neurosurg.,* 1999, 45, 226–246.
118. Lang, F.F., Bruner, J.M., Fuller, G.N. et al., Phase I trial of adenovirus-mediated p53 gene therapy for recurrent glioma: biological and clinical results, *J. Clin. Oncol.,* 2003, 21(13), 2508–2518.
119. Trask, T.W., Trask, R.P., Aguilar-Cordova, E. et al., Phase I study of adenoviral delivery of the HSV-tk gene and ganciclovir administration in patients with current malignant brain tumors, *J. Am. Soc. Gene Ther.,* 2000, 1(2), 195–203.
120. Shand, N., Weber, F., Mariani, L. et al., A phase 1–2 clinical trial of gene therapy for recurrent glioblastoma multiforme by tumor transduction with the herpes simplex thymidine kinase gene followed by ganciclovir GLI328: European-Canadian Study Group, *Hum. Gene Ther.,* 1999, 10(14), 2325–2335.
121. Klatzmann, D., Valery, C.A., Bensimon, G. et al., A phase I/II study of herpes simplex virus type 1 thymidine kinase "suicide" gene therapy for recurrent glioblastoma: Study Group on Gene Therapy for Glioblastoma, *Hum. Gene Ther.,* 1998, 9(17), 2595–2604.

122. Ram, Z., Culver, K.W., Oshiro, E.M. et al., Therapy of malignant brain tumors by intratumoral implantation of retroviral vector-producing cells [comment], *Nature Med.,* 1997, 3(12), 1354–1361.
123. Rainov, N.G., A phase III clinical evaluation of herpes simplex virus type 1 thymidine kinase and ganciclovir gene therapy as an adjuvant to surgical resection and radiation in adults with previously untreated glioblastoma multiforme, *Hum. Gene Ther.,* 2000, 11(17), 2389–2401.
124. Rosenblum, M.L., Wheeler, K.T., Wilson, C.B. et al., *In vitro* evaluation of *in vivo* brain tumor chemotherapy with 1,3-bis(2-chloroethyl)-1-nitrosourea, *Cancer Res.,* 1975, 35(6), 1387–1391.
125. Tator, C.H. and Wassenaar, W., Intraneoplastic injection of methotrexate for experimental brain-tumor chemotherapy, *J. Neurosurg.,* 1977, 46(2), 165–174.
126. Dommasch, D., Przuntek, H., Gruninger, W. et al., Intrathecal cytostatic chemotherapy of meningitis carcinomatosa: clinical manifestation and cerebrospinal fluid cytology in a case of metastatic carcinoma of the breast, *Eur. Neurol.,* 1976, 14(3), 178–191.
127. Garfield, J., Dayan, A.D., and Weller, R.O. Postoperative intracavitary chemotherapy of malignant supratentorial astrocytomas using BCNU. *Clinical Oncology.* 1975, 1(3), 213–222.
128. Garfield, J. and Dayan, A.D., Postoperative intracavitary chemotherapy of malignant gliomas: a preliminary study using methotrexate, *J. Neurosurg.,* 1973, 39(3), 315–322.
129. Tator, C.H., Intraneoplastic injection of CCNU for experimental brain tumor chemotherapy, *Surg. Neurol.,* 1977, 7(2), 73–77.
130. Tator, C.H., Day, A., Ng, R. et al., Chemotherapy of an experimental glioma with nitrosoureas, *Cancer Res.,* 1977, 37(2), 476–481.
131. Zovickian, J., Johnson, V.G., and Youle, R.J., Potent and specific killing of human malignant brain tumor cells by an anti-transferrin receptor antibody: ricin immunotoxin, *J. Neurosurg.,* 1987, 66(6), 850–861.
132. Jain, R.K., Delivery of novel therapeutic agents in tumors, physiological barriers and strategies, *J. Natl. Cancer Inst.,* 1989, 81(8), 570–576.
133. Hall, W.A., Immunotoxin treatment of brain tumors, *Methods Mol. Biol.,* 2001, 166, 139–154.
134. Oldendorf, W.H., Lipid solubility and drug penetration of the blood-brain barrier, *Proc. Soc. Exp. Biol. Med.,* 1974, 147(3), 813–815.
135. Hicks, J.T., Albrecht, P., and Rapoport, S.I., Entry of neutralizing antibody to measles into brain and cerebrospinal fluid of immunized monkeys after osmotic opening of the blood–brain barrier, *Exp. Neurol.,* 1976, 53(3), 768–779.
136. Blasberg, R.G., Patlak, C., and Fenstermacher, J.D., Intrathecal chemotherapy: brain tissue profiles after ventriculo-cisternal perfusion, *J. Pharmacol. Exp. Ther.*, 1975, 195, 73–83.
137. Bouvier, G., Penn, R.D., Kroin, J.S. et al., Direct delivery of medication into a brain tumor through multiple chronically implanted catheters, *Neurosurgery,* 1987, 20(2), 286–291.
138. Morrison, P.F., Laske, D.W., Bobo, R.H. et al., High-flow microinfusion, tissue penetration and pharmacodynamics, *Am. J. Physiol.*, 1994, 266, R292–R305.
139. Bobo, R.H., Laske, D.W., Akbasak, A. et al., Convection-enhanced delivery of macromolecules in the brain, *PNAS,* 1994, 91(6), 2076–2080.
140. Heimberger, A.B., Archer, G.E., McLendon, R.E. et al., Temozolomide delivered by intracerebral microinfusion is safe and efficacious against malignant gliomas in rats, *Clin. Cancer Res.,* 2000, 6(10), 4148–4153.

141. Lieberman, D.M., Laske, D.W., Morrison, P.F. et al., Convection-enhanced distribution of large molecules in gray matter during interstitial drug infusion, *J. Neurosurg.*, 1995, 82(6), 1021–1029.
142. Laske, D.W., Youle, R.J., and Oldfield, E.H., Tumor regression with regional distribution of the targeted toxin TF-CRM107 in patients with malignant brain tumors, *Nature Med.*, 1997, 3(12), 1362–1368.
143. Rand, R.W., Kreitman, R.J., Patronas, N. et al., Intratumoral administration of recombinant circularly permuted interleukin-4–*Pseudomonas* exotoxin in patients with high-grade glioma, *Clin. Cancer Res.*, 2000, 6(6), 2157–2165.
144. Pastan, I.H., Archer, G.E., McLendon, R.E. et al., Intrathecal administration of single-chain immunotoxin, LMB-7 [B3(Fv)-PE38], produces cures of carcinomatous meningitis in a rat model, *PNAS*, 1995, 92(7), 2765–2769.
145. Kreitman, R.J., Wilson, W.H., Bergeron, K. et al., Efficacy of the anti-CD22 recombinant immunotoxin BL22 in chemotherapy-resistant hairy-cell leukemia, *New Engl. J. Med.*, 2001, 345(4), 241–247.
146. Kreitman, R.J., Wilson, W.H., White, J.D. et al., Phase I trial of recombinant immunotoxin anti-Tac(Fv)-PE38 (LMB-2) in patients with hematologic malignancies, *J. Clin. Oncol.*, 2000, 18(8), 1622–1636.
147. Joshi, B.H., Kawakami, K., Leland, P. et al., Heterogeneity in interleukin-13 receptor expression and subunit structure in squamous cell carcinoma of head and neck: differential sensitivity to chimeric fusion proteins comprised of interleukin-13 and a mutated form of *Pseudomonas* exotoxin, *Clin. Cancer Res.*, 2002, 8(6), 1948–1956.
148. Pai, L.H., Bookman, M.A., Ozols, R.F. et al., Clinical evaluation of intraperitoneal *Pseudomonas* exotoxin immunoconjugate OVB3-PE in patients with ovarian cancer [comment], *J. Clin. Oncol.*, 1991, 9(12), 2095–2103.
149. Pai, L.H., Wittes, R., Setser, A. et al., Treatment of advanced solid tumors with immunotoxin LMB-1, an antibody linked to *Pseudomonas* exotoxin, *Nature Med.*, 1996, 2(3), 350–353.
150. Wong, A.J., Bigner, S.H., Bigner, D.D. et al., Increased expression of the epidermal growth factor receptor gene in malignant gliomas is invariably associated with gene amplification, *PNAS*, 1987, 84(19), 6899–6903.
151. Wikstrand, C.J., Stenzel, T., McLendon, R.E. et al., Correlation of EGFRwt RNA transcript levels and patient biopsy protein expression by immunohistochemistry, *Neuro-Oncol.*, 2003, (in press).
152. Libermann, T.A., Razon, N., Bartal, A.D. et al., Expression of epidermal growth factor receptors in human brain tumors, *Cancer Res.*, 1984, 44(2), 753–760.
153. Sampson, J.H. et al., Progress report of Phase I study of the intracerebral microinfusion of a recombinant chimeric protein for the treatment of malignant brain tumors, *J. Neuro-Oncol.*, 65, 27–35, 2003.
154. Puri, R.K., Cytotoxins directed at interleukin-4 receptors as therapy for human brain tumors, *Methods Mol. Biol.*, 2001, 166, 155–176.
155. Debinski, W., Obiri, N.I., Powers, S.K. et al., Human glioma cells overexpress receptors for interleukin-13 and are extremely sensitive to a novel chimeric protein composed of interleukin-13 and *Pseudomonas* exotoxin, *Clin. Cancer Res.*, 1995, 1(11), 1253–1258.

10 Spinal Dysraphism: The Search For Magic

Timothy M. George and David Corey Adamson

CONTENTS

10.1 INTRODUCTION: WHY MAGIC?

Although pediatric neurosurgery is relatively young as a formal subspecialty of general neurosurgery (the first meeting of the Section of Pediatric Neurological Surgery was held in 1972 and the American Society of Pediatric Neurosurgery first met in 1978), it has been practiced for millennia. Trephined pediatric skulls were excavated in Peru and at other ancient sites.[1] The father of neurosurgery, Sir Victor Horsley, performed his first epilepsy surgery on a child in 1886.[2] Harvey Cushing wrote extensively about the unique disorders of childhood.[3,4] Many other notables followed these icons, ensuring the momentum for further progress and refinement in the surgical care of children with disorders of the nervous system.[5]

Congenital spinal nervous system abnormalities continue to be the mainstays and also the pitfalls of pediatric neurosurgery. Paralysis, incontinence, obesity, endocrinopathy, hydrocephalus, short stature, social stigmata, and shortened lifespans are still the norms for children with open neural tube defects (NTDs).[6,7] The number of children born with myelomeningocele has decreased over the past several decades and 3 of every 10,000 children born in the U.S. are handicapped by open spinal NTDs.[8] Additionally, improved imaging techniques have diagnosed even more children who suffer from spinal cord dysfunctions secondary to closed dysraphisms.[9] The future treatment objectives are clear: congenital spinal defects must be prevented or their neurological sequelae must be cured. Imagine that a pill or procedure could

0-8493-1482-8/05/$0.00+$1.50

prevent or cure neurological deficits. Attaining such a goal often seems impossible. It would seem to require magic — but what if magic did exist?

10.2 WHY MAGIC IS NEEDED: CAUSATIVE FACTORS OF NTDS

Current hypotheses suggest that NTDs are caused by complex interactions between extrinsic (drugs, environmental toxins, temperature, etc.) and intrinsic (genetic, metabolic, etc.) variables. Clinical and epidemiological studies in humans have implicated maternal illnesses, medications, environmental toxins, and dietary factors such as folic acid that play causative or at least contributing roles in NTD development.[10] Evidence that mutant genes cause NTDs has been supported by epidemiological studies revealing an increased incidence of NTDs in certain families. In familial cases, the trait for a NTD is semi-dominant, with apparent maternal inheritance.[8,11] Thus, NTDs represent examples of complex genetic disorders in which genes and the environment interact through an unknown relationship.

Extrinsic factors causally related to NTDs have been studied extensively. Vitamins in general and folates in particular have been shown to significantly reduce children's risks of NTD,[12] particularly when siblings have NTDs.[13] The protective mechanism of folate is unknown. Mothers have not been shown to be folate-deficient or have defective intestinal uptakes of folate.[11] Studies of mutant enzymes in the folate metabolic pathway, particularly, methylenetetrahydrofolate reductase (MTHFR), suggest a possible association with NTDs.[14–16] However, Speer et al. could not demonstrate MTHFR as a major risk factor.[17] Other cellular interactions, such as cellular transport mechanisms, are currently under investigation.[18] Folate can act as a methyl donor, permanently altering gene function via an epigenetic mechanism[19] or interfering with a metabolic pathway such as homocysteine conversion to methionine.[20–25] Although folic acid is irrelevant to the predominant basic mechanism of action of folate, supplementation with folic acid has reduced the risk of NTDs worldwide.[26–32]

Other teratogenic agents and maternal diseases have been identified as causal factors for NTD development.[10] Maternal diabetics have greater risks of having children with diabetic embryopathies consisting of NTD and other organ system anomalies.[33–35] Epileptic mothers using valproic acid as an anticonvulsant have a 1 to 2% risk.[36,37] Those exposed to carbamazepine face approximately 0.5% risk of having children with NTDs.[38] Obesity has been associated with increased risk of having a child born with a NTD.[39–41] A twofold increase in NTD incidence was also found in obese versus non-obese mothers, regardless of use of vitamin, folate, and other nutritional supplements. Febrile illnesses and hyperthermia produced by the use of a sauna or hot tub early in pregnancy have been also suggested as causes of NTDs.[10] The exact risk of occurrence due to maternal hyperthermia is not known.

Although strongly implicated, the specific genetic factors that cause NTD are not known. It is proposed that many different genes are involved in neural tube development. Some genes may confer strong genetic components and others may only exert minimal direct effects or require interaction with other genes. Environmental factors may act as triggers to genetic susceptibility.

Several lines of evidence point to a genetic component. Empiric studies have shown that the recurrence risk for NTD is greatest among first-degree relatives of an affected patient and decreases for more distant relatives. The recurrence risk for siblings of an affected patient is 2 to 5%, representing a 25- to 50-fold increase in recurrence risk compared to the general population.[27] Techniques for identification of specific genes are based on identifying populations at high risk, such as twins, investigating the recurrence risks of NTD, and identifying mutated genes.

Mouse mutants have provided many of the genes investigated as candidates for human NTDs. More than 40 mouse species have been described,[42] and the specific gene identified in only 6 species. The six well-known mutations are splotch (Sp),[43–45] extra toes (Xt),[46] short tails (T),[47] patch (Ph),[48] and targeted mutations in apolipoprotein B (ApoB)[49] and Hox-a1.[50] These mouse mutants provided clues to the embryopathies of NTDs and identified potential candidate genes for human investigation. For example, the Pax-3 mutation in splotch mice mirrored the mutated Pax-3 human homologue in Waardenburg's syndrome.[51] Furthermore, several Waardenburg's patients have been reported to have spina bifida.[23] Brachyury, when mutated, is responsible for short-tailed mice, and has been shown to have an association with human spina bifida.[52] Greig's cephalopolysyndactyly corresponds to mouse Xt, with patients revealing mutations in the Gli3 gene.[53] Pax-3, brachyury, and Gli3 have not been shown to be major candidates for human NTDs.[54]

10.3 WHERE MAGIC HAPPENS: DEVELOPMENT OF THE EMBRYO

Normal nervous system development of an embryo requires proper formation of embryonic axes. Determination of dorsoventral (DV) and anteroposterior (AP) domains during gastrulation appears critical for normal neural development. Axis patterning is reliant upon positional signals that provide DV and AP specifications.[55] Furthermore, positional signals appear essential to neural tube induction and patterning.[56–64]

Early embryonic axis determination is dependent on specification of anterior axial mesoderm followed by posterior axial mesodermal induction.[55] A specified group of cells (organizers) are known to function to organize the AP domains of the embryonic axis.[65–68] Several genes (brachyury, goosecoid, noggin, XLIM-1, Not-1) and diffusible morphogens (retinoids, activins, fibroblast growth factors) appear to be important in the regulation of organizer activity, specifically in posterior development of the axis.[56,58–60,62,69,70] The anterior axial mesoderm (chordamesoderm) induces competent ectoderm to form archencephalic structures (telencephalon, diencephalon, optic rudiment). The posterior mesoderm (notochord) induces competent ectoderm to form the deuterencephalon (metencephalon, myelencephalon, cerebellum) and spinal cord. Induced ectoderm forms the brain, hindbrain, and spinal cord by the process of neurulation.

Neurulation follows two stages: primary and secondary. Primary neurulation[71] begins after gastrulation when the primitive ectoderm is induced by the axial mesoderm to form a neural plate. The neural plate undergoes further elevation, folding, and fusion to form the neural tube. Neural crest cells migrate from the dorsal aspect of the neural tube. Primary neurulation forms all functional levels of the brain and spinal cord to the second sacral level in humans.

The caudal elements of the spinal cord, conus medullaris and filum terminale, are formed by secondary neurulation,[72–78] which begins at a transitional zone where the dorsally located primary neural tube overlaps the more ventral mesenchymal cells of the tail bud in the future lumbosacral area. In this overlap zone, randomly arranged mesenchymal cells condense to form the medullary cord. Radially oriented peripheral cells surround a cellular central core in the medullary cord. Cavitation occurs centrally, forming multiple lumina that coalesce to form a secondary neural tube.

The source of secondary neural tube cells is under scrutiny. Recent evidence in chick embryos suggests that cells may migrate from more rostral neural plates to attain their proper positions in the secondary neural tubes.[79,80] Normal caudal spinal cord patterning in humans has been described[81] and abnormal patterning has been demonstrated in dysraphic states.[82,83] Aberrant positional identity of caudal spinal cord cells may be a consequence of disrupted positional signals, faulty differentiation, or improper migration. Governing factors in the caudal neural tube pattern such as the brachyury and Pax-3 patterning genes have not been identified as major factors in spinal dysraphism.[54]

10.4 MAGIC PILLS

Exciting and provocative evidence demonstrates that some manifestations of NTDs are preventable or reversible at any one of numerous steps along the pathway from preconception to childhood, and possibly even into adulthood (Figure 10.1). Several different therapeutic interventions (or "magic pills") may be developed to treat the remaining types of NTDs. These pills may target genetic loci, proteins, or any of several metabolites involved in NTD development.

We now understand a great deal about the development of the neural tube, and are quickly approaching a more complete genetic characterization of the process. Ideally, NTDs could be detected early enough in development to target the defects before any permanent manifestations occurred. The epidemiological studies described definitively implicate maternal risk factors as well as inheritable and/or acquired genetic influences that may be targeted. The combination of genetic, epigenetic, and environmental factors offers numerous targets for interventions.

Preconception would be the optimal time for prevention. Mothers with modifiable risk factors should be identified and counseled. Perhaps one of the most remarkable advances in NTD treatment has been the introduction of periconceptional folic acid supplementation for the prevention of myelodysplasias. Whether taken in pill form or supplemented in dietary flour, this simple and inexpensive measure has cut the incidence and devastating sequelae of myelomeningocele by more than half. Despite this extraordinary achievement, it is still a challenge to prevent this unfortunate disorder of aberrant neural tube closure.

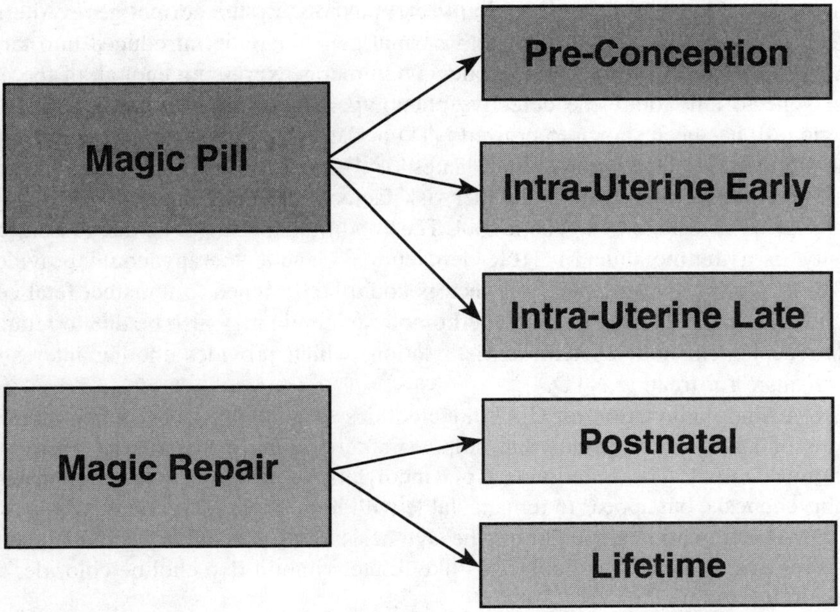

FIGURE 10.1 The magic phases of spinal dysraphism.

Other maternal risk factors that may prove important include good control of diabetes, reduction of obesity and infections, vitamin supplementation (folate, inositol, and vitamin B_{12}), and avoidance of over-heated environments like saunas. Additionally, mothers taking valproate and carbamazapine antiepileptic medications should discontinue use or take other medications if possible to eliminate the increased risk.

It may be possible in some cases to identify mothers with inheritable genetic predispositions and counsel them during the preconception period in preparation for possible treatment during pregnancy. Several possible medications could be developed to provide genetic targeting during early fetal development. Tools for targeting candidate genes at the DNA, RNA, or protein level are all plausible possibilities. These tools could target defects in genes involved in proper neural tube patterning, folate-dependent and -independent mechanisms, or healing mechanisms. The next decade certainly will see attempts at *in vitro* correction of genetic defects during the blastocyst stage or manipulation of these genes *in utero* via delivery systems like viral vectors.

Several studies with animal models have elucidated some of the genes involved in the induction of proper neural tube development, for example, Wnt-1, Gnot1 (a notochord family homeobox gene), HOX-1, and activin.[50,56,60] Activin and retinoic acid regulate Gnot1 expression prior to gastrulation. The neural tube-inducing properties of sonic and bone morphogenic protein genes are also under intense investigation. The Sp mouse model has defects in neural tube closure due to mutations in the Pax-3 paired box gene.[44,45] When genes are deleted or mutated, the fetal cells may be transfected *in utero* with viral vectors expressing the normal gene. Alternatively, embryonic stem cell lines with normal genes may be introduced into target embryos by blastocyst injection, producing chimeras expressing enough of the normal gene to ameliorate the defective phenotype. Interestingly, folate, the earliest magic pill, has been shown to prevent NTD in the Sp and other mouse models with mutations in Cart1 and crooked tail genes.[67,84,85]

Hyperhomocysteinemia is another risk factor linked to increased risk of NTD that may be amenable to a genetic tool. The condition appears to be due to homozygosity of a thermolabile MTHFR deficiency.[20] Genetic therapy could provide a solution. Currently available viral vectors could be designed to transfect fetal cells with the normal MTHFR gene. Hyperhomocysteinemia may also be due to reduced folate-dependent homocysteine remethylation, which provides another interesting mechanism for treating NTD.

Cytosine methylation on CpG dinucleotides of genomic DNA is one of many forms of DNA modifications that help maintain stability of numerous regions of genomic DNA.[86] These heritable CpG methylation sites may be altered in early embryogenesis, but appear to remain stable with high fidelity afterward.[87] This form of DNA methylation depends on the synthesis of S-adenosylmethionine, which requires methyl donors and cofactors like folate, vitamin B_{12}, choline chloride, and anhydrous betaine.[88]

Maternal nutrition may affect fetal phenotype via DNA methylation. The areas of methylation that change during embryogenesis are at transposable element insertion sites in the genome that underlie epigenetic-induced phenotypic variability.[89] Transient exposure to methyl donors *in utero* has been demonstrated to shift an epigenotype via CpG methylation of genomic DNA in mice.[19] This experimentally altered phenotype persisted into adulthood. It is hypothesized that such a mechanism may underlie the corrected NTD phenotype in folate supplementation. Other methyl donors may also serve as magic pills.

Another compound that prevented folate resistance NTD in the curly tail mouse and recently in humans is inositol.[90,91] The mechanism may occur via upregulation of the retinoic acid receptor beta.[91,92] Inositol is also important in glucose metabolism and may play a role in hyperglycemic or obesity-related causes of NTD. All these therapeutic measures are meant to prevent or correct defects early enough in development to prevent NTDs. However, efforts to correct defects are still needed. Most forms of what we can designate as "magic repairs" are applied during intrauterine development or after birth.

10.5 MAGIC REPAIRS

In a typical scenario, a child born with a NTD undergoes repair of the defect in the first few days after birth (as with myelomeningocele) or when neurological deterioration or substantial neurological risk is determined (as with closed dysraphism). Both paradigms are designed to minimize further risk, prevent progressive functional loss, and possibly reverse neurological deterioration. Clearly, in the case of an open NTD, reversal of paralysis or sacral dysfunction is not expected or attained. Novel repair strategies should be aimed at restoration of neurological function.

10.5.1 Fetal Surgery

Recent evidence suggests that the neurological deterioration associated with open NTDs may have resulted from progressive intrauterine injury alone or in concert with the primary defect of neurulation. For example, fetal ultrasonography revealed that human fetuses with myelomeningoceles retained lower extremity movements early in gestation and that the movements were lost by term.[93] These data and maternal reports that describe losses of fetal movements suggest that an event occurring during gestation damaged fetal function.[94]

In the event of intrauterine injury, intrauterine intervention such as a surgical repair may protect against progressive neurological deterioration. Animal models designed with spina bifida were tested after intrauterine repair. Neurological function was preserved in repaired animals.[95] This result led to intrauterine repairs of open, exposed spinal cords in humans.[96,97]

To determine the outcomes of fetal myelomeningocele repairs, the National Institute of Child Health and Human Development (NICHD) sponsored the Management of Myelomeningocele Study (MOMS), a continuing clinical trial [http://www.nichd.nih.gov]. Parameters undergoing study include optimal timing, neurological recovery, and effects of repairs on associated hydrocephalus and Chiari II malformations. The study is comparing two approaches to the treatment of babies with spina bifida: surgery before birth (prenatal surgery) and the standard closure surgery after birth (postnatal surgery). Preliminary results of human surgery show failure to preserve fetal neurological function. Furthermore, when it appeared that spinal cord function was present to a degree, it was less than predicted based on data from the animal models.[98] Improvements in the degree of hindbrain herniation noted in the associated Chiari II malformation have also been demonstrated.[99] Additionally, a reduction in the need for CSF shunting for hydrocephalus has been reported.[97,100]

Reported complications of fetal myelomeningocele surgery have been few; the most common complication is preterm delivery.[94] Major complications of intrauterine intervention such as maternal death from uterine rupture have been reported for other types of fetal surgery.[101] No uterine rupture resulting in maternal or fetal demise has been reported to date for fetal myelomeningocele repair.[94,97] Technical advancements, such as less invasive endoscopic procedures, have been proposed to avert this severe complication.[102,103]

One key to predicting optimal outcomes of novel fetal surgery treatments for myelomeningocele is understanding the structure of the placode. If the placode retains normal patterning and is simply un-neurulated, a repair may be effective in preventing secondary injury. There are mixed reports on whether placodes are normal in animal models.[104,105] Similar controversies surround human studies. Meuli et al. characterized the human placode as having partial loss of tissue, containing hemorrhages and abrasions, while preserving developed elements of dorsal and ventral parts of the spinal cord with nerve roots and ganglia.[106] The abnormalities were attributed to intrauterine injury.

Conversely, George and Cummings characterized the placode as having abnormal patterning along the dorsoventral and rostrocaudal axes indicative of a change in pattern determination and a paucity of maturing neurons with evidence of significant inflammatory infiltrate, gliosis, and fibrosis consistent with secondary injury.[83] These data suggest that the myelomeningocele placode shows abnormal development along with evidence of injury.

Reexamination of the animal model is needed to help clarify this controversy. George and Fuh made several observations in a review.[107] Two definitions of NTDs were used to describe the surgical models: spina bifida or spina bifida-like and surgical NTDs. All mammals except mice had spina bifida lesions in which the skin, muscle, lamina, and dura were opened, but the spinal cord itself was not disturbed.[108–112] Surgical NTDs were developed in avian species and mice; the dorsal elements of the spinal cord were opened and splayed apart, and exposed the central elements of the spinal cord to the surrounding environment.[113–118] The surgical models uncovered three mechanisms of injury:

1. Toxicity of the amniotic fluid
2. Direct intrauterine trauma
3. Developmental and growth distortion from laminectomy defect

Timing of lesions was critical. Spontaneous healing resulted if lesions occurred early in gestation instead of later.[111,114,117] Subsequent functional outcomes were virtually indistinguishable between groups lesioned early in gestation and spontaneously healed and repaired fetuses lesioned later in gestation.[83] Last, the surgical animal models used were not the products of abnormal primary neurulation, and could not directly address questions concerning the placode. These surgical models represent a reopening mechanism of a closed neural tube that has not been shown to appear in humans, but was reported in curtailed mouse mutants.[119]

The future of fetal surgery may rest in uncovering the mechanisms of fetal healing and directly reconstituting the spinal cord. In the study of fetal wounds, healing was demonstrated to occur rapidly and without scarring. The exact mechanisms of fetal scarless healing remain unknown. However, transforming growth factor-beta and hyaluronic acid-rich wound matrix play pivotal roles in scarless repair.[120]

The mechanism of annealing or healing that can lead to protection of the neural tube has also not been defined. The fusion of reapproximated dorsal neural elements in chicks has been suggested.[118] A preliminary study in our laboratory utilizing surgical NTDs in chicks and adding inhibitors of primary neurulation failed to prevent reclosure of the neural tube (unpublished data). Therefore, reclosure in chicks does not appear to be a recapitulation of primary neurulation. The underlying molecular and cellular mechanisms that regulated the repair remain unclear, but the ability of spinal cord cells to proliferate appeared important.[118] These data suggest that fetal interventions should be targeted at reinstituting mechanisms of fetal healing that were turned off after a critical developmental phase.

10.5.2 Spinal Cord Regeneration

Current work on restoration of spinal cord function has focused on regeneration after a spinal cord injury. If the precept from the fetal surgery is true, that the neurological sequelae in open NTDs are caused by intrauterine injuries, restoration of cord function should be attainable. In fact, the majority of research has revealed that an injured spinal cord can be restored by reconstituting or reestablishing molecular or cellular developmental mechanisms.[121] Therefore, the developing spinal cord appears to be the ideal substrate for regeneration of specific cell types and functional connections as long as the milieu can be properly manipulated.

Paramount for the regeneration of the spinal cord is that the neuron becomes "regeneration-capable" — it can restore the ability to demonstrate axonal growth and proper targeting. A number of genes have been shown to be constitutively expressed or upregulated in response to axonal growth. They have been termed "regeneration-associated genes" and their products include transcription factors such as c-jun, cytoskeleton components such as alpha tubulin, cytoplasmic growth cone proteins such as GAP-43 and CAP-23, and cell adhesion molecules such as NCAM and L1 that are important for growth cone guidance.[122]

The rate-limiting factor impacting regeneration is the inhibitory environment of the mature CNS. CNS inhibition to axonal growth is broadly divided into nonpermissive factors related to myelin and the inhibitory nature of the gliotic scar. Proteins identified in CNS myelin (NI-35 and NI-250) have been shown to function as neurite inhibitory factors.[123] At the injury site, dead cells, inflammation, and degraded tissue are present. They contain reactive astrocytes, microglia, oligodendrocytes, and meningeal cells that form gliotic scars that function as three-dimensional barriers to axonal growth.[124]

As noted earlier, George and Cummings demonstrated that the myelomeningocele placode may have abnormal patterning along the dorsoventral and rostrocaudal axes. This finding is indicative of a change in pattern determination, along with a paucity of maturing neurons with evidence of significant inflammatory infiltrate, gliosis, and fibrosis consistent with secondary injury.[82] The impact that aberrant development plays on the ability of the injured placode to regenerate and overcome the inhibitory environment is unclear and remains a goal of future research.

Regenerative strategies in spinal cord injury include administration of trophic factors, gene therapy, and cell transplantation. Intrathecal administration of trophic factors such as neurotropin, nerve growth factor and glial-derived neurotrophic factor upregulated growth cone proteins such as GAP-43 and CAP-23, propagated axonal regrowth across an area of crush injury, and established functional connections.[125] Interestingly, the administration of folate has been reported to assist in regenerating axons in a spinal cord injury model via intraperitoneal administration (personal communication). The mechanism of folate-assisted regeneration remains unknown.

Gene therapy strategies provide a way for longer lasting delivery of important trophic factors. Trophic genes can be supplied *ex vivo* to an injured spinal cord by inserting genetically altered cells that produce trophic factors.[126] Another method is applied *in vivo*: the neurotrophic gene is tranfected into the native spinal cord, usually via a viral vector.[127] Trophic factors listed above also serve as candidates for gene therapy. Other classes of gene candidates are endogenous receptors or morphogens important in embryonic development. For example, retinoic acid (RA) is important in embryonic neural development[69] and has been shown to stimulate embryonic neurite outgrowth.[128] RA administration failed to induce neurite growth in an injured adult spinal cord, presumably due to the lack of retinoic acid receptor-beta 2 (RAR β2) upregulation.[129] However, when the RAR β2 is upregulated, neurite outgrowth can occur.[130] Transfection into the adult spinal cord of RAR β2 alone was shown to stimulate neurite outgrowth.[130] Therefore, reinstitution of developmental mechanisms may be another methodology of cord regeneration.

Cellular transplantation strategies are aimed at circumventing the inhibitory surround created by the gliotic scar. Candidates for transplantation are neural stem cells and fetal cells that have the potential to develop into mature neurons or glia and restore function by replacing or repairing axons and synaptic relays.[131] Mature cells such as Schwann cells or olfactory ensheathing cells also provide neurotrophic support and myelination, thereby enhancing the regenerative environment. How the myelomeningocele placode would respond to cellular transplantation remains unclear. The lack of understanding of cell connectivity and patterning and the way that environment responds to injury makes outcomes unpredictable, but unveils a focal point for future study.

A final challenge to spinal cord regeneration of a NTD is that most of the studies examined models of acute injury. Spinal cord dysfunction secondary in the congenital setting is more likely to be chronic in nature. An important consideration in studies of chronic injury is the survival of the injured neurons. Reports indicate that 25 to 50% of neurons die as early as 4 weeks postaxotomy,[132] while the remaining cells become atrophic. There is some evidence that trophic factors[133] and fetal cell transplants[134] can enhance survival, even if applied 1 year after injury. Since many patients with open and closed defects will present with neurological dysfunction within this time frame, attempts at spinal cord regeneration remain viable techniques to pursue.

10.6 SHOULD WE BELIEVE IN MAGIC?

The short answer is "yes." Recent advances in genomics, proteomics, developmental cell biology, biochemistry, embryology, neurobiology and neuroimaging have created the potential for a "golden age" in the cure of NTDs. Until then, NTDs remain physically debilitating and are socioeconomic burdens. The time to advance neurosurgical management from supportive to restorative is now. It will be like magic!

REFERENCES

1. Hrdlicka, A., Trephination among prehistoric people especially in America, *Ciba Foundation Symposium,* 1:170–177, 1939.
2. Horsley, V., Brain surgery, *Br. Med. J.,* 2:670–675, 1886.
3. Cushing, H., Intracranial tumors of preadolescence, *Am. J. Dis. Children*, 33:551–584, 1927.
4. Fulton, J., *Harvey Cushing*, Oxford, Blackwell Scientific, 1946.
5. Winston, K., Pediatric neurosurgery: pride and prejudice, *Ped. Neurosurg.,* 32:58–68, 2000.
6. Begeer, I.H., Staal-Schreinemachers AL: the benefits of team treatment and control of adult patients with spinal dysraphism, *Eur. J. Ped. Surg.,* 6:15–16, 1996.

7. Bowman, R.M., McLone D.G., Grant, J.A., et al., Spina bifida outcome: a 25-year prospective, *Ped. Neurosurg.,* 34:114–120, 2001.
8. Centers for Disease Control, Spina bifida incidence at birth: USA, 1983–1990, *WMMR,* 41:497–500, 1992.
9. Reigel, D. and McLone, D.G., The tethered spinal cord, in *Pediatric Neurosurgery,* Philadelphia, W.B. Saunders, 1994.
10. Copp, A., Brook, F.A., Estibeiro, J.P., Shum, A.S.W., and Cockroft, D.L., The embryonic development of mammalian NTDs, *Prog. Neurobiol.,* 35:363–403, 1990.
11. Centers for Disease Control, Recommendations for the use of folic acid to reduce the number of cases of spina bifida and other NTDs, *WMMR,* 41:1–7, 1992.
12. Holmes, L., Harris J., Oakley G.P., Jr. et al., Teratology Society consensus statement on use of folic acid to reduce the risk of birth defects, *Teratology,* 55:381, 1997.
13. Milunsky, A., Jick, H., Jick, S. et al., Multivitamin/folic acid supplementation in early pregnancy reduces the prevalence of NTDs, *JAMA,* 262:2847–2852, 1991.
14. Armstrong, D.C., Congenital malformations of the spine, *Topics in Mag. Res. Imaging,* 5:131–140, 1993.
15. Kirke, P.N., Mills, J.L., Whitehead, A.S. et al., Methylenetetrahydrofolate reductase mutation and NTDs, *Lancet,* 348:1037–1038, 1996.
16. Van der Put, N.M., Gabreels, F., Stevens, E.M. et al., A second common mutation in the methylenetetrahydrofolate reductase gene: an additional risk factor for neural tube defects? [comment], *Am. J. Human Gen.,* 62:1044–1051, 1998.
17. Speer, M.C., Worley, G., Mackey, J.F. et al., The thermolabile variant of methylenetetrahydrofolate reductase (MTHFR) is not a major risk factor for neural tube defect in American Caucasians: NTD Collaborative Group, *Neurogenetics,* 1:149–150, 1997.
18. Anderson, R., Caveolae: where incoming and outgoing messengers meet, *PNAS,* 90:10909–10913, 1993.
19. Waterland, R.A. and Jirtle, R.L., Transposable elements: targets for early nutritional effects on epigenetic gene regulation, *Mol. Cell. Biol.,* 23:5293–5300, 2003.
20. Bakker, R.C. and Brandjes, D.P., Hyperhomocysteinaemia and associated disease, *Pharm. World Sci.,* 19:126–132, 1997.
21. Blom, H.J., Mutated 5,10-methylenetetrahydrofolate reductase and moderate hyperhomocysteinaemia, *Eur. J. Ped.,*157:S131–134, 1998.
22. Burgoon, J.M., Selhub, J., Nadeau, M. et al., Investigation of the effects of folate deficiency on embryonic development through the establishment of a folate deficient mouse model, *Teratology,* 65:219–227, 2002.
23. Chatkupt, S., Chatkupt, S., and Johnson, W., Waardenburg syndrome and myelomeningocele in a family, *J. Med. Gen.,* 30:83–84, 1993.
24. Epeldegui, M., Pena-Melian, A., Varela-Moreiras, G. et al., Homocysteine modifies development of neurulation and dorsal root ganglia in chick embryos, *Teratology,* 65:171–179, 2002.
25. Gos, M., Jr. and Szpecht-Potocka, A., Genetic basis of NTDs II: genes correlated with folate and methionine metabolism, *J. Appl. Gen.,* 43:511–524, 2002.
26. Castilla, E.E. and Orioli, I.M., Epidemiology of NTDs in South America, *Am. J. Med. Gen.,*22:695–702, 1985.
27. Elwood, J., Little, J., and Elwood, J., Eds., *Epidemiology and Control of NTDs,* Oxford, Oxford University Press, 1992.
28. Friel, J.K., Frecker, M., and Fraser, F.C., Nutritional patterns of mothers of children with NTDs in Newfoundland, *Am. J. Med. Gen.,* 55:195–199, 1995.
29. Jorde, L.B., Fineman, R.M., and Martin, R.A., Epidemiology of NTDs in Utah, 1940–1979, *Am. J. Epidemiol.,* 119:487–495, 1984.

30. Koch, M. and Fuhrmann, W., Epidemiology of NTDs in Germany, *Hum. Gen.*, 68:97–103, 1984.
31. McDonnell, R.J., Johnson, Z., Delaney, V. et al., East Ireland, 1980–1994: epidemiology of NTDs, *J. Epidemiol. Commun. Health*, 53:782–788, 1999.
32. Xiao, K.A., Epidemiology of NTDs in China, *Chung-Hua i Hsueh Tsa Chih* (*Chin. Med. J.,*) 69:189–191, 114, 1989.
33. Main, D. and Mennuti, M., NTDs: Issues in prenatal diagnosis and counseling, *Obstetr. Gynecol.,* 67:1–16, 1986.
34. Seller, M., Risk in spina bifida, *Dev. Med. Child. Neurol.,* 36:1021–1025, 1994.
35. Seller, M.J., Vitamins, folic acid and the cause and prevention of NTDs, *Ciba Found. Symp.,* 181:161–173, 1994.
36. Lammer, E., Sever, L., and Oakley, G., Teratogen update: valproic acid, *Teratology*, 35:465–473, 1987.
37. Robert, E. and Guibaud, P., Maternal valproic acid and congenital NTDs, *Lancet*, 937:2, 1982.
38. Rosa, F., Spina bifida in infants of women treated with carbamazepine during pregnancy. *New Engl. J. Med.,* 324:674–677, 1991.
39. Shaw, G., Velie, E., and Schaffer, D., Risk of neural tube defect-affected pregnancies among obese women, *JAMA*, 275, 1996.
40. Watkins, M., Scanlon, K., Mulinare, J. et al., Is maternal obesity a risk factor for anencephaly and spina bifida? *Epidemiology*, 7, 507–512, 1996.
41. Werler, M., Louik, C., Shapiro, S. et al., Prepregnant weight in relation to risk of neural tube defects, *JAMA*, 275:1089–1092, 1996.
42. Harris, M. and Juriloff, D., Genetic landmarks for defects in mouse neural tube closure, *Teratology*, 56:177–187, 1997.
43. Epstein, D.J., Malo, D., Vekemans, M. et al., Molecular characterization of a deletion encompassing the splotch mutation on mouse chromosome 1, *Genomics*, 10:89–93, 1991.
44. Epstein, D.J., Vekemans, M., and Gros, P., Splotch (Sp2H), a mutation affecting development of the mouse neural tube, shows a deletion within the paired homeodomain of Pax-3, *Cell*, 67:767–774, 1991.
45. Epstein, D.J., Vogan, K.J., Trasler, D.G. et al., A mutation within intron 3 of the Pax-3 gene produces aberrantly spliced mRNA transcripts in the splotch (Sp) mouse mutant, *PNAS,* 90:532–536, 1993.
46. Schimmang, T., Lemaistre, M., Vortkamp, A. et al., Expression of the zinc finger gene Gli3 is affected in the morphogenetic mouse mutant extra-toes (Xt), *Development*, 116:799–804, 1992.
47. Wilson, V., Rashbass, P., and Beddington, R., Chimeric analysis of T (brachyury) gene function, *Development*, 117:1321–1331, 1993.
48. Morrison-Graham, K., Schatteman, G., Bork, T. et al., A PDGF receptor mutation in the mouse (patch) perturbs the development of a non-neuronal subset of neural crest-derived cells, *Development*, 115:133–142, 1992.
49. Homanics, G., Smith, T., Zhang, S. et al., Targeted modification of the apolipoprotein B gene results in hypobetalipoproteinemia and developmental abnormalities in mice, *PNAS,* 90:2389–2393, 1993.
50. Lufkin, T., Dierich, A., LeMeur, M. et al., Disruption of the Hox-1.6 homeobox gene results in defects in a region corresponding to its rostral domain of expression, *Cell*, 66:1105–1119, 1991.
51. Moase, C.E. and Trasler, D.G., Splotch locus mouse mutants: models for NTDs and Waardenburg syndrome type I in humans, *J. Med. Gen.,* 29:145–151, 1992.

52. Hui, C. and Joyner, A., A mouse of Greig cephalopolysyndactyly syndrome: the extra toes mutation contains an intragenic delection of the Gli3 gene, *Nature Gen.*, 3:241–246, 1993.
53. Morrison, K., Papapetrou, C., Attwood, J. et al., Genetic mapping of the human homologue (T) of mouse T (brachyury) and a search for allele association between human T and spina bifida, *Hum. Mol. Gen.*, 5:669–674, 1996.
54. Melvin, E.C., George, T.M., Worley, G. et al., Genetic studies in NTDs: NTD Collaborative Group, *Ped. Neurosurg.*, 32:1–9, 2000.
55. Yamada, T., Caudalization by the amphibian organizer: brachyury, convergent extension and retinoic acid, *Development*, 120:3051–3062, 1994.
56. Cooke, J., Takada, S., and McMahon, A., Experimental control of axial pattern in the chick bastoderm by local expression of Wnt and activin: role of HNK-1-positive cells, *Dev. Biol.*, 164:513–527, 1994.
57. Jacobson, A., Inductive processes in embryonic development, *Science*, 152:25–34, 1966.
58. Jessell, T. and Melton, D., Diffusible factors in vertebrate embryonic induction, *Cell*, 68:257–270, 1992.
59. Kispert, A., Ortner, H., Cooke, J., and Herrmann, B.G., The chick brachyury gene: developmental expression pattern and response to axial induction by localized activin, *Dev. Biol.*, 168:406–415, 1995.
60. Knezevic, V., Ranson, M., and Mackem, S., The organizer-associated chick homeobox gene, Gnot1, is expressed before gastrulation and regulated synergistically by activin and retinoic acid, *Dev. Biol.*, 171:458–470, 1995.
61. Menkes, B. and Sandor, S., Researches on the development of axial organs, *Roum Rev. Embryol. Cytol.*, 6:65–88, 1969.
62. Mitrani, E. and Shimoni, Y., Retinoic acid inhibits growth in agarose of early chick embryonic cells and may be involved in regulation of axis formation, *Development*, 107:275–280, 1989.
63. Mitrani, E.G.Y., Shohat, H., and Ziv, T., Fibroblast growth factor during mesoderm induction in the early chick embryo, *Development*, 109:387–393, 1990.
64. Ziv, T., Shimoni, Y., and Mitrani, E., Activin can generate ectopic axial structures in chick blastoderm explants, *Development*, 115:689–694, 1992.
65. Belting, H.G., Hauptmann, G., Meyer, D. et al., Spiel ohne grenzen/pou2 is required during establishment of the zebrafish midbrain-hindbrain boundary organizer, *Development*, 128:4165–4176, 2001.
66. Deardorff, M.A., Tan, C., Conrad, L.J., et al., Frizzled-8 is expressed in the Spemann organizer and plays a role in early morphogenesis, *Development*, 125:2687–2700, 1998.
67. Zhao, Q., Behringer, R.R., and de Crombrugghe, B., Prenatal folic acid treatment suppresses acrania and meroanencephaly in mice mutant for the Cart1 homeobox gene, *Nature Gen.*, 13:275–283, 1996.
68. Zinyk, D.L., Mercer, E.H., Harris, E. et al., Fate mapping of the mouse midbrain–hindbrain constriction using a site-specific recombination system, *Curr. Biol.*, 8:665–668, 1998.
69. Maden, M. and Holder, N., The involvement of retinoic acid in the development of the vertebrate central nervous system, *Dev. Suppl.*, 2:87–94, 1991.
70. Yamada, T., Pfaff, S.L., Edlund, T., and Jessell, T.M., Control of cell pattern in the neural tube: motor neuron induction by diffusible factors from notochord and floor plate, *Cell*, 73:673–686, 1993.

71. Schoenwolf, G.C. and Smith, J.L., Mechanisms of neurulation: traditional viewpoint and recent advances, *Development,*109:243–270, 1990.
72. Griffith, C.M., Wiley, M.J., and Sanders, E.J., The vertebrate tail bud: three germ layers from one tissue, *Anat. Embryol.,* 185:101–113, 1992.
73. Muller, F. and O'Rahilly, R., The development of the human brain, the closure of the caudal neuropore, and the beginning of secondary neurulation at stage 12, *Anat. Embryol.,* 176:413–430, 1987.
74. Nievelstein, R.A., Hartwig, N.G., Vermeij-Keers, C. et al., Embryonic development of the mammalian caudal neural tube, *Teratology,* 48:21–31, 1993.
75. O'Rahilly, R. and Muller, F., Neurulation in the normal human embryo, *Ciba Found. Symp.,* 181:70–82, 1994.
76. Schoenwolf, G.C., Histological and ultrastructural studies of secondary neurulation in mouse embryos, *Am. J. Anat.,* 169:361–376, 1984.
77. Schoenwolf, G.C. and Delongo, J., Ultrastructure of secondary neurulation in the chick embryo, *American J. Anat.,* 158:43–63, 1980.
78. Yang, H.J., Wang, K.C., Chi, J.G. et al., Neural differentiation of caudal cell mass (secondary neurulation) in chick embryos: Hamburger and Hamilton Stages 16–45, *Brain Res. Dev. Brain Res.,* 142:31–36, 2003.
79. Catala, M., Teillet, M.A., De Robertis, E.M. et al., A spinal cord fate map in the avian embryo: while regressing, Hensen's node lays down the notochord and floor plate thus joining the spinal cord lateral walls, *Development,* 122:2599–2610, 1996.
80. Le Douarin, N.M., Teillet, M.A., and Catala, M., Neurulation in amniote vertebrates: a novel view deduced from the use of quail-chick chimeras, *Int. J. Dev. Biol.,* 42:909–916, 1998.
81. Cummings, T. and George, T., The immunohistochemical profile of the normal conus medullaris and the filum terminale, *Neuroembryology* (in press), 2003.
82. George, T.M., Bulsara, K.R., and Cummings, T.J., The immunohistochemical profile of the tethered filum terminale, *Ped. Neurosurg.,* 39:227–233, 2003.
83. George, T.M. and Cummings, T.J., Immunohistochemical profile of the myelomeningocele placode: is the placode normal? *Ped. Neurosurg.,* 39:234–239, 2003.
84. Carter, M., Ulrich, S., Oofuji, Y. et al., Crooked tail (Cd) models human folate-responsive NTDs, *Hum. Mol. Gen.,* 8:2199–2204, 1999.
85. Fleming, A. and Copp, A.J., Embryonic folate metabolism and mouse NTDs, *Science,* 280:2107–2109, 1998.
86. Kowal, M. and Wojcierowski, J., Role and significance of DNA methylation, *Polski Merkur/ Lekarsk.,* 14:364–368, 2003.
87. Reik, W., Dean, W., and Walter, J., Epigenetic reprogramming in mammalian development, *Science,* 293:1089–1093, 2001.
88. Van den Veyver, I.B., Genetic effects of methylation diets, *Annu. Rev. Nutr.,* 2:255–282, 2002.
89. Rakyan, V.K., Blewitt, M.D., Druker, R. et al., Metastable epialleles in mammals, *Trends Gen.,* 18:348–351, 2002.
90. Cavalli, P. and Copp, A.J., Inositol and folate-resistant NTDs, *J. Med. Gen.,* 39:E5, 2002.
91. Greene, N.D. and Copp, A.J., Inositol prevents folate-resistant NTDs in the mouse, *Nature Med.,* 3:60–66, 1997.
92. Corcoran, J., What are the molecular mechanisms of NTDs? *Bioessays,* 20:6–8, 1998.
93. Korenromp, M.J., van Gool, J.D., Bruinese, H.W. et al., Early fetal leg movements in myelomeningocele, *Lancet,* 1:917–918, 1986.

94. Sutton, L.N., Sun, P., and Adzick, N.S., Fetal neurosurgery, *Neurosurgery*, 48:124–142, 2001.
95. Meuli, M., Meuli-Simmen, C., Yingling, C. et al., *In utero* repair of experimental myelomeningocele saves neurological function at birth, *J. Pediatr. Surg.*, 31:397–402, 1996.
96. Adzick, N., Sutton, L., Crombleholme, T. et al., Successful fetal surgery for spina bifida, *Lancet*, 352:1675–1676, 1998.
97. Tulipan, N., Bruner, J.P., Hernanz-Schulman, M. et al., Effect of intrauterine myelomeningocele repair on central nervous system structure and function, *Ped. Neurosurg.*, 31:183–188, 1999.
98. Hirose, S., Farmer, D.L., and Albanese, C.T., Fetal surgery for myelomeningocele, *Curr. Opin. Obstetr. Gynecol.*, 13:215–222, 2001.
99. Tulipan, N., Hernanz-Schulman, M., and Bruner, J.P., Reduced hindbrain herniation after intrauterine myelomeningocele repair: report of four cases, *Ped. Neurosurg.*, 29:274–278, 1998.
100. Bruner, J.P., Tulipan, N., Paschall, R.L. et al., Fetal surgery for myelomeningocele and the incidence of shunt-dependent hydrocephalus, *JAMA*, 282:1819–1825, 1999.
101. Ranzini, A.C., White, M., Guzman, E.R. et al., Prenatal sonographic diagnosis of uterine rupture following open fetal surgery, *Obstetr. Gynecol.*, 93:826–827, 1999.
102. Bruner, J.P., Boehm, F.H., and Tulipan, N., The Tulipan–Bruner trocar for uterine entry during fetal surgery, *Am. J. Obstetr. Gynecol.*, 181:1188–1191, 1999.
103. Oberg, K.C., Robles, A.E., Ducsay, C.A. et al., Endoscopic intrauterine surgery in primates: overcoming technical obstacles, *Surg. Endos.*, 13:420–426, 1999.
104. McLone, D.G., Dias, M.S., Goossens, W. et al., Pathological changes in exposed neural tissue of fetal delayed splotch (Spd) mice, *Child Nerv. Syst.*, 13:1–7, 1997.
105. Selcuki, M., Manning, S., and Bernfield, M., The curly tail mouse model of human NTDs demonstrates normal spinal cord differentiation at the level of the meningomyelocele: implications for fetal surgery, *Child Nerv. Syst.*, 17:19–23, 2001.
106. Meuli, M, Meuli-Simmen, C., Hutchins, G.M. et al., The spinal cord lesion in human fetuses with myelomeningocele: implications for fetal surgery, *J. Ped. Surg.*, 32:448–452, 1997.
107. George, T.M. and Fuh, E., Review of animal models of surgically induced spinal NTDs: implications for fetal surgery, *Ped. Neurosurg.*, 39:81–90, 2003.
108. Calvano, C.J., Moran, M.E., Mehlhaff, B.A. et al., Minimally traumatic techniques for *in utero* access and fetal surgery, *J. Soc. Laparoendoscopic Surg.*, 2:227–233, 1998.
109. Heffez, D.S., Aryanpur, J., Rotellini, N.A. et al., Intrauterine repair of experimental surgically created dysraphism, *Neurosurgery*, 32:1005–1010, 1993.
110. Housley, H.T., Graf, J.L., Lipshultz, G.S. et al., Creation of myelomeningocele in the fetal rabbit, *Fetal Diagn. Ther.*, 15:275–279, 2000.
111. Meuli, M., Meuli-Simmen, C., Yingling, C.D., Hutchins, G.M. et al., Creation of myelomeningocele *in utero*: a model of functional damage from spinal cord exposure in fetal sheep, *J. Pediatr. Surg.* 30:1028–1033, 1995.
112. Michejda, M., Intrauterine treatment of spina bifida: primate model, *Zeitschr. Kinderchirurg.*, 39:259–261, 1984.
113. Campbell, L.R. and Sohal, G.S., Pattern of NTDs created by secondary reopening of the neural tube, *J. Child Neurol.*, 5:336–340, 1990.
114. Clark, B.J. and Scothorne, R.J., Variation in the response of chick embryos to incision of the roof plate of the neural tube at different developmental stages, *J. Anat.*, 168:167–184, 1990.

115. Zhang, X.M., Lin, E., Yang, X.J., Sonic hedgehog-mediated ventralization disrupts formation of the midbrain-hindbrain junction in the chick embryo, *Dev. Neurosci.*, 22:207–216, 2000.
116. Inagaki, T.S.G. and Walker, M.L., Experimental model: change in the posterior fossa with surgically induced spina bifida aperta in mouse, *Pediatr. Neurosurg.* 26:185–189, 1997.
117. Sim, K.B., Cho, B.K., Chi, J.G. et al., Morphological study of surgically induced open neural tube defect in old (14 and 21 days) chick embryos, *Neurosci. Lett.*, 192:61–64, 1995.
118. Sim, K.B., Cho, B.K., Lee, Y.J. et al., Chronological changes of re-closure capacity in surgically induced spinal open NTDs of chick embryos, *Neurosci. Lett.*, 292:151–154, 2000.
119. Park, C., Pruitt, J.H., and Bennett, D., A mouse model for NTDs: curtailed (Tc) mutation produces spina bifida occulta in Tc/+ animals and spina bifida with meningomyelocele in Tc/t, *Teratology*, 39:303–312, 1989.
120. Samuels, P. and Tan, A.K., Fetal scarless wound healing, *J. Otolaryng.*, 28:296–302, 1999.
121. Chernoff, E.A., Sato, K., Corn, A. et al., Spinal cord regeneration: intrinsic properties and emerging mechanisms, *Sem. Cell Dev. Biol.*, 13:361–368, 2002.
122. Bulsara, K.R., Iskandar, B.J., Villavicencio, A.T. et al., A new millenium for spinal cord regeneration: growth-associated genes, *Spine,* 27:1946–1949, 2002.
123. Caroni, P. and Schwab, M.E., Oligodendrocyte- and myelin-associated inhibitors of neurite growth in the adult nervous system, *Adv. Neurol.,* 61:175–179, 1993.
124. Davies, S.J., Fitch, M.T., Memberg, S.P. et al., Regeneration of adult axons in white matter tracts of the central nervous system, *Nature*, 390:680–683, 1997.
125. Ramer, M., Priestley, J., and McMahon, S., Functional regeneration of sensory axons into the adult spinal cord [comment], *Nature,* 403:312–316, 2000.
126. Tuszynski, M., Peterson, D., Ray, J. et al., Fibroblasts genetically modified to produce nerve growth factor induce robust neuritic ingrowth after grafting to the spinal cord, *Exp. Neurol.,*126:1–14, 1994.
127. Huber, A., Ehrengruber, M., Schwab, M. et al., Adenoviral gene transfer to the injured spinal cord of the adult rat, *Eur. J. Neurosci.,*12:3437–3442, 2000.
128. Maden, M., Role of retinoic acid in embryonic and post-embryonic development, *Proc. Nutr. Soc.,* 59:65–73, 2000.
129. Corcoran, J., So, P.L., Barber, R.D. et al., Retinoic acid receptor beta-2 and neurite outgrowth in the adult mouse spinal cord *in vitro*, *J. Cell Sci.,* 115:3779–3786, 2002.
130. Maden, M. and Hind, M., Retinoic acid, a regeneration-inducing molecule, *Dev. Dynamics*, 226:237–244, 2003.
131. Gage, F., Mammalian neural stem cells, *Science,* 287:1433–1438, 2000.
132. Feringa, E., McBride, R., and Pruitt, J.N., Loss of neurons in the red nucleus after spinal cord transection, *Exp. Neurol.,* 100:112–120, 1988.
133. Houle, J. and Ye, J., Survival of chronically-injured neurons can be prolonged by treatment with neurotrophic factors, *Neuroscience*, 94:929–936, 1999.
134. Bregman, B. and Reier, P., Neural tissue transplants rescue axotomized rubrospinal cells from retrograde death, *J. Comp. Neurol.,* 244:86–95, 1986.

11 Delayed Cerebral Vasospasm: Current Hypotheses and Future Treatments

Kent C. New, Cheryl Smith, Laura Niklason, and Dennis A. Turner

CONTENTS

11.1 INTRODUCTION

Delayed cerebral vasospasm (DCV) is the leading cause of morbidity and mortality in patients who have ruptured intracranial aneurysms and are admitted to tertiary care centers.[1,2] Thick focal collections of blood visualized on a CT scan are highly predictive of the risk of DCV.[3,4] The time course of the event is well established,[5] although the pathophysiology has remained a puzzle for many years. Recent advances in cellular and molecular biology techniques have led to the development of new hypotheses regarding this very important clinical problem.

In some ways *vasospasm* is a misnomer because it implies a reactive vascular tone increase with secondary vessel narrowing. However, a critical difference between an ordinary vasospasm and the vasospasm of DCV is that vessels lose their

0-8493-1482-8/05/$0.00+$1.50

sensitivity to most agents acting directly on vessel walls in DCV. For example, nitric oxide and nitroprusside, among other equivalent agents, normally act directly to significantly dilate smooth muscles in vessel walls, but have little effect in DCV.[6,7] An active tone increase implies a significant decrease in elasticity due to the contraction; whereas the vessel can be easily dilated with angiographic balloons (balloon angioplasty; see Chapter 12 for details).

This contrary phenomenon occurs because the tension in the vessel wall is proportional to the luminal radius; therefore, as the radius of the spastic vessel decreases, the tension in the wall also decreases. Thus, human DCV occurs in a delayed fashion, shows severe luminal narrowing which is not a vasospasm in the usual sense of active muscle contraction, and cannot be relaxed except with mechanical dilatation via angioplasty from the luminal side.

While the time course of DCV is well known, the reason the onset occurs several days (typically 5 to 7) after the initial subarachnoid hemorrhage has not been explained adequately. We will construct a presumed chain of events leading to the peculiar time course, together with a brief review of pertinent hypotheses and possible new treatment mechanisms.

11.2 TIME COURSE, DIAGNOSIS, AND MANAGEMENT OF DCV

DCV occurs only after a precipitating event. The most commen event is a subarachnoid hemorrhage (SAH) occurring in the basal cisterns secondary to rupture of a berry aneurysm. The volume of the SAH as determined by computerized tomography (CT) scan (Fisher grade) clearly relates to the probability of DCV.[3] Other causes of SAH, for example, head injury, may also precipitate DCV if the SAH is sufficiently dense.[8,9]

It is not known whether a rapid and immediate reactive vasospasm occurs directly after SAH due to the multiple vasospastic mediators present in blood, platelets, and serum that completely surround the vessel adventitia after the hemorrhage in the place of normal cerebrospinal fluid (CSF). When cerebral angiograms are performed relatively soon (within a few hours) after the onset of SAH, approximately 10% reveal angiographic evidence of immediate vasospasm.[5,10] These data suggest that an immediate reactive vasospasm occurs in a minority of SAH patients and implies a different mechanism for DCV. However, the later occurring DCV may in some way be linked to a transient immediate vasospasm.

The immediate morbidity from aneurysmal SAH is frequently secondary to increased intracranial pressure (ICP), as evidenced by the observation that 78% of SAH patients with acute symptomatic hydrocephalus improved after ventricular drainage.[11] ICP elevation after SAH appears to result from subarachnoid blood that causes malfunction of the arachnoid villi through an acute blockage, preventing normal resorption of CSF.[12] Additionally, because the blood leakage from the vessel at the time of the hemorrhage is directed into the CSF space at arterial pressure, the hemorrhage only stops after the local CSF pressure rises to arterial level. This high pressure precludes cerebral perfusion. The rapidity of restoration of cerebral perfusion dictates in many ways the resulting morbidity and mortality from the hemorrhage.

Clinical observations suggest that the cerebral vessels are still reactive during the initial time period after a SAH that precedes the onset of DCV. During this period, usually within 48 hours of the hemorrhage, most direct clip ligation treatments of berry aneurysms are performed. Some manipulation of the parent cerebral vessels is usually done during surgery and this often produces a direct visible vasospastic response of the vessels. This vasospastic response is apparent on the exteriors of the vessels and can be relieved by direct application of papaverine or similar agents that facilitate smooth muscle relaxation.[13] This vascular reactivity clearly implies that the smooth muscle in the region of the SAH (where the later DCV will occur) functions relatively normally in the early period after the SAH.

Clinical symptoms of DCV usually appear between the fifth and twelfth days following the hemorrhage.[14] Angiography performed during this time may reveal a diffuse constriction of major vessels, often including the internal carotid artery.[15] It is presumed that smaller vessels including arterioles are equally or more constricted although they are more difficult to visualize via angiography. Vascular constriction may be only radiographically apparent (radiographic vasospasm) or also clinically apparent, often resulting in focal neurological signs or permanent deficits or infarcts.[2] Various methods to diagnose DCV include transcranial Doppler to detect increased blood flow velocity,[16] computed tomographic angiography (CTA), and direct cerebral angiography — at present the gold standard diagnostic test.[14]

One of the clinical treatments shown to be most effective in prevention of DCV is early administration of the relatively specific cerebral smooth muscle calcium channel blocker, nimodipine.[17,18] After DCV has begun, the mainstay of treatment is hyperdynamic therapy (enhanced blood volume and relative hypertension).[19–21] This therapy promotes as much blood flow past the relatively constricted regions as possible by raising the pressure head in the systemic arteries leading to the brain. The most clinically effective treatment for severe vasospasm is therapeutic angioplasty. When possible, it is used to dilate proximal spastic vessels that may enhance blood flow into smaller arterioles that may also be involved.[22]

Antifibrinolytic agents have been used to decrease rebleeding after aneurysmal SAH by preventing lysis of the blood clot tamponading the rent in the aneurysm.[23] A meta-analysis of several trials of antifibrinolytic agents in SAH patients demonstrated that these agents significantly decrease the rebleeding rate. Unfortunately, they also significantly increase cerebral ischemia secondary to DCV, and therefore have no significant effect on outcomes compared to control patients.[24] Failed trials of antifibrinolytic therapies in SAH patients confirm the importance of blood products in the subarachnoid space as critical in the pathogenesis of DCV. If blood products remain longer due to decreased lysis, then the DCV rate (along with secondary symptoms such as stroke and death) is considerably higher.

11.3 CRITICAL QUESTIONS ABOUT DCV

The clinical knowledge accumulated over many years raised a large number of questions about DCV and prompted multitudes of research studies of the human condition and animal models. Unfortunately, DCV is very difficult to duplicate in animal models for assessing the time course, histology, and other parameters of treatments. Research studies have partially answered several critical questions and the results will be discussed next.

11.3.1 Why Are Subarachnoid Arteries Susceptible to DCV?

It is unclear why cerebral arteries located in the subarachnoid space are susceptible to DCV. One hypothesis links interference in the nutrition of the intracerebral vessels to their susceptibility to vasospasm.[25] Intracerebral vessels appear to lack the common nutrient-penetrating abilities present in other systemic arteries (vaso vasorum). Instead, it appears that pores or communication channels within the adventitia of intracerebral vessels allow access of CSF to the vessel media for critical glucose and nutrient supply. Thus, a thick coating of blood directly adjacent to the outer vessel wall after SAH may prevent nutrient access to the media and this eventually leads to media necrosis.

This does not appear to be a problem for the first few days after the SAH because the vessels remain externally reactive at operative exposure at least up to 72 hours after SAH. Eventually the lack of nutrient supply (combined with the enhanced tendency toward contraction due to the vasoconstrictive environment surrounding the vessel) leads to pathological changes within the vessel and at least partial necrosis of the media. Medial fibrosis and necrosis have been consistent findings in human specimens with symptomatic DCV.[26]

Another hypothesis suggests DCV after SAH is due to vascular mitogens released by activated platelets inducing vascular cell proliferation in the arterial walls.[27] Platelet-derived growth factor-AB is a powerful mitogenic growth factor for vascular smooth muscle cells[28,29] and also promotes cell migration.[30] Smooth muscle proliferation is stimulated within hours after injury and may increase wall thickness producing vessel stiffening that contributes to cerebral vasospasm. During the first week after SAH, it has been found that platelet-derived growth factor (PDGF) levels in the CSF of SAH patients are significantly higher than levels of nonSAH patients.[27,28]

The time course of DCV is consistent with that of a cellular proliferation process (Figure 11.1). In animal models, immunohistochemical labeling using proliferating cell nuclear antigen (PCNA) shows smooth muscle replication in the vascular wall[27] and significant changes in vascular mechanical properties. Consequently, in the days and weeks following SAH, small changes in arterial wall dimensions could theoretically thicken the vessel walls, which would dramatically decrease arterial compliance. Thus, vessel wall thickness may be a function of both media necrosis and smooth muscle proliferation, partly in response to the necrosis (to renew the vessel wall) and to mitogens readily available from blood products (Figure 11.1).

11.3.2 What Are the Likely Spasmogens Contributing to DCV?

It is well established that the proximity of dissolving blood in the subarachnoid space to the outer vessel wall leads to a large array of vasoactive substances that

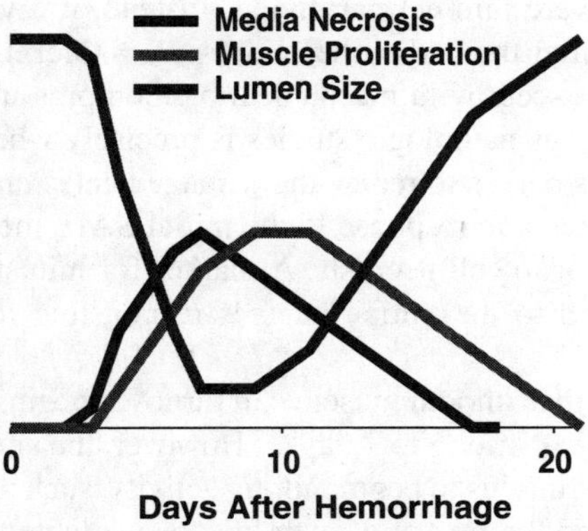

FIGURE 11.1 (See color insert following page 146.) Time course of delayed cerebral vasospasm. Two critical processes contribute to delayed cerebral vasospasm. The first is media necrosis that likely begins soon after subarachnoid hemorrhage and peaks at 5 to 7 days. The start of media necrosis acts as a signal to begin smooth muscle cell proliferation for eventual replacement of the smooth muscle cells in the media. However, the additional cells created by new dividing myoblasts and fibroblasts further increase the width of the media. The processes of media necrosis and media cellular proliferation significantly narrow the lumen, beginning early, but peaking at the 5- to 10-day range.

maintain continuous contact with the outer surfaces of the blood vessels.[31] It is postulated that the presence of these vasoactive substances around the walls of intracerebral vessels, which have at least partial wall necrosis, contributes to post-SAH DCV. Incubation of cerebral vessels in clotted blood followed by administration of blood products can lead to vasoconstriction.[32] It has been difficult, however, to identify the spasmogen most responsible for DCV — the mechanism by which the effect occurs — and the linkage between short-term muscle contraction and the subsequent DCV.

Several agents have been hypothesized to be responsible for DCV, all of which are present in blood products, including serotonin, catecholamines, eicosanoids, and others.[33] Convincing evidence suggests, however, that the vasoactive substance likely to be responsible for initiation of DCV is oxyhemoglobin.[33] Oxyhemoglobin has several mechanisms of action that may be important in vasospasm including the release of free radicals, the initiation and propagation of lipid peroxidation, metabolism to the vasoactive substance bilirubin, release of eicosanoids and endothelin from the vessel walls, perivascular nerve damage, inhibition of endothelium-dependent relaxation, and induction of structural damage to the vessel wall.[33] The precise role of these processes in the pathogenesis of DCV remains to be elucidated.

11.3.3 Why is the Onset of DCV Delayed?

With the combination of relative ischemia of the vessel wall due to lack of CSF nutrients and the intense vasoactive presence maintained against the outer arterial wall, eventually the arterial wall becomes thickened. A combination of necrotic smooth cells fills most of the media, together with proliferating smooth cell precursors, all leading to severe luminal narrowing. Instead of a vasospastic response at this time (5 to 7 days after the SAH), the vessel wall is thickened, has a small lumen, and cannot be dilated except with mechanical balloon pressure (angioplasty). What is not clear from previous pathologic studies is precisely when the mitotic turnover of smooth muscle cells begins to renew the damaged cells, and whether this smooth muscle cell proliferation is in response to the initial SAH, media necrosis, or earlier factors that appear prior to cell necrosis. A marker for mitosis could indicate when the SAH insult has led to the initial changes responsible for vessel necrosis and thickening.

One hypothesis is that smooth muscle cell turnover begins rapidly after the SAH insult, and reaches a peak after 5 to 7 days.[27] However, the smooth muscle cells may require a more potent stimulus to begin mitotic activity, such as the later combination of relative ischemia and the mix of growth factors available from the blood coating the outer wall. The vessel thickening would then correspond to a combination of vessel necrosis of smooth muscle cells in the media and mitosis and hypertrophy of an underlying population of cells, which would lead to smooth muscle renewal and proliferation. The smooth muscle cell proliferation would presumably then proceed over days to a few weeks, leading to a repopulation of the media and resumption of normal vessel reactivity and caliber.

Thus, the time course of DCV is presumably delayed due to the slow onset of smooth muscle necrosis over several days. This, together with the combination of mitotic activity and hypertrophy of remaining cells, markedly increases the width of the media, leading to shrinkage of the vessel lumen. The 5-day period may be an unfortunate superimposition of these two processes of necrosis with associated cell swelling and the secondary hypertrophy and mitotic activity of smooth muscle cell turnover. This time period is compounded by the slow lysis of blood products by CSF and a correspondingly slow resumption of adequate vessel nutrition, presumably as CSF adventitial pores are reopened or reconstituted.

Cerebral vessels may show luminal narrowing for reasons other than media thickening and direct changes in smooth muscle cells. For example, there may be an infiltrative component suggestive of inflammation within the vessel wall in response to the SAH that may be separately treatable. The possible role of inflammation in vasospasm should be the focus of a search to determine the exact cellular content (other than smooth muscle precursor cells and mature or dying smooth muscle cells) within the thickened media. If inflammatory cells are specifically identified as significant components of thickened vessel walls, new therapeutic options for vasospasm may be developed in the future.

11.3.4 Why Does SAH Density Correlate with Risk of DCV?

The most probable explanation for the correlation of thickness of SAH on CT scans with the risk of DCV is that blood deposition adjacent to the vessel induces vascular wall necrosis by interfering with vessel nutrition and releasing spasmogens such as oxyhemoglobin. Theoretically, enhancing blood lysis in the CSF early after SAH could lead to decreased risk of and faster recovery from DCV (but promote rebleeding if early aneurysm clipping is not performed). This approach is advocated by those attempting to treat vasospasm with infusion of urokinase or tissue plasminogen activator (tPA) into the subarachnoid space after SAH.

Several trials have demonstrated the potential benefits of intracisternal urokinase or tPA infusion after SAH in the reduction of DCV.[34–37] These results led to a multicenter, randomized, blinded, placebo-controlled trial of intracisternally administered tPA in attempts to prevent DCV after aneurysmal SAH.[38] Unfortunately, although the trial revealed a significant decrease in incidence of severe vasospasm in patients with thick subarachnoid clots treated with tPA, all other outcome measures, including overall incidence of angiographic vasospasm, incidence of clinical vasospasm, and outcome at 3 months were not significantly affected. Interestingly, overall bleeding complication rates did not increase with tPA. Although the benefits of tPA could potentially reach statistical significance in a larger trial, the results of this trial have dampened enthusiasm for fibrinolytic agents in SAH patients.

11.4 CAN ADVANCES IN SMOOTH MUSCLE CELL BIOLOGY FACILITATE UNDERSTANDING?

Cerebral blood vessels are composed primarily of smooth muscle cells (long, tapering, single nuclei cells with thick-to-thin filaments aligned with the long axis) within the media. Smooth muscle contraction is involuntarily triggered by the autonomic system or by hormones, and is designed for slow, long-lasting contraction. Smooth muscle cells are specifically designed to maintain tension for prolonged periods (passive maintenance) while hydrolyzing five- to tenfold less ATP than skeletal muscle cells performing the same task. Like other muscle cells, contraction occurs because of myosin and actin. The actin in smooth muscle cells has a different amino acid sequence than that of cardiac or skeletal muscle cells, but there appears to be no known functional significance.

Smooth muscle myosin resembles skeletal myosin; functionally, the level of ATPase activity is tenfold lower, which allows more direct calcium regulation of contraction. Also, smooth muscle myosin can interact with actin filaments and cause contraction only when its light chains are phosphorylated. When the myosin is dephosphorylated, it cannot interact with actin and the muscle relaxes. Specific enzymes accomplish this calcium-dependent phosphorylation and dephosphorylation of the myosin light chain.

Arteries have thick walls of connective tissue and vascular smooth muscle cells (VSMCs) lined by monolayers of endothelial cells. The endothelial cells are separated from the smooth muscle cells by a basal lamina and then the elastic fibers of the internal elastic lamina. The arterial wall morphology can change by both smooth muscle hypertrophy and hyperplasia. Hypertrophy occurs by adding cytoplasmic elements, but is reversible because the cells enlarge without changes in DNA. Unlike skeletal and cardiac muscle, smooth muscle can divide and may recruit undifferentiated cells (pericytes) to become smooth muscle cells. This mitotic behavior is stimulated by various growth factors.

The predominant growth regulators of VSMCs and pericytes are fibroblast growth factors (FGFs), platelet-derived growth factors (PDGFs), transforming growth factor-beta 1 (TGF-β1), and epidermal growth factor (EGF). When stimulated by any of these growth factors at appropriate concentrations, VSMCs can begin execution of the mitosis program within hours. For vascular smooth muscle cells, PDGF-BB precipitates the greatest degree of growth, with PDGF-AA stimulating small but significant growth, and PDGF-AB causing an intermediate amount of growth.

PDGF-AB is the predominant form of growth factor released from activated platelets. Depending on dose, TGF-β1 is inhibitory to SMCs but not to pericytes. Both acidic and basic FGFs are strong mitogens to pericytes and SMC proliferation.[39] Also, tumor necrosis factor-alpha (TNF-α), a ubiquitous cytokine involved in inflammatory states, has been reported to stimulate SMC growth in culture. TNF receptor activation is known to induce SMC apoptosis more in rapidly proliferating neointimal cells than in more slowly replicating medial cells.[40]

Although SMC proliferation likely occurs as part of the media replacement during and after DCV, little direct evidence for this has been shown in human arterial samples to this point. However, multiple mitogens leading to such proliferation are clearly present in the SAH mix around cerebral vessel walls and other factors, such as hypoxia, can induce mitogens.

11.5 SUGGESTED RESEARCH AVENUES AND TREATMENT OPTIONS

The development of a suitable model system for the study of DCV has been difficult. In general, three types of model systems have been used to investigate cerebral vasospasm: cell culture, isolated cerebral vessels, and whole animals. Whole animal models of vasospasm range from intracisternal injection of autologous blood to craniotomy for exposure of cerebral arteries and direct application of blood clot to their surfaces.[41,42] Although the craniotomy model replicates well the human disease process and even its response to nimodipine,[43] it involves primates that are very expensive and ethically troubling, and the model itself is technically difficult.

Less challenging and expensive *in vitro* models of DCV have used isolated cerebral vessels or cultured cells. The difficulty with isolated cerebral vessels is that they survive at best only a few days in culture, and are only beneficial in the study of early immediate vasospasm rather than the entire DCV process.[44,45] Interestingly, SMC isolated from rat aortas and exposed to hemoglobin *in vitro* have been found to develop changes similar to those seen in DCV, suggesting that some mechanistic features of the disease process may be investigated in cell culture systems.[46] Of course, the ability to study pharmacologic and other therapies in a vessel-free system is limited.

An alternative experimental approach is available as a result of the recent development of the ability to grow blood vessels entirely *in vitro*.[47,48] This system has the advantage of allowing *in vitro* study of vessels of the size desired and over a longer period than isolated vessels are able to survive. Additionally, since the growth media can be changed as desired, these model vessels offer a novel way to investigate changes in the vascular SMC on a detailed time schedule in an ischemic or vasoconstrictive environment.

Treatment options can also be directly demonstrated in this model because it allows easy access to both the luminal and adventitial sides of the vessel. In many ways, these vessels grown *in vitro* are similar to human cerebral vessels. They are of the same size (a few millimeters) and both lack vaso vasorum or nutrient feeding vessels to the media. The *in vitro* cultured vessels are surrounded by a culture growth medium that can be altered to be like CSF, and then the vessels can be deprived of substrates or surrounded by blood to imitate in many ways the SAH process that underlies DCV.

Unfortunately, short of animal models that fully duplicate the sequence of events present in the human situation, further human tissue may be the most valuable study source and clearly the most valid in terms of predicting human treatment. Studies focusing on muscle cell turnover and mitotic activity in human specimens will be critical for mapping out the full sequence of events of DCV beyond the limits of ordinary pathological examination. This type of analysis could include assessing proliferation of smooth muscle cell precursors, hypertrophy, mitotic activity, and in particular assessing the relative contributions to the media enlargement of SMC necrosis, SMC hypertrophy, and inflammation.

Further treatment efforts could be directed at early or late phase. Early intervention could be performed to enhance CSF lysis of blood products in an effort to restore appropriate nutrition levels to the media. If an early proliferative phase exists and if it can be safely slowed or postponed to await the resolution of necrosis, less reduction of the vessel caliber may occur. The danger of slowing down reactive smooth muscle changes is that SMC growth may be insufficient by the time of resolution of the necrosis for vessel strength, which could lead to spontaneous vessel necrosis and possibly rupture. Other interventions may reduce necrosis or enhance tolerance of SMC to the relative ischemic conditions present after SAH. Thus, preventing or delaying necrosis may obviate the need for delayed SMC proliferation.

Many ischemic effects observed in clinical DCV are results of vasospasm in small vessels that are not amenable to current vascular interventional treatment (therapeutic angioplasty). Thus, further systemic or local medical treatment may be very helpful for treating or forestalling cerebral ischemic changes observed in DCV. Vasospasm has been most intensively studied in larger vessels, but the pathogenesis in small vessels (i.e., arterioles) may differ due to the different mixtures of vessel wall components compared to the larger more proximal vessels. Thus, an *in vitro* model that duplicates some features of small vessels may also be of significance. The smaller arterioles share many features of the larger cerebral vessels, in that vaso vasorum is also absent and the vessels are also located within the subarachnoid space, susceptible to SAH and its secondary effects.

11.6 CONCLUSIONS

DCV is a complex and time-dependent phenomenon that is not completely understood, partly due to the lack of a suitable experimental model that clearly reproduces the changes observed in human vessels in DCV. Further, more effective clinical treatments will likely come from enhanced understanding of the pathophysiology of the disease, particularly the biology of smooth muscle cells because the majority of empiric treatments over the past 30 years have not demonstrated substantial efficacy.

Short-term animal models of DCV seem to have little relevance or validity — a conclusion echoed in 1985 by Wellum et al.[49] Thus, development of new animal models and understanding mechanisms involved in both necrosis and proliferation may be the key to future translational treatments.

REFERENCES

1. Kassell, N.F. et al., Cerebral vasospasm following aneurysmal subarachnoid hemorrhage, *Stroke,* 16, 562–572, 1985.
2. Kassell, N.F. et al., The International Cooperative Study on the Timing of Aneurysm Surgery. Part 1: Overall management results, *J. Neurosurg.,* 73, 18–36, 1990.
3. Fisher, C.M., Kistler, J.P., and Davis, J.M., Relation of cerebral vasospasm to subarachnoid hemorrhage visualized by computerized tomographic scanning, *Neurosurgery,* 6, 1–9, 1990.
4. Adams, H.P. et al., Predicting cerebral ischemia after aneurysmal subarachnoid hemorrhage: influences of clinical condition, CT results, and antifibrinolytic therapy. Report of the Cooperative Aneurysm Study, *Neurology,* 37, 1586–1591, 1987.
5. Heros, R.C., Zervas, N.T., and Varsos, V., Cerebral vasospasm after subarachnoid hemorrhage: an update, *Ann. Neurol.,* 14, 599–608, 1983.
6. Hatake, K. et al., Impairment of endothelium-dependent relaxation in human basilar artery after subarachnoid hemorrhage, *Stroke,* 23, 1111–1116, 1992.
7. Onoue, H. et al., Altered reactivity of human cerebral arteries after subarachnoid hemorrhage, *J. Neurosurg.,* 83, 510–515, 1995.
8. Zubkov, A.Y. et al., Risk factors for the development of post-traumatic cerebral vasospasm, *Surg. Neurol.,* 53, 126–130, 2000.
9. Soustiel, J.F., Shik, V., and Feinsod, M., Basilar vasospasm following spontaneous and traumatic subarachnoid haemorrhage: clinical implications, *Acta Neurochir.,* 144, 137–144, 2002.
10. Qureshi, A.I. et al., Prognostic value and determinants of ultra-early angiographic vasospasm after aneurysmal subarachnoid hemorrhage, *Neurosurgery,* 44, 967–973, 1999.
11. Hasan, D. et al., Management problems in acute hydrocephalus after subarachnoid hemorrhage, *Stroke,* 20, 747–753, 1989.
12. Johnson, R.N. et al., Mechanism for intracranial hypertension during experimental subarachnoid hemorrhage: acute malfunction of arachnoid villi by components of plasma, *Trans. Am. Neurol. Assn.,*103, 38–42, 1978.
13. Heffez, D.S. and Leong, K.W., Sustained release of papaverine for the treatment of cerebral vasospasm: *in vitro* evaluation of release kinetics and biological activity, *J. Neurosurg.,* 77, 783–787, 1992.
14. Weir, B. et al., Time course of vasospasm in man, *J. Neurosurg.,*48, 173–178, 1978.
15. Burch, C.M. et al., Detection of intracranial internal carotid artery and middle cerebral artery vasospasm following subarachnoid hemorrhage, *J. Neuroimag.,* 6, 8–15, 1996.

16. Aaslid, R., Huber, P., and Nornes, H., A transcranial Doppler method in the evaluation of cerebrovascular spasm, *Neuroradiology,* 28, 11–16, 1986.
17. Haley, E.C., Kassell, N.F., and Torner, J.C., A randomized controlled trial of high-dose intravenous nicardipine in aneurysmal subarachnoid hemorrhage: report of the Cooperative Aneurysm Study, *J. Neurosurg.,* 78, 537–547, 1993.
18. Haley, E.C., Jr., Kassell, N.F., and Torner, J.C., A randomized trial of nicardipine in subarachnoid hemorrhage: angiographic and transcranial Doppler ultrasound results: report of the Cooperative Aneurysm Study, *J. Neurosurg.,* 78, 548–553, 1993.
19. Kosnik, E.J. and Hunt, W.E., Postoperative hypertension in the management of patients with intracranial arterial aneurysms, *J. Neurosurg*., 45, 148–154, 1976.
20. Kassell, N.F. et al., Treatment of ischemic deficits from vasospasm with intravascular volume expansion and induced arterial hypertension, *Neurosurgery,* 11, 337–343, 1982.
21. Awad, I.A. et al., Clinical vasospasm after subarachnoid hemorrhage: response to hypervolemic hemodilution and arterial hypertension, *Stroke,* 18, 365–367, 1987.
22. Zubkov, Y.N., Nikiforov, B.M., and Shustin, V.A., Balloon catheter technique for dilatation of constricted cerebral arteries after aneurysmal SAH, *Acta Neurochir.,* 70, 65–79, 1984.
23. Vermeulen, M. et al., Antifibrinolytic treatment in subarachnoid hemorrhage, *New Engl. J. Med.,* 311, 432–437, 1984.
24. Roos, Y.B. et al., Systematic review of antifibrinolytic treatment in aneurysmal subarachnoid haemorrhage, *J. Neurol. Neurosurg. Psychiatr.,* 65, 942–943, 1988.
25. Zervas, N.T. et al., Cerebrospinal fluid may nourish cerebral vessels through pathways in the adventitia that may be analogous to systemic vaca vasorum, *J. Neurosug.,* 56, 475–481, 1982.
26. Smith, R.R. et al., Arterial wall changes in early human vasospasm, *Neurosurgery,* 16, 171–176, 1985.
27. Borel, C.O. et al., Possible role for vascular cell proliferation in cerebral vasospasm after subarachnoid hemorrhage, *Stroke,* 34, 427–433, 2003.
28. Gaetani, P. et al., Platelet derived growth factor and subarachnoid haemorrhage: a study on cisternal cerebrospinal fluid, *Acta Neurochir.,* 139, 319–324, 1997.
29. Braun-Dullaeus, R.C., Mann, M.J. and Dzau, V.J., Cell cycle progression: new therapeutic target for vascular proliferative disease, *Cardiovasc. Res.,* 98, 82–89, 1998.
30. Boehm, M. and Nabel, E.F., Cell cycle and cell migration: new pieces to the puzzle, *Circulation,* 130, 2879–2881, 2001.
31. Sonobe, M. and Suzuki, J., Vasospasmogenic substances produced following subarachnoid haemorrhage, and its fate, *Acta Neurochir.,* 44, 97–106, 1978.
32. Osaka, K., Prolonged vasospasm produced by the breakdown products of erythrocytes, *J. Neurosurg.,* 47, 403–411, 1977.
33. Macdonald, R.L. and Weir, B.K., A review of hemoglobin and the pathogenesis of cerebral vasospasm, *Stroke,* 22, 971–982, 1991.
34. Kodama, N. et al., Cisternal irrigation therapy with urokinase and ascorbic acid for prevention of vasospasm after aneurysmal subarachnoid hemorrhage: outcomes in 217 patients, *Surg. Neurol.,* 53, 110–117, 2000.
35. Moriyama, E. et al., Combined cisternal drainage and intrathecal urokinase injection therapy for prevention of vasospasm in patients with aneurysmal subarachnoid hemorrhage, *Neurol. Med. Chirurg.,* 35, 732–736, 1995.
36. Sasaki, T. et al., A phase II clinical trial of recombinant human tissue-type plasminogen activator against cerebral vasospasm after aneurysmal subarachnoid hemorrhage, *Neurosurgery,* 35, 597–604, 1994.

37. Usui, M. et al., Vasospasm prevention with postoperative intrathecal thrombolytic therapy: a retrospective comparison of urokinase, tissue plasminogen activator, and cisternal drainage alone, *Neurosurgery,* 34, 235–244, 1994.
38. Findlay, J.M. et al., A randomized trial of intraoperative, intracisternal tissue plasminogen activator for the prevention of vasospasm, *Neurosurgery,* 37, 168–176, 1995.
39. D'Amore, P.A. and Smith, S.R., Growth factor effects on cells of the vascular wall: a survey, *Growth Factors,* 8, 61–75, 1993.
40. Niemann-Jonsson, A. et al., Increased rate of apoptosis in intimal arterial smooth muscle cells through endogenous activation of TNF receptors, *Arterioscler. Thromb. Vasc. Biol.*, 21, 1909–1914, 2001.
41. Espinosa, F. et al., Chronic cerebral vasospasm after large subarachnoid hemorrhage in monkeys, *J. Neurosurg.,* 57, 224–232, 1982.
42. Espinosa, F., Weir, B., and Noseworthy, T., Rupture of an experimentally induced aneurysm in a primate, *Can. J. Neurol. Sci.,* 11, 64–68, 1984.
43. Nosko, M. et al., Nimodipine and chronic vasospasm in monkeys. Part 1: clinical and radiological findings, *Neurosurgery,* 16, 129–136, 1985.
44. Peerless, S.J. et al., Angiographic study of vasospasm following subarachnoid hemorrhage in monkeys, *Stroke,* 13, 473–479, 1982.
45. Macdonald, R.L. et al., Morphometric analysis of monkey cerebral arteries exposed *in vivo* to whole blood, oxyhemoglobin, methemoglobin, and bilirubin, *Blood Vessels,* 28, 498–510, 1991.
46. Fujii, S. and Fujitsu, K., Experimental vasospasm in cultured arterial smooth-muscle cells. Part 1: Contractile and ultrastructural changes caused by oxyhemoglobin, *J. Neurosurg.,* 69, 92–97, 1988.
47. Niklason, L.E. et al., Functional arteries grown *in vitro*, *Science,* 284, 489–493, 1999.
48. Niklason, L.E. and Langer, R.S., Advances in tissue engineering of blood vessels and other tissues, *Transplant Immunol.,* 5, 303–306, 1997.
49. Wellum, G.R., Peterson, J.W., and Zervas, N.T., The relevance of *in vitro* smooth muscle experiments to cerebral vasospasm, *Stroke*, 16, 573–581, 1985.

12 Future Directions of Endovascular Neurosurgery

Osama O. Zaidat and Michael J. Alexander

CONTENTS

12.1 INTRODUCTION

The major advances in endovascular neurosurgery over the past 15 years are reflections of the pioneering work of previous generations of enthusiastic, persistent, and optimistic physicians. Since the introduction of cerebral angiography by António

0-8493-1482-8/05/$0.00+$1.50

Egas Moniz in animal models in 1926 and subsequently in humans, the possibility of using a less invasive endovascular approach to treat cerebrovascular diseases was pursued.[1–5] History unraveled many attempts by neurosurgeons searching for less invasive approaches to treating cerebrovascular diseases[1–5] until Guido Guglielmi, an Italian neurosurgeon at the Medical Center of the University of California at Los Angeles, developed the platinum detachable coil as an effective and feasible treatment alternative to cerebral aneurysm clipping.[6,7] This technology initiated the great advancements we see in endovascular neurosurgery today.

A second historical landmark was the publication in 2002 of the randomized International Subarachnoid Aneurysm Trial (ISAT) study comparing aneurysmal coiling and clipping.[8] The study showed that the endovascular coil embolization of cerebral aneurysm is safe, feasible, and effective in comparison with open craniotomy and clipping. Despite the wide controversy surrounding the trial and its findings, it constituted another step in better defining the role of coiling and its long-term efficacy in aneurysm therapy based on scientific background.

Unfortunately, endovascular therapy for arteriovenous malformation (AVM) has not undergone a similar rapid pace of achievement. Newer flow directed microcatheters to facilitate closer access to the AVM nidus were introduced[9] and AVM embolic agents are under investigation in the search for the ideal embolic agent.[10–13] These advances may provide a more effective adjunctive or even curative role for endovascular AVM therapy.

The catheter-based treatment of atherosclerotic carotid disease is rapidly evolving despite the disappointing early clinical trials.[14] The results may be related to the use of older generation stents. Biodegradable and biocompatible materials and drug coatings have been incorporated in contemporary stent designs to provide better trackability, flexibility, conformability, and compressibility and prevent restenosis.[14–16]

Interventional acute stroke therapy is rapidly evolving, with newer pharmacological agents that may be administered locally into clots via the intra-arterial route in combination with intravenous administration or with mechanical microdevices to achieve clot disruption.[17–20]

The pace and the dynamic evolution of endovascular therapy advances have escalated rapidly over the past 10 years. The potential exists for further advances and the only limitation is imagination. The minimally invasive treatment of vascular neurosurgical diseases is the desired approach of the future. Keeping up with clinical advancements in interventional and catheter-based technologies is the key factor for improving clinical outcomes. The future is going to be marked by constant changes and the development of more minimally invasive techniques to treat central nervous system (CNS) diseases.

The core of advances in treating different CNS vascular diseases lies in refining existing techniques and tools and developing more biocompatible ones. The current management strategies and approaches may also evolve over time and be replaced by techniques tailored to specific vascular anomalies.

Materials are becoming more bioactive and less irritating or toxic to vascular structures. They are designed with the different anatomical and structural variations of the vascular diseases in mind. New tools are being designed to lessen the complication rates during or following endovascular interventions. Future developments involve all aspects of endovascular neurosurgery. This chapter will provide an overview of recent developments and future directions of endovascular neurosurgical approaches for treating various CNS diseases.

12.2 CEREBRAL ANEURYSMS

The rate of cerebral aneurysm treatment via nonconventional endovascular therapy (rather than surgical clipping) is steadily increasing. The endovascular approach is accomplished by filling aneurysm lumens with balloons, Guglielmi detachable coils (GDC), or liquid polymers.[6–8,21–26] The recent publication of the prospective and randomized ISAT is considered a significant step in providing clinical evidence of endovascular aneurysm coiling. The study showed a 6.9% absolute risk reduction in functional outcomes in patients with ruptured aneurysms treated with GDC when compared to surgical clipping.[8]

Although the study results provoked wide controversies, they influenced the treatment approach to ruptured cerebral aneurysms in many institutions worldwide. Further studies in North America are on the way and may define better the exact future role of endovascular therapy.

In addition to clinical advances, technology is constantly evolving and the pace of improvement may be hastened by the spread of endovascular approaches. The standard platinum-based GDC has been improved by the addition of 3-dimensional shapes and the use of new biologically compatible polymer-coated Matrix® detachable coils (Boston Scientific, Fremont, CA).[28–31] Different detachable coil shapes, materials, coatings, and designs under development may allow better conformation to the shape of an aneurysm and improve healing, fibrosis, endothelialization, and obliteration of the aneurysm lumen.

Several technical and design aspects of endovascular treatment of cerebral aneurysms are being refined to enhance trackability, ease of deployment, and biological activity in promoting aneurysmal neck neoendothelialization. The aneurysm coil-coating compositions are made mainly from biodegradable lactose or cellulose copolymer derivatives.[28–31] Molecular biology and translational basic research are becoming the bases for developing newer coil designs.[32,33] Those advances are summarized in Table 12.1 and discussed next.

12.2.1 Surface Modification of GDC Embolization

Biological material-based coatings intended to achieve complete luminal filling are being developed to enhance the efficacy of the newer generation of aneurysm coils. Coating with bioactive materials would make GDC more biocompatible and could stimulate clot organization with aneurysm fibrosis and neck endothelialization. Matrix 2- and 3-dimensional coils composed of 75% bioabsorbable polyglycolic/poly-L-lactic acid copolymer outer coats and 25% platinum cores are currently in use.[28] In preclinical studies, Matrix detachable coils were shown to accelerate the formation of intra-aneurysmal connective tissue, fibrosis, endothelialization with increased aneurysmal neck thickness, and reduced aneurysm size over a shorter

TABLE 12.1
Overview of Future Aneurysmal Endovascular Neurosurgical Therapy

Advances	Comments
Surface-Modified Aneurysmal Coils	
Polyglycolic acid bioabsorable polymeric coating Polyglycolic acid Polyglycolic/poly-L-lactic acid copolymer Fibroblast tissue allograft	Theoretical advantage: improves intra-aneurysmal fibrosis Neointima formation across aneurysm neck
Others	
Smooth muscle, growth factor, and ion implantation Collagen coating Polyurethane, Gelfoam, Dacron, and Fibronectin implants	Promotes tissue reaction and fibrosis; fibroblast proliferation and aneurysm obliteration; fibroblasts rapidly proliferate, are immunologically inert
New Techniques	
Balloon remodeling, stent-assisted coiling, double catheters, 3-D coils, liquid polymers (Onyx, cyanoacrylate, cellulose acetate, etc.)	

period of follow-up when compared to standard GDC.[28] The copolymer was hypothesized to enhance the inflammatory response to coil deployment and hasten aneurysm healing. As a stronger neck forms, the biological material is degraded and absorbed, leaving the platinum core and promoting shrinkage of the aneurysm.[28]

Fibroblast cells and growth factors are some of the other materials used to coat GDC in aneurysm animal models in an aim to improve the rate of luminal fibrosis and neck closure.[32,33] Fibroblast allograft delivery via deployed coils was shown to be safe and feasible in animal studies.[33] After 2 weeks of the fibroblast allograft coated coil deployment, the fibroblasts remained viable and contained within the aneurysm lumen, with cellular proliferation and fibrosis versus unorganized thrombi in the control aneurysms.[33] Other radiological and histological studies involving collagen, protein, ion-implantation, and polyester coating showed variable success rates of aneurysm occlusion and neck endothelialization.[32,33] Current modifications to GDC coils are in their early phases. The Matrix coated detachable coils are the first available for clinical use. Follow-up data on long-term efficacy of the Matrix may be available in the next few years. Other modified GDC or complete biologically active coils are expected to become available in the near future.

The Hydrocoil® (Microvention, Aliso Viejo, CA) is another recently released polymer-coated coil. The product has a pH-activated hydrogel coating that expands over a period of several minutes after hydration. Modification of the polymer coating allows for a high percentage of aneurysm filling with coils and may ultimately reduce the recanalization rate due to coil compaction. Other companies are also attempting to develop polymer-based coils that can completely fill an aneurysm and increase the healing response to seal the aneurysm neck.

12.2.2 New Techniques: 3-D Coils, Balloons, and Stent-Assisted Coiling

Aneurysm coiling is best suited for small aneurysms with narrow necks (dome-to-neck ratio greater than 1.5). Aneurysms with irregular shapes and wide necks and those that have been partially clipped are usually approached with innovative endovascular techniques.[30,31,34–37] Balloon remodeling was devised to protect the parent vessel lumen from coil herniation by inflating the balloon across the aneurysm neck after placing a microcatheter into the aneurysm lumen.[30,31] More recently, flexible and small intracranial stents (e.g., Boston Scientific's Neuroform®) have become available. They can be used to protect parent vessels with wide neck or irregularly shaped aneurysms.[31,34–37]

Other techniques include the double-catheter approach.[31] Two microcatheters are placed in the aneurysm lumen prior to detaching the two coils. The interventionalist can then test the positioning of the two coils. If deployment is satisfactory, the result is basket protection formed by deploying two coils simultaneously.[31] Finally, the availability of 3-D coils has made it safe and feasible to deploy and detach coils in irregularly shaped aneurysms with wider necks.[30]

Flexible stents covered with polyethylene terephthalate may be used to treat spontaneous, traumatic, or postsurgical pseudoaneurysms, where no significant branches of the parent vessels exist. The Symbiot-covered stent (Boston Scientific/SCIMED, Minneapolis, MN) has been used in a few reported cases to cross a pseudoaneurysm neck and effectively treat pseudoaneurysms of the internal carotid arteries.[37]

12.2.3 Liquid Polymers

The use of liquid materials to embolize cerebral aneurysms has emerged again despite the controversy, increased risk of procedural complications, and past failures to obtain perfect luminal obliteration.[22–26] Liquid material may indeed provide an advantage in that it can conform to the shape of an aneurysm and may enhance and promote intraluminal fibrosis. However, it also carries high potential risks of protrusion into the parent vessel lumen with distal migration and inability to stimulate endothelialization across the aneurysm neck. The latter risk makes the currently available liquid embolic agents unfavorable first choices to embolize cerebral aneurysms. Such limitations may be lessened with the use of protective devices such as balloons or stents or in combinations with coils.[22–26]

One liquid polymer currently undergoing clinical trials and in use in humans is Onyx® (Micro Therapeutic Inc., Irvine, CA).[22–26] Onyx is an ethylene–vinyl alcohol (EVGOH) copolymer dissolved in dimethyl sulfoxide (DMSO) and mixed with micronized tantalum powder to achieve the appropriate radio-opacity. Onyx is a nonadhesive biocompatible polymer that allows slow delivery and complete filling of an aneurysm, but requires a balloon or other protective device to contain the delivered material in place.

A study by Muramaya et al. revealed that despite the improvement in liquid leakage to the parent vessel lumen, such a complication remained a difficult challenge even with the use of protective devices to contain the liquid polymer inside the aneurysm lumen.[23] They used 12% Onyx: (1) alone, (2) in combination with GDC, (3) proximal, (4) across the aneurysm protective balloon, and (5) with neck stenting. The study was limited in the design and sample size (five to ten patients per group), but showed that the use of Onyx combined with protective devices provided more complete filling, with migration rates into the parent vessels ranging from 9 to 33%, but with no significant differences among the groups.

Cyanoacrylate embolization with GDC coil protection in an animal model of carotid bifurcation aneurysm revealed a better filling rate at 3-month follow-up.[24] Unfortunately, the cyanoacrylate escape rate to the parent blood vessel remained high (25%).[25] Other liquid polymers used *in vitro* and *in vivo* are iodinized cellulose mixed esters that provided good results in aneurysm models in sheep.[26] These embolic agents need further modifications of solvent concentrations to become less toxic prior to use on a larger scale in human subjects.[26]

12.3 ARTERIOVENOUS MALFORMATION AND ARTERIOVENOUS FISTULA

Arteriovenous malformations (AVM) and arteriovenous fistulae (AVF) are congenital lesions that can present at any age, although they are most common in the third and fourth decades of life. The main presenting symptoms are related to headache, seizure, and intracranial bleeding. AVM and AVF may be also found via magnetic resonance imaging (MRI) and angiography. The risk of bleeding may range from 3 to 4% per year. The risk of chronic neurological disability or death following intracranial bleeding ranges from 20 to 50%. The mainstay of AVM therapy depends on the clinical and imaging grade of the condition that closely correlates with postoperative complications and predicts surgical outcome. To reduce surgical risk, presurgical endovascular embolization is usually attempted. In addition, endovascular therapy may be implemented to reduce AVM volume prior to surgical resection or radiosurgery treatment. Recent endovascular advances may define better the adjunctive and occasionally curative role of local AVM embolization (Table 12.2).

12.3.1 Development of New Embolization Materials: Glues and Polymers

Endovascular AVM therapy is progressing at a slower rate than aneurysm therapy because it involves different obstacles. The current endovascular neurosurgical therapeutic approach to AVM remains adjunctive rather than curative. It is used to aid gamma knife radiosurgery of eloquent and large AVM and conventional surgical resection for smaller noneloquent AVM.

Obstacles in achieving an important milestone in treating AVM via endovascular therapy include the lack of ideal materials for embolization of the AVM, the need to access and embolize all the feeders, and the ability to deal with hemodynamic changes upon abruptly occluding large amounts of inflow or outflow to the AVM. The current complications rate remains around 2 to 15% due to inadvertent embolization of an arterial branch supplying an important functional brain region.

TABLE 12.2
Advances in Arteriovenous Malformation Endovascular Therapy

Advances	Comments
Embolic Agents	
n-BCA	Immediate solidification; risk of escape of the agent; microcatheter adhesions; microcatheters can be used only once
Ethibloc	Slow solidification; microcatheters can be used up to four times; risk of rupture of microcatheter
PVA	Several particle sizes: 150 to 1000 μmm; temporary occlusion; mainly used preoperatively
Others	
Onyx, GDC, gelfoam, silastic or latex balloon, fibrin silicon, hydrogel glue, calcium alginate, etc.	

Embolizing materials include Onyx, cyanoacrylate (n-butyl cyanoacrylate or n-BCA), Ethibloc, polyvinyl alcohol (PVA), GDC, and other agents including silastic or latex balloons, gelfoam powder, cellulose, fibrin glue, silicone spheres, calcium alginate, surgical silk thread, and microhydrogel spheres.[9–13,39–41] The most common embolic agent is n-BCA,[10] which solidifies immediately on contact with free hydrogen ions in the blood; the casting effect is permanent.[10] The n-BCA is dissolved in lipiodol, and injected via a flow-directed microcatheter. The risks with n-BCA include escape of the agent to the venous side or normal arteries, possibly leading to cerebral infarction, and adhesion of the catheter to the wall of the blood vessel due to back-reflux of the n-BCA. Adhesion prevents reuse of microcatheters. Each catheter should be withdrawn immediately as soon as the injection is completed.[10]

Ethibloc is a solution of ethanol and zein, a corn protein (210 mg zein/ml ethanol) in an aqueous solution. The ethanol dissolves and the zein precipitates.[9] Ethibloc provides an advantage over n-BCA in that it is less adhesive and allows re-use of flow-directed microcatheters rather than removing the catheters immediately after a single use, as required with n-BCA. Unfortunately, Ethibloc must be infused via a microcatheter with an outer diameter of 1.8 French; otherwise it could cause rupture of the microcatheter.

Both ethibloc and n-BCA are mixed with lipid-based oil before injection.[9] Other agents used include PVA with different particle sizes: small (50 to 150 μmm), medium (250 to 450 μmm), and large (500 to 750 μmm). The PVA is used mainly for preoperative embolization of tumors and AVMs due to the increased incidence of recanalization.[39–41] Another treatment strategy is use of an embolic agent in combination with GDC coils or using the coils alone to close some of the AVM feeders.[13,42] Balloon protection and assisted closure with trispan devices may also be used.[42]

Technical advances in AVM endovascular therapy involve reemergence of the transvenous approach and induced systemic hypotension during such therapy.[43–45] The rationale is to lessen the hemodynamic effect of abrupt occlusion of the AVM outflow. Induced systemic hypotension to 70 to 80 mm torr mean arterial blood pressure during the AVM transvenous embolization procedure via systemic vasodilators or adenosine-induced cardiac pause could be performed successfully without complications.[44,45] These studies may reopen the door to AVM therapy via the transvenous approach in combination with induced hypotension.[44,45]

12.3.2 Arteriovenous Fistula

Carotid-cavernous fistulae (CCF) are of four types: type A (fast flow) and slow flow types B, C, and D. Treatment is achieved via venous or arterial approaches. Success has been variable, ranging from 50 to 80%.[42,44] The current methods include embolizing the fistulae with liquid embolic agents such as n-BCA or microparticles, balloon occlusions, GDC, and hydrocoils. Inflatable detachable balloons and hydrocoils are usually used for type A fast flow post-traumatic CCF via arterial or venous approaches through the inferior petrosal sinus. Type B dural shunts receive slow flows via meningeal branches of the internal carotid arteries and endovascular surgery usually is not feasible. Type D dural arteriovenous shunts receive contributions via the meningeal branches of the external or internal carotid arteries. Types C and D are usually treated via embolization with liquid agents to external carotid artery feeders.[42,44]

12.4 STENTING FOR ATHEROSCLEROTIC DISEASE

The carotid endarterectomy remains the gold standard for treating patients with carotid disease to prevent future neurological deficit and stroke.[46] Carotid endarterectomy is more effective in preventing stroke and death than medical therapy alone in symptomatic patients with carotid artery stenosis measuring more than 50%, if the surgical complication rate is less than or equal to 1.5% according to several studies.[46,47]

Earlier studies comparing endovascular therapy with endarterectomy failed to show significant differences in favor of carotid angioplasty or stenting (CAS) due to high periprocedural complication rates reaching 10%.[48] The early Wallstent® study was terminated because of a high event rate in the CAS arm.[49] Results of further studies designed with advanced techniques and protective embolic devices (Carotid Revascularization Endarterectomy versus Stent Trial [CREST][50]); Stenting and Angioplasty with Protection in Patients at High Risk for Endarterectomy [SAPHIRE][51]) are still pending. Despite recent advances in endovascular stenting and angioplasty, several hurdles must be overcome. Increased rates of restenosis and periprocedural complications due to distal embolization and stroke following CAS remained important factors that limit wide clinical application.

Current and future technical advances are focused on improving the restenosis and neurological complication rates after CAS. Recent developments in carotid stent design and clinical CAS trials are making use of drug-coated stents, embolic

TABLE 12.3
Areas of Focus and Advances in Stent Development

Advances	Comments
Stent design	Different links and shapes; tight versus loose cells; covered versus uncovered; porous versus nonporous
Stent composition to decrease restenosis rate and thrombogenecity	Traditional stainless steel, nitinol, gold, titanium, tantalum, cobalt chromium alloy stents, etc.; total biodegradable or combined metallics and copolymers
Drug-coated and drug-eluting stents	Heparin, hirudin, iloprost, abciximab, prednisone, methotrexate, rapamycin, paclitaxel, collagen or polylactic acid incorporating adenovirus gene-loaded vector, radioactive stent, polytetrafuoroethylene
Embolic protection devices	PercuSurge, Angioguard filter, PAEC, and FilterWire-EX EPD
Adjunctive medical therapy	Clopidogrel, aspirin, heparin, parenteral GP IIb/IIIb inhibitors (abciximab, eptifibatide, tirofiban, lamifiban), oral GP IIb/IIIa inhibitors (orbofiban, sibrafiban), neuroprotective agents (nimodipine, citicholine, labulazole, etc.)

protection devices, and adjunctive medical therapy to lessen these periprocedural and long-term CAS complications. These advances are summarized in Table 12.3.

Intracranial angioplasty and stenting present additional technical challenges. The flexibility of the stent and its interference with the small arterial perforators originating at stenotic lesions are of paramount significance in designing an intracranial stent that will attain good results and achieve low morbidity and mortality.

12.4.1 Drug-Coated and Drug-Eluting Stents

The pathophysiology of atherosclerotic disease involves inflammatory changes incited by intimal wall injury, with leukocyte activation and secretion of prostacyclins and cytokines. Platelet activation follows, with subsequent adhesion, aggregation, and thrombus formation. In addition to the atherosclerotic process when deploying an intraluminal stent, the tension and stress on the intima may stimulate growth factors and lead to intimal hyperplasia and subsequent peri- or within-stent restenosis.[14,15,52–54] These pathophysiological changes now serve as the basis for developing a newer generation of drug-coated stents.[50,51] Impregnating stents with heparin, iloprost, hirudin, or a combination of these drugs may help decrease thrombotic responses and prevent intimal hyperplasia.[14,15,52–54]

Drug-eluting stents with abciximab,[55] prednisone, methotrexate, paclitaxel, rapamycin, or angiopeptin may also provide some protection against restenosis by inhibiting inflammatory response, intimal proliferation, and thrombosis. Current data on drug-coated stents for coronary intervention indicate that antimitotic agents such as sirolimus and paclitaxel are promising in preventing neointimal hyperplasia and restenosis.[15,52,53] These studies should be duplicated in CAS, particularly because it involves a different pathophysiology and vessel caliber from coronary atherosclerosis. Coating the stents with radioactive substances provides an additional method for preventing restenosis by inhibiting intimal hyperplasia. One drawback of the radioactive approach may be related to less radiation delivery at the periphery of the stent due to penumbral effects that may lead to peri-stent restenosis.[15,52,53]

12.4.2 Surface-Modified and Biocompatible Stents

Surface modifications to stents are performed to achieve increased biocompatibility, better conformation to blood vessel shape by wall opposition, and ability to withstand external crush forces. The new expandable stents provide better physical fit to a narrowed arterial lumen. These stents are mainly made of nitinol — a four-atom biocompatible composite of nickel and titanium. Nitinol changes its crystal structure upon contact with human blood and reverts to its original austenite crystals with high elastic properties at body temperature. This provides a stent with better hemodynamic resilience, conformability, and wall opposition.[15]

Other biocompatible stent modifications include collagen polymer coatings to reduce neointimal hyperplasia and electrochemical polishing of the stent surface that decreases thrombogenic reactions when stents are deployed in the arterial lumen.[14,15] More recently, biochemical and tissue engineering techniques are being applied in animal studies seeking the ideal stent design, particularly a "living stent." Fibrin-, nitric oxide-, and phosphorylcholine-coated stents try to mimic living vascular wall tissue (with its physiological function and antiproliferative feedback mechanism) to lessen atherosclerosis and intimal hyperplasia.[14,15,56,57] Surface modification of stents is still in its early phases, and further studies promise to delineate better their role in managing patients with high risks of carotid artery peri-stent and in-stent restenosis.

12.4.3 Vector-Coated Stents

While the Human Genome Project unravels the genetic code sequences for humans, gene therapy to inhibit local atherosclerosis and plaque formation presents a real potential therapeutic alternative. Coating metallic stents with biologically active materials and hydrogels, such as lactic acid derivatives and gelatin macromers, allows incorporation of different drugs and gene therapies in stents and local delivery to nearby arterial walls.[56,57] Coating stents with specific gene-carrying vectors that may inhibit expression of known growth factors or act on platelet aggregation and smooth muscle growth factors are also undergoing testing.

Theoretically, delivering these agents to arterial walls would prevent neointimal formation and proliferation and halt the restenosis process and progression of carotid atherosclerosis.[56,57] Vector-coated stents were tested in rabbit carotid arteries *in vivo*.[56] After the stents were coated with adenovirus vectors expressing bacterial beta-galactosidase, the genetic material was transmitted to the vascular wall. Gene expression was altered within 3 weeks of the bioactive stent deployment and the adenovirus vector indeed induced production of beta-galactosidase in the vascular wall near the stent.[56] This study indicates the feasibility and potential of this technique to be applied with an effective gene therapy to halt the progression of or reverse carotid and intracranial atherosclerosis.

12.4.4 Embolic Protective Strategies

The risks associated with carotid stent deployment from transcranial ultrasound studies are known to occur early; they are associated with microembolic phenomena, often leading to neurological deficits. To minimize such risks, several studies are investigating embolic protective devices (EPDs) during CAS, including flow reversal devices, filters, umbrellas, and other membranous devices.[58–60]

In one study, three types of EPDs were used on 30 high-risk surgical patients during CAS. The CAS with pre-EPD placement was found to be safe and feasible. Only one patient suffered a major periprocedural stroke, making the complication rate equal to 3%.[59] Another small study of 36 CAS procedures used a FilterWire-EX EPD consisting of a 0.014-inch guide wire with an integrated expandable distal nitinol loop attached to a thin microporous polyurethane filter. The procedures were performed successfully without any permanent neurological deficits at 30 days.[60]

Transient neurological deficits lasting 30 minutes without residual effects were noted in two patients (5.7%).[60] In the cardiac literature and in a multicenter study, a total of 801 stents have been deployed. Cases were randomly assigned stents with PercuSurge EPDs or stents without EPDs. The study showed a significant reduction in periprocedural cardiac complications, decreasing from 14.7% in the control group to 8.6% in the PercuSurge EPD group ($p = 0.008$).[58]

Large, randomized studies to better delineate optimal EPDs and proper patient selection are underway (CREST and SAPHIRE), and results should be forthcoming in the next few years.[50,51] The risk of distal embolization persists due to lack of protection from EPD malfunction during the diagnostic segment of the CAS or during placement of the EPD.

12.4.5 Adjunctive Medical Therapy

The routine use of antiplatelet therapy before or during CAS is mainly derived from cardiac literature protocols. The evidence for such therapy during CAS is based solely on anecdotal evidence or case series.[61] Current practice is to load patients with 300 mg of clopidogrel on the day preceding the CAS, use heparin to extend activated clotting time above 250 seconds during the CAS, and continue both aspirin and clopidogrel for 6 weeks followed by use of one agent only.

The use of different types of intravenous glycoproteins IIb and IIIa has not been well delineated. They are currently used in cases of emergency CAS that do not involve large areas of cerebral infarction, as documented by neurological examination or imaging studies, to avoid the risk of intracerebral hemorrhage.[61] The glycoprotein IIb and IIIa inhibitors available in the United States are abciximab (ReoPro®, Centocor Inc., Malvern, PA; Eli Lilly & Company, Indianapolis, IN), eptifibatide (Integrilin®, Cor Therapeutics Inc., South San Francisco, CA; Key Pharmaceuticals Inc., Kenilworth, NJ), and tirofiban (Aggrestat®, Merck & Co. Inc., Whitehouse Station, NJ).

These agents have been used sporadically in CAS and the rationale for their use is derived from anecdotal experience or small published case series.[61,62] Randomized clinical trials to define their exact role in CAS are needed, although the systemic glycoprotein may be used in urgent or emergency CAS and oral versions may be used for elective CAS. Meanwhile, additional data can be expected from ongoing studies comparing CAS to endarterectomy.

The area of adjunctive medical therapy in CAS is still in its early stages: defining the role of periprocedural medications to prevent neurological complications and stent restenosis. Another potential adjunctive medical therapy is the use of neuroprotective agents to halt ischemic cascades in acute stroke patients.[14] Because stroke risk may be high during CAS, a neuroprotective agent with a high safety profile may be administered before and during the CAS or even immediately after the onset of neurological symptoms.

12.5 Endovascular Stroke Treatment

In June 1996, the U.S. Food and Drug Administration (FDA) approved intravenous (IV) therapy of recombinant tissue plasminogen activator (rtPA) for use in acute ischemic stroke patients within 180 minutes of symptom onset.[63] This approval followed publication by the National Institute of Neurological Disorders and Stroke (NINDS) of improved 3-month outcomes in patients who received 0.9 mg/kg within 3 hours of symptom onset, compared with a placebo, based on the modified Rankin disability scale.[63] The narrow time window for treatment (within 3 hours) and lack of public awareness preclude offering this therapy to large number of patients.

Questions remain about the effectiveness of IV therapy and how the proportion of treated patients can be increased. Moreover, IV rtPA efficacy may be marginal because of low-drug concentration delivered to the clot, given the stagnation and slow blood flow surrounding the blocked artery. The risk of symptomatic intracerebral hemorrhage (ICH) is about 6.4% in the NINDS group, although 40% had no disabilities at 3 months. Although this is better than placebo results, 60% retained different degrees of disabilities at 90 days.

The ideal goal of future intervention would be to improve the proportion of patients with better outcomes and have fewer patients with ICHs. Hence, endovascular, local administration of thrombolytics or mechanical clot retrieval devices is appealing.[17,18,64–73] Several studies have shown the efficacy of intra-arterial administration of rtPA in various types of vessel occlusions.[17,18,65,66] In the prolyse in acute cerebral thromboembolism (PROACT) study, pro-urokinase was administered to the horizontal portion of the middle cerebral artery with good recanalization rate and improved clinical outcomes at 3 months, but at the expense of an increase in ICH rate to about 10%.[17]

The main obstacles to intra-arterial thrombolytic therapy for acute ischemic stroke are the narrow therapeutic time window (6 hours from symptom onset) and the lack of public awareness of the emergency nature of stroke treatment. In acute ischemic stroke, the marginal benefit of thrombolytic agents more than 6 hours after symptom onset is outweighed by the incremental risk of ICH as time passes. An additional obstacle is the efficacy of clot lysis following administration of currently available thrombolytic agents. The complete recanalization rate is modest, even when treatment is administered early. Intra-arterial therapy may provide a higher recanalization rate, but at the expense of increased risk of bleeding. To try to improve the rate of recanalization, a combination strategy of administering IV followed by intra-arterial thrombolytics has been implemented in many tertiary care centers.[65,66] The vessel patency rate improved slightly, but the risk of ICH was as high as or higher (10 to 15%) than intra-arterial therapy alone.

To improve patency rates after administration of thrombolytics, second, third, and fourth generations of rtPA have been introduced. The newer generations were developed by altering the terminal N units of rtPA and include tenecteplase (TNK), reteplase, alteplase, monteplase, lanoteplase, and pamiteplase.[67] The modification may have improved the plasma half-life from 4 minutes to 40 minutes on average, but clinical trials in cardiology showed only marginal benefits over rtPA in achieving vessel patency. TNK is a mutant rtPA with higher fibrin specificity and longer plasma half-life due to slower clearance. Staphylokinase, a non-rtPA derivative produced by *Staphylococcus aureus*, has extreme fibrin specificity and a 6-minute plasma half-life, in comparison to 3 to 4 minutes for rtPA. Specificity to fibrin is thought to correspond to drug efficacy and lower incidence of hemorrhagic complications.[67]

In addition to fibrinolytic agents, the availability of a new generation of parenteral glycoprotein IIb and IIIa antagonists will provide stroke victims with alternative therapeutic options.[68–70] The preliminary results of the study of abciximab in acute ischemic stroke are encouraging, and the risk of hemorrhage does not seem to be higher than the risk with IV rtPA.[69] A Phase I safety study of rtPA plus tirofiban showed that the combination is safe and feasible.[70]

Although the pharmacological advances for acute interventional stroke therapy are still improving, several conclusions may be drawn:

1. The recanalization rate using IV rtPA is less than intra-arterial therapy, and the latter seems to be less with combined therapy.
2. Even with the best strategy, the current pharmacological agents provide modest vessel patency rates and are time consuming to administer.
3. Increasing the doses of therapeutic agents or combining different anti-platelet and fibrinolytic drugs may only lead to increased risk of ICH in stroke patients.

This leads us to contemporary microendovascular device designs and innovative techniques that may provide significant advantages over pharmacological approaches.[19,71–73] Endovascular approaches with mechanical devices ideally would offer stroke patients faster recanalization and more effective flow reconstitution, possibly make blood clots more amenable to lower doses of thrombolytics or anti-platelet agents, and reduce the risk of ICH.[19,20,71–73] Several devices are available and have undergone Phase I trials or have been reported in case series format in the literature and await large-scale Phase III trials.[19,20,71–73] Clot retrieval devices have been developed to physically capture clots and remove them from the body via a microcatheter.

The Microsnare is a simple primitive design reported to capture or disrupt blood clots, but it can be associated with vessel dissection, perforation, or distal clot migration.[71] The Concentric Thrombus Retriever (Concentric Medical, Mountain View, CA) is a more advanced design to retrieve clots from the intracranial circulation. The nitinol corkscrew-like tip on the microwire can be pulled back to an inflated balloon at the tip of a microcatheter when the clot is captured. Ideally, the blood clot, the tips of the microwire, and the microcatheter should be engulfed by the end balloon and should be pulled out as one unit. Initial studies of nine vessels in swine models showed good retrieval and no dissection or perforation.[19] A Phase I trial of mechanical embolus removal in cerebral ischemia (MERCI) within 3 to 8 hours of symptom onset using this device is ongoing.[19]

Another device also in Phase I trials is the new generation basket-like Neuronet endovascular snare (Guidant Corporation, Indianapolis, IN). A European trial known as the Neuronet evaluation in embolic stroke disease (NEED) is currently being conducted.[19] Thrombus obliteration devices including the AngioJet (Possis Medical, Inc., Minneapolis, MN) and the X-ciser (Endicor Medical, Inc., San Clemente, CA), are being tested in pilot safety and feasibility studies.[19,72] The AngioJet uses a vacuum created by a high-pressure saline solution jet to aspirate clots.[19,72] It is currently being tested within 6 hours of stroke onset in a trial known as TIME (thrombectomy in middle cerebral artery embolism). The X-ciser uses a dual lumen microcatheter with rotating blades within a central hollow core and vacuum simultaneously to aspirate the debris of a clot.[19,72] Several other devices are in development, including a catheter with several wires that form a basket when the catheter tip is placed in the clot.

One other contemporary design is the EKOS catheter (EKOS Corporation, Bothell, WA) — a 2.5-French drug infusion catheter with a 2.1-mm distal ultrasound transducer.[73] The rationale behind this design is that the use of transcranial and endovascular ultrasound has been shown to intensify the effects of thrombolytic therapy in animal models and early human studies.[19,73] The catheter is placed proximal to the clot and ideally the transducer is embedded in the clot. A total of 14 ischemic stroke patients were treated in the North American EKOS trial without any complications and with a 57% recanalization rate using the thrombolysis-in-myocardial infarction (TIMI 2–3) scale.[19,73]

Lasers are also thought to produce clot emulsification by transforming photoenergy into acoustic energy. Two laser emitting catheters are being tested in a Phase I trial. The intra-arterial endovascular photo acoustic recanalization (EPAR) laser system (Endovasix, Inc., San Francisco, CA) trial enrolled 26 patients within 6 hours of symptoms onset. A total of 31 vessels were treated with 48% recanalization rate (TIMI 2–3), although 2 vessels were perforated during microcatheter placement and before laser therapy.[19] Endovascular stroke therapy is summarized in Table 12.4.

12.6 OTHER ENDOVASCULAR APPLICATIONS

12.6.1 Neoplastic Diseases

Vascular tumors such as meningiomas and hemangioblastomas are currently treated preoperatively with embolization of the vascular tumor bed. Embolic materials similar to those used for AVM and AVF therapies may be used to embolize tumor feeders. With enhancement of microbioengineering technology, new microcatheters, wires, and particles will lead to more effective adjunctive tumor embolotherapy.

TABLE 12.4
Endovascular Therapies for Acute Ischemic Stroke

Advances	Comments
Pharmacological	
New generation fibrinolytic agents such as alteplase, reteplase, TNK, staphylokinase, ancrod	Higher fibrin specificity; longer plasma half-life; may be given intravenously, intra-arterially, or in combination
Glycoprotein IIb/IIIa; parenteral and oral formulations will be available	May play role in local intra-arterial therapy alone or in combination with thrombolytics
Neuroprotective agents	May reemerge in combination with thrombolytics, antiplatelets, or endovascular devices
Mechanical	
Mechanical devices: Thrombus Clot Retriever, Neurosnare, AngioJet, X-Ciser	Several human safety trials with enhanced newer catheters and device designs are ongoing
Endovascular ultrasonification, laser clot lysis (e.g., EKOS and EPAR microcatheters)	Early stages of human studies are complete; devices appear safe and feasible. Larger Phase II studies planned

Endovascular neurosurgery would be an effective and direct means of administering chemotherapeutic agents to brain tumors locally with fewer unwanted side effects from systemic administration. As the molecular biologies and bases of neoplastic diseases are being uncovered, the endovascular approach may be the choice in some cases to deliver gene therapy or newer and more effective antimitotic agents.

12.6.2 Degenerative Diseases

Newer disease-modifying drugs for both genetic and degenerative neurological diseases would be probably safer, less toxic, and more effective when applied directly to the affected areas rather than systemically. The current microcatheter technology allows selective catheterization of small arterial branches to deliver higher concentrations of therapeutic agents directly to the affected neuronal tissues. As the genetic codes unfold, the endovascular approach may become the preferred method of administering gene therapy to combat various genetic and degenerative diseases.

12.7 CONCLUSIONS

The field of endovascular neurosurgery is evolving rapidly. Newer and more biocompatible devices are becoming increasingly available and the interests of physicians and industry will hasten the progress even further. In the field of endovascular intracranial aneurysm therapy, the first and largest randomized control study (ISAT) comparing coiling to clipping was completed and published with positive results. North American trials are being designed and funded. When completed, they are expected to clarify the exact role of aneurysm coiling, taking into account the fact that U.S. practice strategies are different from those in Europe where ISAT was initiated.[74]

Newer coil materials and compositions, with more emphasis on biocompatibility and bioactive substances, are currently available. Advances in microwire, microcatheter, and guide catheter technology are also imperative to safer and more successful coiling. Angiography equipment in many endovascular suites includes three-dimensional rotational capability that allows better delineation of small vessels, aneurysmal origins, and aneurysmal neck and provides better endovascular guidance.[75,76] Newer, more flexible stents for intracranial deployment now allow endovascular neurosurgeons to secure wide neck aneurysms with better coil packing and less residual filling.

Endovascular microangioscopy or aneurysmoscopy is still in its infancy, but may become the future imaging technique for cerebral aneurysms and AVMs.[77] AVM endovascular therapy is now more effective. Flow-directed microcatheters allow access to more feeders and a user can get closer to the nidus. New nonadhesive embolic agents such as Onyx and Ethibloc allow reuse of microcatheters with fewer complications. The field of CAS expanded further with the development of specific, self-expandable extracranial and intracranial flexible stents. Several embolic protection devices that may reduce periprocedural neurological complications are available. Drug-coating and drug-eluting technologies to reduce rates of stenosis are in development. For example, rapamycin-eluting stents are associated with remarkable reductions in rates of restenosis in coronary vessels and may prove as effective in the carotid and intracranial blood vessels.[78]

New stroke therapies with intra-arterial thrombolytic agents and combination therapies are also available. Endovascular nonpharmacological means of clot removal and recanalization including the AngioJet, Microsnare, and ultrasound and laser catheters with or without thrombolytic therapies are currently in Phase I trials. The role of near real-time magnetic resonance angiography (MRA) is not well defined yet, and remains to be explored when instant and fluoroscopic real-time MR capability with its compatible catheters, devices, and patient accessibility becomes available.[79,80] The future of endovascular neurosurgery will continue to see dynamic and constant changes over the next decade, with wider applications and enhanced techniques, devices, and imaging capabilities.

REFERENCES

1. Hopkins, N.L., Giuseppe, L., and Guterman L.R., Treating complex nervous system vascular disorders through a "needle stick": origins, evolution, and future of neuroendovascular therapy, *Neurosurgery,* 48, 463–475, 2001.
2. Rosenwasser, R.H., Endovascular tools for the neurosurgeon, *Clin. Neurosurg.,* 49, 115–135, 2002.
3. Katzen, B.T., The future of catheter-based angiography: implications for the vascular interventionalist, *Radiol. Clin. N. Am.,* 40, 689–692, 2002.
4. Levy, E.I. et al., Endovascular surgery: the future without limits, *Clin. Neurosurg.,* 49, 229–246, 2002.

5. Pelz, D.M., Advances in interventional neuroradiology, *Stroke,* 34, 357–358, 2002.
6. Guglielmi, G., Viñuela, F., Sepetka, I., and Macellari, V., Electrothrombosis of saccular aneurysm via endovascular approach. Part I: basis, technique, and experimental results, *J. Neurosurg.,* 75, 1–7, 1991.
7. Guglielmi, G., Viñuela, F., Sepetka, I., and Macellari, V., Electrothrombosis of saccular aneurysm via endovascular approach. Part II: preliminary clinical experience, *J. Neurosurg.,* 75, 8–14, 1991.
8. Molyneux, A. et al., International Subarachnoid Aneurysm Trial (ISAT) of neurosurgical clipping versus endovascular coiling in 2143 patients with ruptured intracranial aneurysms: a randomized trial, *Lancet,* 360, 1267–1274, 2002.
9. Doerfer, A. et al., Flow-directed microcatheters for cerebral embolization using Ethibloc: an *in vitro* study, *Neuroradiology,* 43, 1112–1117, 2001.
10. Debrun, G.M. et al., Embolization of the nidus of brain arteriovenous malformation with n-butyl cyanoacrylate, *Neurosurgery,* 40, 112–121, 1997.
11. Becker, T.A. et al., *In vivo* assessment of calcium alginate gel for endovascular embolization of a cerebral arteriovenous malformation model using the swine *Rete mirabile*, *Neurosurgery,* 51, 453–459, 2002.
12. Phadke, R.V. et al., Embolization of cranial/spinal tumours and vascular malformations with hydrogel microsphere: an experience of 69 cases, *Acta Radiol.,*43, 15–20, 2002.
13. Jahan, R. et al., Embolization of arterio-venous malformation with Onyx: clinicopathological experience in 23 patients, *Neurosurgery,* 48, 984–997, 2001.
14. Tan, W.A., Jarmolowski, C.R., Wechsler, L.R., and Wholey, M.H., New development in endovascular interventions for extracranial carotid stenosis, *Tex. Heart Inst. J.,* 27, 273–280, 2000.
15. Kandzari, D.E., Tcheng, J.E., and Zidar, J.P., Coronary artery stents: evaluating new design for contemporary percutaneous intervention, *Cath. Cardiovasc. Intervention,* 56, 562–576, 2002.
16. Stendel, R., Krischek, B., and Pietilä, A., Biodegradable implants in neurosurgery, *Acta Neurochir.,* 143, 237–243, 2001.
17. Furlan, A. et al., Intra-arterial prourokinase for acute ischemic stroke: the PROACT II study: a randomized controlled trial: prolyse in acute cerebral thromboembolism, *JAMA,* 282, 2003–2011, 1999.
18. Zaidat, O.O. et al., Response to intra-arterial and combined intravenous and intra-arterial thrombolytic therapy in patients with distal internal carotid artery occlusion, *Stroke,* 33, 1821–1826, 2002.
19. Leary, M.C. et al., Beyond tissue plasminogen activator: mechanical intervention in acute stroke, *Ann. Emerg. Med.,* 41, 838–846, 2003.
20. Qureshi, A.I. et al., Aggressive mechanical clot disruption and low dose intra-arterial third generation thrombolytic agent for ischemic stroke: a prospective study, *Neurosurgery,* 51, 1319–1329, 2002.
21. Dovey, Z. et al., Guglielmi detachable coiling for the treatment for intracranial aneurysms: the story so far, *Arch. Neurol.,* 58, 559–564, 2001.
22. Macdonald, R.L., Mojtahedi, S., Johns, L., and Kowalczuk, A., Randomized comparison of Guglielmi detachable coils and cellulose acetate polymer for treatment of aneurysms in dogs, *Stroke,* 29, 478–486, 1998.
23. Muramaya, Y. et al., Endovascular treatment of experimental aneurysm by use of a combination of liquid embolic agents and protective devices, *AJNR,* 21, 1726–1735, 2000.

24. Raymond, J. et al., Cyanoacrylate embolization of experimental aneurysms, *AJNR,* 23, 129–138, 2002.
25. Mottu, F., Rufenacht, D.A., Laurent, A., and Doelker, E., Iodine–containing cellulose mixed esters as radiopaque polymers for direct embolization of cerebral aneurysms and arteriovenous malformations, *Biomaterials,* 23, 121–131, 2002.
26. Matsumaru, Y. et al., Embolic materials for endovascular treatment of cerebral lesions, *J. Biomater. Sci. Polym. Ed.,* 8, 555–569, 1997.
27. Nakahara, I. et al., Endovascular treatment of aneurysms on the feeding arteries of intra-cranial arteriovenous malformations, *Neuroradiology,* 41, 60–66, 1999.
28. Murayama, Y. et al. Bioabsorbable polymeric material coils for embolization of intracranial aneurysms: a preliminary experimental study, *J. Neurosurg.,* 94, 454–463, 2001.
29. Abraham, J.M. et al., Surface modifications enhancing biological activity of Guglielmi detachable coils in treating intracranial aneurysms, *Surg. Neurol.,* 54, 34–41, 2000.
30. Cloft, H.J. et al., Use of three-dimensional Guglielmi detachable coils in the treatment of wide-necked cerebral aneurysms, *AJNR,* 21, 1312–1314, 2000.
31. Tong, F.C., Cloft, H.J., and Dion, J.E., Endovascular treatment of intracranial aneurysms with Guglielmi detachable coils: emphasis on new techniques, *J. Clin. Neurosci.,* 7, 244–253, 2000.
32. Amar, A.P., Zlokovic, B.V., and Apuzzo, M.L.J., Endovascular restorative neurosurgery: a novel concept for molecular and cellular therapy of the nervous system, *Neurosurgery,* 52, 402–413, 2003.
33. Marx, W.F. et al., Endovascular treatment of experimental aneurysms by use of biologically modified embolic devices: coil-mediated intra-aneurysmal delivery of fibroblast tissue allografts, *AJNR,* 22, 323–333, 2001.
34. Han, P.P. et al., Percutaneous intracranial stent placement for aneurysm, *J. Neurosurg.,* 99, 23–30, 2003.
35. Irie, K., Kawanishi, M., and Nagao, S., Balloon-assisted coil placement in wide-necked cerebral aneurysms: preliminary clinical experience, *Neurol. Med. Chir.,* 40, 603–608, 2000.
36. Lieber, B.B. and Gounis, M.J., The physics of endoluminal stenting in the treatment of cerebrovascular aneurysms, *Neurol. Res.,* 24, S33–S42, 2002.
37. Alexander, M.J., Smith, T.P., and Tucci, D.L., Treatment of an iatrogenic petrous carotid artery pseudoaneurysm with a Symbiot-covered stent: technical case report, *Neurosurgery,* 50, 658–662, 2002.
38. Schellhammer, F. et al., Polyethylene terephthalate and polyurethane coatings for endovascular stents: preliminary results in canine experimental arterio-venous fistulas, *Radiology,* 211, 169–175, 1999.
39. Kazekawa, K. et al., Nontoxic embolic liquids for treatment of arteriovenous malformations, *Appl. Biomater.,* 38, 79-86, 1997.
40. Hamada, J. et al., A non-adhesive liquid embolic agent composed of ethylene vinyl alcohol copolymer and ethanol mixture for the treatment of cerebral arteriovenous malformations: experimental study, *J. Neurosurg.,* 97, 889–895, 2002.
41. Goto, K., Uda, K., and Ogata, N., Embolization of cerebral arteriovenous malformations (AVMs): material selection, improved technique, and tactics in the initial therapy of cerebral AVMs, *Neurol. Med. Chir.,* 38, S193–S199, 1998.
42. Weill, A. et al., Use of the trispan device to assist coil embolization of high-flow arteriovenous fistulas, *AJNR,* 23, 1149–1152, 2002.

43. Massoud, T.F. and Hademenos, G.J., Transvenous retrograde nidus sclerotherapy under contolled hypotension (TRENSH): a newly proposed treatment for brain arteriovenous malformation: concepts and rationale, *Neurosurgery,* 45, 351–370, 1999.
44. Hara, T., Hamada, J., Kai, Y, and Ushio, Y., Surgical trans-venous embolization of a carotid-cavernous dural fistula with cortical drainage via a petrosal vein: two technical case reports, *Neurosurgery,* 50, 1380–1384, 2002.
45. Pile-Spellman, J. et al. Adenosine-induced cardiac pause for endovascular embolization of cerebral arterio-venous malformation: technical case report, *Neurosurgery,* 44, 881–886, 1999.
46. North American Symptomatic Carotid Endarterectomy Trial Collaborators, Beneficial effect of carotid endarterectomy in symptomatic patients with high-grade carotid stenosis, *NEJM,* 325, 445–453, 1991.
47. Asymptomatic Carotid Atherosclerosis Study Executive Committee, Endarterectomy for asymptomatic carotid stenosis, *JAMA,* 273, 1421-1428, 1995.
48. Wholey, M.H. and Wholey, M., Current status of carotid artery stent placement, *Cardiovasc. Surg.,* 44, 331–339, 2003.
49. Connors, J.J., III et al., Treatment of atherosclerotic disease at the cervical carotid bifurcation: current status and review of the literature, *AJNR,* 21, 444–450, 2000.
50. Hobson, R.W., II, Update on the Carotid Revascularization Endarterectomy versus Stent Trial (CREST) protocol, *J. Am. Coll. Surg.,* 194, S9–S14, 2002.
51. Mukherjee, D. and Yadav, J.S., Percutaneous treatment for carotid stenosis, *Cardiol. Clin.,* 20, 589–597, 2002.
52. Babapulle, M.N. and Eisenberg, M.J., Coated stents for the prevention of restenosis: part I, *Circulation,* 106, 2734–2740, 2002.
53. Babapulle, M.N. and Eisenberg, M.J., Coated stents for the prevention of restenosis: part II, *Circulation,* 106, 2859–2866, 2002.
54. Marty, B., Leu, A.J., Mucciolo, A., and Von Segesser, L.K., Biological fixation of polyester- versus polyurethane-covered stents in a porcine model, *J. Vasc. Interv. Radiol.,* 13, 601–607, 2002.
55. Fontaine, A.B. et al., Evaluation of local abciximab delivery from the surface of a polymer-coated covered stent: *in vivo* canine studies, *J. Vasc. Interv. Radiol.,* 12, 487–492, 2001.
56. Ye, Y.W. et al., Bioresorbable microporous stents deliver recombinant adenovirus gene transfer vectors to the arterial wall, *Ann. Biomed. Eng.,* 26, 398–408, 1998.
57. Nakayama, Y. et al., Development of high performance stent: gelatenous photogel-coated stent that permits drug delivery and gene transfer, *J. Biomed. Mater. Res.,* 57, 559–566, 2001.
58. Baim, D.S. et al., Randomized trial of a distal embolic protection device during percutaneous intervention of saphenous vein aorto-coronary bypass graft, *Circulation,* 105, 1285–1290, 2002.
59. Ohki, T. et al., Initial experience with cerebral protection devices to prevent embolization during carotid artery stenting, *J. Vasc. Surg.,* 36, 1175–1185, 2002.
60. Grube, E. et al., Initial multicenter experience with a novel distal protection filter during carotid artery stent implantation, *Cath. Cardiovasc. Intervention,* 58, 139–146, 2003.
61. Qureshi, A.I. et al., Intracerebral hemorrhages associated with neurointerventional procedures using a combination of antithrombotic agents including abciximab, *Stroke,* 33, 1916–1919, 2002.
62. Hobson, R.W., II, Carotid angioplasty stent: clinical experience and role for clinical trials, *J. Vasc. Surg.,* 33, S117–S123, 2001.

63. NINDS Study Group, Tissue plasminogen activator for acute ischemic stroke, *NEJM, 83*, 1581–1587, 1995.
64. Fisher, M., Antithrombotic and thrombolytic therapy for ischemic stroke, *J. Thrombosis Thrombolysis,* 7, 165–169, 1999.
65. Lewandowski, C.A. et al., Combined intravenous and intra-arterial rTPA versus intra-arterial therapy of acute ischemic stroke: emergency management of stroke (EMS) bridging trial, *Stroke,* 30, 2598–2605, 1999.
66. Ernst, R. et al., Combined intravenous and intra-arterial recombinant tissue plasminogen activator in acute ischemic stroke, *Stroke,* 31, 2552–2557, 2000.
67. Deitcher, S.R. and Jaff, M.R., Pharmacologic and clinical characteristics of thrombolytic agents, *Rev. Cardiovasc. Med.,* 3, S25–S33, 2002.
68. Fintel, D.J., From bench to bedside: GP IIb-IIIa inhibitors, *Neurology,* 57, S12–S19, 2001.
69. Bogousslavsky, J. and Lecler, J.R., Platelet glycoprotein IIb/IIIa antagonists for acute ischemic stroke, *Neurology,* 57, S53–S57, 2001.
70. Seitz, R.J. et al., Thrombolysis with recombinant tissue plasminogen activator and Tirofiban in stroke: preliminary observations, *Stroke*, 34, 1932–1935, 2003.
71. Chopko, B.W., Kerber, C., Wong, W., and Georgy, B., Transcatheter snare removal of acute middle cerebral artery thromboembolism: technical case report, *Neurosurgery,* 46, 1529–1531, 2000.
72. Lustep, H.L. et al., Intra-arterial suction thrombectomy in acute stroke, *Stroke,* 30, 270, 1999
73. Behrens, S. et al., Potential use of therapeutic ultrasound in ischemic stroke treatment, *Echocardiography,* 18, 359–363, 2001.
74. Kaku, Y., Endovascular aneurysm treatment from the neurosurgeon's point of view, *Acta Neurochir. Suppl.,* 82, 99–103, 2002.
75. Van den Berg, J.C., Overtoom, T.T.C., De Valois, J.C., and Moll, F.L., Using three-dimensional rotational angiography for sizing of covered stents, *AJR,* 178, 149–152, 2002.
76. Ernemann, U. et al., 3-D angiography in planning the treatment of cerebral aneurysms, *Electromedica,* 68, 31–37, 2000.
77. Massoud, T.F., Murayama, Y., Viñuela, F., and Utsumi, A., Laboratory evaluation of a microangioscope for potential percutaneous cerebrovascular application, *ANJR,* 22, 363–365, 2001.
78. Thompson, C.A. and Oesterle, S.N., Biointerventional cardiology: the future interface of interventional cardiovascular medicine and bioengineering, *Vasc. Med.,* 7, 135–140, 2002.
79. Dion, Y.M., DeWailly, G.W., and Moisan, C., Endovascular procedures under near-real-time MRI guidance: present status and future perspectives, *Surg. Technol. Int.,* 10, 161–167, 2002.
80. Lardo, A.C., Real-time magnetic resonance imaging: diagnostic and interventional applications, *Pediatr. Cardiol.,* 21, 80–98, 2000.

13 Neuroscience ICU Therapeutics

Ashutosh A. Pradhan and Dennis A. Turner

CONTENTS

13.1 INTRODUCTION

Neurotrauma, neurosurgical intervention, and cerebrovascular disease all involve combinations of immediate or primary injury and delayed or secondary injury (see Chapter 4 for information on cellular treatments of stroke and hypoxia/ischemia). The primary injury occurs at the time of or a few minutes after a trauma, brain procedure or vascular insult. The secondary injury evolves over time — usually hours to several days.

0-8493-1482-8/05/$0.00+$1.50

The primary injury may be optimally influenced by care provided to prevent the trauma (i.e., seatbelt use in cars and other injury prevention schemes), prevent intra-operative complications, or manage medical problems such as diabetes and hypertension that may predispose a patient to stroke occurrence. Intensive care unit (ICU) therapeutic measures focus on preventing, treating, and reversing the evolving secondary effects of brain damage and its systemic complications following primary events.[1]

In order to improve treatment options, much attention has been directed to more sensitive monitoring methods to prevent further injury in addition to standard comprehensive medical care and prevention of systemic problems.[2] This chapter reviews the current hypotheses in neuroscience ICU therapeutics and discusses promising lines of bench research that will make their way into ICUs within the next 5 years. It also includes a brief review of the modest strides made in management of head injury, postoperative issues, and cerebrovascular problems, including subarachnoid hemorrhage and ischemic stroke.

13.2 CLINICAL PROBLEMS AMENABLE TO ICU MANAGEMENT

13.2.1 Brain Trauma

The two distinct types of severe brain injuries are cerebral contusions or gray matter injuries and diffuse axonal shears or white matter injuries. The evolution over time after injury varies by type of injury, although many patients have combinations of different injury types.[2]

Cerebral contusions are commonly present following a variety of different injury mechanisms, and often the patient is nearly normal or only slightly disoriented initially. Computerized tomography (CT) scans may show little initial evidence of the extent of cerebral contusion although some swelling or mild hemorrhage may often be present. A common pattern is the coup–contrecoup injury in which a contusion appears under the skull at the location of the impact and also opposite the area of injury. The typical evolution is that local swelling and cortical irritation occur, often with focal or punctate hemorrhages within the cortex over 2 to 3 days. Occasionally these small hemorrhages (observed on CT scans as diffuse salt-and-pepper areas of hemorrhage) will coalesce to form a substantial intraparenchymal hemorrhage that may form a mass. The secondary consequences of the injury commonly include seizure, brain swelling and increased intracranial pressure, and hyponatremia. If the injured cortical areas were functioning close to normal, recovery over time is usually excellent when the intermediate problems of swelling, intracranial pressure, and seizure are managed effectively.

Shear injury or diffuse axonal injury (DAI) commonly is caused by more severe trauma. Patients can experience severe deficits from the time of injury. In several ways, DAI is similar to spinal cord injury because both involve primarily severe white matter injury and poor degrees of recovery over time. Rarely do intracranial pressure (ICP) problems consisting primarily of DAI appear because white matter does not show significant swelling after injury. These patients create less of a concern for intracranial swelling or raised pressure, although all the other ICU management issues including preventing infection, treatment of seizures and electrolyte abnormalities, airway protection, and feeding arise. While some patients with severe DAI recover, they commonly require tracheostomy and gastrostomy for long-term care. Many long-term comatose patients in this category are maintained for years, often in low-functioning or vegetative states.

13.2.2 Postoperative Neurosurgical Care

ICU management after neurosurgical intervention has two common bases.[1] The first is preventative management, for rapid detection of postoperative problems such as seizures, brain hemorrhages, airway problems, or systemic concerns such as blood loss replacement. The second basis is active treatment of patients with marginal brain function, immediate brain swelling or hemorrhage after vascular occlusion, and intraoperative complications. Essentially all patients undergoing open brain procedures are monitored at least overnight in an ICU due to the large number of problems that can arise.

13.2.3 Cerebrovascular Problems

Multiple forms of spontaneous cerebrovascular events resulting in brain damage, increased intracranial pressure, hydrocephalus, hypertension, and repeat subarachnoid hemorrhage (SAH) can be optimally managed via a neuroscience ICU approach (see Chapter 4 and Chapter 11). For example, patients with severe SAH may not be immediately amenable for either direct neurosurgical intervention, such as aneurysm clipping or endovascular treatment (see Chapter 12), but may still have severe deficits and require ICP monitoring, blood pressure, and volume management (such as hyperdynamic therapy).

Common forms of stroke include ischemic stroke, such as major coronary artery occlusion, after which patients often develop severe brain swelling. This swelling can in some cases be ameliorated with ICP monitoring and occasionally by decompressive craniectomy. Deep intracerebral hemorrhages due to hypertension often also result in severe neurological deficits requiring intensive management, secondary medical problems, and hydrocephalus. Mortality can be as high as 85%. All of these possibilities represent common reasons to stabilize and treat patients in a neuroscience ICU setting.

13.2.4 Inappropriate Clinical Care Situations

Neuroscience ICU care intuitively is life-saving, and can function as a key bridge between difficult neurological problems and their recovery over time. However, in many instances recovery is very unlikely, in spite of intensive care capabilities. For example, terminal care in patients with poor prognosis or brain death does not require ICU care, nor would patients in a permanent vegetative state who are stable. Also, physiologically stable patients may not require an ICU environment, because no critical treatment issues may exist. In most hospitals ICU are a scarce resource, and such a resource must be managed wisely, considering which patients may optimally benefit from expensive care and resources.

13.3 MODALITIES OF NEUROSCIENCE ICU MANAGEMENT

Neuroscience ICU management includes a number of key modalities of care including frequent and sensitive neurological examination, specific concern for brain or spinal cord-related deterioration, various forms of monitoring of brain and systemic function, and attention to the details of healing, nutrition, airway management, and prevention of complications.[1,2] These various modalities will be discussed individually.

13.3.1 Specific Neurological and General Care

The mainstay of neuroscience ICU management is careful neurological assessment. Most neuroscience ICUs use detailed flowsheets to help monitor neurological function, including assessment on the Glasgow coma scale for brain function, spinal cord function, and particularly detailed responses to environmental stimuli. Intensive neurological exams are usually performed at least every 2 hours by nurses specialized in neurological assessment. Detection of specific neurological events including seizures, herniation syndromes, changes in levels of consciousness and awareness, and spinal cord functions is critical. A key point is that detection of worsening may imply a cascade of pathological alterations that may be prevented by prompt treatment and management.

In addition to neurological assessment, excellent general medical care is also critical and should include assessment of airway problems, need for continued intubation, nutrition, and methods of diet supplementation. In addition, maintenance of normal temperature, or in some cases lower than normal temperature (as in cooling), requires vigilance and possibly cooling blankets. Since most patients with severe brain alterations typically have intact gastrointestinal function, rapid resumption of nutrition is usually possible, initially through an orogastric or nasal tube into the small bowel. In many cases, a gastrostomy can be helpful for long-term nutritional supplementation. Prevention of infection, skin ulceration, and other medical problems is critical because these conditions may significantly slow recovery.

13.3.2 Monitoring Techniques

Many noninvasive and invasive monitoring techniques are currently used in the ICU setting.[2,3] Almost any physiological function can be intermittently or continuously monitored, depending on the need for rapid intervention. Many monitoring functions such as arterial and intracranial pressures are viewed as analog signals and converted to numbers in the case of pulse rate, systolic and diastolic pressures, cerebral perfusion pressures, levels of oxygenation, etc.

Typical systemic monitoring functions include cardiac pulse and blood pressure, temperature, weight, cardiac output, ventricular pressures, oxygenation and arterial blood gases, systemic electrolytes, and blood counts. Additional modalities specific to the neuroscience ICU setting include ICP measurement, cerebral oxygen and substrate levels, cerebral blood flow and transcranial Doppler monitoring, cerebral metabolism, electroencephalographic monitoring of seizure and electrical activity of the brain, and level of sedation.[3]

Other types of monitoring include structural assessments of the brain and spinal cord via CT and magnetic resonance imaging (MRI) scans. While most of these general and neurological modalities of monitoring are common, new methods of monitoring brain function continuously arise. Many monitoring procedures involve an overlap between monitoring and treatment capabilities, for example, ICP monitoring and drainage. The major forms of monitoring are discussed next.

13.4 INTRACRANIAL PRESSURE MEASUREMENTS

Intracranial pressures (ICP) reflect a combination of the brain's pulsatile response to incoming arterial blood with each cardiac cycle and its compliance.[4–7] The shape of the ICP waveform closely resembles that of the arterial blood pressure waveform, but with a delay and smaller amplitude. This fact led to the development of waveform analysis to reveal the compliance of the brain in response to incoming cardiac pulsation and to demonstrate whether pathological changes within the brain alter this compliance and lead to increased ICP.[4,8–10] While many types of ICP measurements are in common clinical use, no data indicate whether they actually improve outcomes, as compared to empirical treatment of presumed ICP elevations.

13.4.1 CSF and Intracranial Pressure

Cerebrospinal fluid (CSF) normally completely surrounds the brain and occupies the subarachnoid space. The freely diffusible CSF normally equilibrates the ICP around the brain rapidly and buffers the brain mechanically. Because CSF drains into veins with positive pressures (4 to 6 mmHg above sagittal sinus pressure), the usual pressure is 10 to 12 mmHg in a lateral horizontal position. As long as CSF can circulate freely, elevated ICP has no deleterious consequences on brain function, assuming that cerebral perfusion pressure remains in the normal range (65 to 70 mmHg typically). However, secondary systemic hypertension may occur as a reflex in order to maintain cerebral perfusion pressure if the ICP is elevated.

This is clearly observed in the example of benign intracranial hypertension (BIH; formerly called pseudotumor cerebri prior to CT and MRI imaging). A CSF absorption deficit is commonly present in BIH, but because normal CSF production continues, CSF pressures can rise. Interestingly, since BIH commonly occurs in young patients with normal ventricular size (rather than hydrocephalus), the ventricles typically do not dilate. The brain typically resists the increased pressure, but papilledema can result. Long-standing papilledema can lead to loss of visual function. Visual loss and headaches are the only discernable abnormalities from high ICP in the absence of a mass, often up to 50 mmHg. However, if a mass is present and CSF circulation around the brain is disturbed because of the mass (from shift, loculation, etc.), the CSF cannot equilibrate the pressure. CSF can accumulate on the side of the mass, enhancing the brain shift and mass effect, creating a vicious cycle of increasing mass. In such cases, ICP monitoring can be very helpful for discerning whether such brain shift and ICP buildup are occurring. Additionally, CSF drainage through an intraventricular catheter can effectively treat the mass effect and prevent additional untoward consequences in many cases.

13.4.2 Types of Intracranial Pressure Monitors

ICP monitoring catheters include those that have drainage channels (and hence are placed in locations where CSF can be drained, for example, in lateral ventricles), and those that only monitor pressure. Potential locations for catheters include epidural, subdural, intraparenchymal, and intraventricular, as well as lumbar subarachnoid spaces. Epidural and subdural catheters are not as popular due to the dearth of reliable data. Intraventricular catheters remain the most popular because of their ability to measure pressure and drain CSF. The intraparenchymal catheter has changed little in the past 20 years and includes a device such as a moveable diaphragm on the tip of the catheter that will transduce brain pressure. The fiberoptics that relay signals through the catheter have improved and less drift occurs over time when these catheters are used.

Intraventricular catheters have also been combined with fiberoptic catheters that allow continuous monitoring when draining. Most intraventricular catheters can be tunneled under the skin away from the site of insertion. This decreases infection rates and allows longer catheter use time *in vivo*.[11] However, most parenchymal catheters are stiff and cannot be tunneled and are thus more subject to damage or shear.

Because the brain is soft and does not transmit pressure well, ICP may vary from location to location, as demonstrated by several studies with multiple catheter locations and the resulting disparate pressure measurements. This is particularly true when CSF circulation is impeded by mass or shift, leading to pressure gradients in the brain that are then not equalized by CSF movement. Thus, intracranial pressure measurement in a distal location (such as contralateral to a mass) may be misleading and may show apparent low value that could potentially misinform a clinician about a patient's true status.

13.4.3 Intracranial Pressure and Waveform Monitoring

Although ICP recordings are commonly used to monitor increased pressure, clinical interpretation of increased mean ICP has major limitations because it is an indirect measure of potential neurological deterioration; a high degree of variability of mean ICP levels exists among patients.[4,8–10] Consistently increased mean ICP values (>40 mmHg) correlate well with poor clinical status and outcome in compromised patients, but this relationship is less predictable in patients with moderately increased values between 20 and 30 mmHg. Theoretically, the volume–pressure relationship provides a measure of the compensatory reserve and the likelihood of neurological deterioration. However, no reliable and safe direct clinical method of calculating intracranial compliance or elastance (the inverse of compliance) allows full reconstruction of the volume–pressure curve.

One technique proposed to provide additional information about neurological status is spectral analysis of arterial blood pressure (ABP) in the intracranial cavity as an input function and analysis of the ICP waveform as an output function.[8–10] The ratio of the frequency components of the ICP and ABP spectra yields a transfer function that includes both cerebrovascular and brain compliance components. ICP waveforms have been analyzed in this manner using a variety of models and techniques, generally across multiple cardiac cycles. The influence of respiratory cycles and central venous pressures (CVP) and their relationship to ICP may also be important in understanding ICP.

A normal ICP recording consists of a pulsatile waveform with two components: one corresponding to arterial pulsations, the other corresponding to the much slower respiratory excursions, related most closely to changes in CVP.[9] The pulsatile waveform has been analyzed by the inflections and components as well as by fractionating the frequency components using the Fourier transform. This can done on a cycle-by-cycle basis to provide an almost instantaneous measure of cerebral compliance for each cardiac pulse.[8] This type of analysis will require further clinical studies to assess its overall usefulness and predictive value.

13.5 ADDITIONAL MONITORING MODALITIES

13.5.1 Cerebral Blood Flow

Cerebral blood flow (CBF) is the velocity of blood through the cerebral circulation,[1] together with estimates of the total blood volume in the various arterial and venous compartments. Many metabolic parameters are dependent on knowing the CBF. Changing the CBF at different times can help treat patients, particularly in low-flow situations, such as after occlusion of a major trunk artery. CBF follows Poiseulle's law that essentially identifies three variables the clinician can affect: (1) perfusion pressure, (2) vascular radius, and (3) blood viscosity.

During different physiologic states such as vasospasm or after an ischemic infarct, perfusion pressure may be increased for greater blood flow and enhancement of collateral formation. In addition, blood may be diluted to decrease viscosity to an optimal hematocrit in the range of 30 to 33%. Cerebral autoregulation primarily functions to maintain constant CBF during fluctuations in cerebral perfusion pressure within a wide normal-range systemic blood pressure (approximately 60 through 180 mmHg).

Diminished CBF can indicate ischemia, which may lead to damage to regions of the brain. Thus, measurements of CBF can allow a physician to change treatment paradigms. Multiple techniques of direct or indirect measurement of CBF have been developed. A simple method such as the Kety–Schmidt nitrous oxide technique can be used at the bedside. Only arterial and venous samples are needed to measure nitrous oxide differences. Radiological imaging can also be used to determine blood flow via many modalities, classically by using radioactive monitors over the skull to measure the amount of radioactive xenon coursing through blood vessels of the brain after inhalation. Because the number of surface monitors is limited, this type of crude blood-flow assessment is rarely done.

Recently, MR diffusion and perfusion were used to measure the diffusion coefficient of water, which relates areas of low blood flow and/or evolving ischemic infarct. Perfusion can be used with contrast to determine areas of low-blood volume. CT can also be used to measure flow by looking at specific tracers that provide quantitative measurements. The most common form of this CT blood-flow approach is the use of inhaled xenon (up to 40% by mask). Because xenon is a heavier molecule than iodine (the most common CT contrast agent), xenon provides excellent visualization in vessels for measuring both blood volume and flow. Unfortunately, the dose of xenon needed by inhalation for this technique is at the level for achieving anesthesia, leading to confusion and sedation. Single photon emission computed tomography (SPECT) can also be used for blood-flow studies in stroke, brain death determination, and epilepsy. All these techniques can be used to determine areas of decreased blood flow.

13.5.2 Brain Oximetry

Two modalities of brain oximetry are currently in limited use for assessing regional oxygen levels.[12] Near-infrared spectroscopy (NIRS) is a noninvasive method that can be used in the operating room and ICU to observe changes in brain oxygen demand. This technique is based on using oxyhemoglobin concentration as a tracer to determine CBF. However, since NIRS depends on the ability of infrared light to cross the scalp and dura to reach the brain when the skull is closed, it can only be used in infants with thin skulls or through the fontanelle. Even then, controlled studies to determine the validity and accuracy of this indirect measure are needed, particularly because no method clearly distinguishes scalp blood oxygen levels from blood oxygen levels in the brain.

Intraparenchymal oxygen tension catheters for human use based on the Clark style oxygen electrode (Licox, Integra Neurosciences, Plainsboro, NJ) have become available recently.[13–16] The principle of this electrode that dates back to the 1930s is the use of a gold sensor sensitive to oxygen and diffusion of oxygen through a dialysis membrane into an internal electrolyte solution. The impermeable dialysis membrane allows a small (1 mm in diameter) catheter to be sterilized. Despite some preliminary experience, the indications for use are not yet clear, but the device seems promising. The device has been used to ensure adequate oxygenation of injured tissue in traumatic injury.[2,17–19] Several studies suggest that normal brain oxygen levels in the extracellular space range from 30 to 40 mmHg, and that brain levels rise considerably with systemic oxygen challenges to 100% inspired oxygen.[19] Levels below 20 mmHg are considered hypoxic and increased oxygen or increased cerebral perfusion may be used to reverse a trend toward hypoxia in brain regions.[15]

Our institution has used the Licox catheter in a small number of patients. The catheter is placed similarly to a frontal intraparenchymal bolt. Continuous recording of brain oxygen tension is performed. When values fall below a certain level, the percent of inhaled oxygen is increased. The thought is that damaged tissue may be more sensitive to lowered levels of oxygen, with permanent damage caused by periods of hypoxia. The overall hypothesis for Licox use is that by ensuring adequate oxygenation of the brain, marginal areas may be prevented from cell death. In an observational study in the Netherlands, patients had catheters placed without any major complications. The patients were observed for partial oxygen pressure and outcome. The study determined that the depth and duration of brain tissue hypoxia correlated well with outcome and they proved to be independent predictors of unfavorable outcomes.[17]

In addition, the catheter can also be used to manage partial pressure of carbon dioxide and ventilation. Hyperventilation is a useful tool in the armamentarium of ICP management. However, cerebral autoregulation is usually disturbed in injured states. Because blood vessels may severely constrict at a pCO_2 level below 25 to 28 mmHg, the possible risk of causing ischemia with excessive hypocarbia is real. With a partial oxygen pressure reading, ventilation can be titrated so as not to cause ischemia while controlling ICP. This relationship of brain oxygen tension and blood gas carbon dioxide levels was confirmed in the laboratory using swine and Licox catheters.[13]

Jugular bulb oximetry is a recent method of assessing oxygen extraction that has fallen out of favor due to difficulty in its application and unreliable data. It is used to estimate the brain's metabolic needs because arterio–venous oxygen differences (oxygen extraction in a global sense) can be roughly determined. Because the jugular vein has a highly variable distribution of brain blood drainage, this method only hints at global brain metabolism and is highly nonspecific. Normal values range from 60 to 80%. Low levels can signify ischemia secondary to hyperventilation, increased metabolic demand, agitation, or seizure, suggesting increased oxygen extraction due to demand. Conversely, high levels can signify hypercarbia, hyperemia, late ischemia, or cerebral blood flow cessation. Difficulties with the catheters include migration, extracerebral contribution to jugular venous blood causing contamination, and low resolution to identifying areas of decreased metabolism.[1,12]

Overall, the tissue levels of oxygen in the brain are now well defined through both preclinical and human studies. In several early human studies, the low levels of oxygen were thought to be due to ischemia, but it has since become clear that oxygen is tightly regulated within the brain at fairly low levels. Supply is coupled to demand through vascular control and autoregulation.[19] However, a clear hypoxic threshold (oxygen level below which local ischemia or cell death occurs) has not yet been determined.[15] Rather, a loose clinical correlation between brain oxygen levels and survival exists, but these two disparate factors may or may not be correlated.

13.5.3 Cerebral Microdialysis

Lactate, glutamate, pyruvate, glucose, and other critical metabolites within the brain play a large role in secondary injuries that occur before and after neuronal damage. Much interest now focuses on measuring levels of these compounds and correlating them to ischemic or detrimental events. The Licox catheter can also be used for microdialysis. The bolt placed into the skull for fixation is large enough to allow a two- or three-way manifold to be placed into the bolt to accommodate up to three different catheters. Often, the simplest combination is the oxygen tension monitor, ICP parenchymal monitor, and brain temperature monitor. However, one or two of these can be exchanged for a microdialysis catheter. A group in Germany recently showed that prior to a hypoxic period, glucose decreased significantly and glutamate increased three- to fourfold.[18] This suggests that either a reduction in hyperventilation therapy or an increase in FiO_2 was indicated. Further research is needed to determine whether changes in treatment modalities affect outcome.

Determination of resting and stress levels of metabolites within the brain has proven very enlightening for understanding CNS metabolism and correlating human values with those obtained in preclinical studies. A popular preclinical hypothesis presented in many studies since 1997 has been the "lactate shuttle" concept, borrowed from muscle metabolism analysis. Briefly, concept suggests that one of the primary glial functions in the brain is to produce lactate from glycolytic metabolism of glucose and then excrete this lactate into the extracellular space. Neurons, according to this hypothesis, preferentially use lactate (rather than glucose) for much of the production of ATP and energy. An excellent review article critically discusses this concept and concludes that neurons may use lactate if available, but glucose is a critical fuel, particularly for membrane pumping of ions.[20]

The levels of lactate measured within the human brain partially support the lactate shuttle concept. In a head injury study in which the Licox catheters were placed contralateral to brain lesions (in the most normal areas possible), the measured brain lactate levels exceeded 3 m*M*.[19] These values were much higher than systemic lactate values near 1 m*M*, suggesting a complete dissociation of brain and systemic lactate due to the loss of monocarboxylate transporter activity (particularly MCT1) with maturity in the blood–brain barrier.

The relatively high level of brain lactate was initially thought to represent a high degree of anaerobic glycolysis within the brain (in other words, hypoxia), but on oxygen challenge, the lactate did not change. This finding of persistent lactate, even in a highly enriched oxygen environment, suggests a high degree of aerobic glycolysis, presumably partly within glial cells, as proposed by the lactate shuttle concept. The level of glucose measured was near 2 m*M*, suggesting highly limited transport into the brain, and/or high utilization by glial cells. Interestingly, the levels of pyruvate were very low ($<$0.2 m*M*), indicating that pyruvate is rapidly transported into cells and mitochondria when available, and is only a transient molecule in the extracellular space.

These results suggest an intricate interplay of metabolism between neurons and glia. Presumably neurons are the primary consumers of lactate, along with glucose for membrane pumping and other needs, whereas glia are net lactate producers, particularly because the glial citric acid cycle is primarily used for glutamine generation. The basic mechanisms of CNS metabolism appear in many ways to be radically different from those of systemic circulation, so it is not necessarily correct to borrow systemic concepts, such as, for example, lactate indicates anaerobic metabolism.[19,20]

13.5.4 Serum and CSF Markers

Blood tests that could signal impending intracranial hypertension or vasospasm would be very useful and innocuous to patients in intensive care settings. Panels of serum values currently under development may help intensive care specialists discuss prognosis or heighten concern regarding vasospasm. For example, S-100 is a cytosolic calcium-binding protein normally found in striated muscle, heart, kidney, astroglial, and Schwann cells. In adults, the levels of S-100β are elevated in multiple sclerosis, intracranial tumors, subarachnoid hemorrhage, and cerebral infarction.[21–25]

Similar correlations were recently validated for children. If blood was drawn within an hour of injury, a high level demonstrated 95% specificity and 86% sensitivity for predicting a poor outcome. In our own institution, McGirt et al. assessed serum markers that became elevated prior to clinical vasospasm. These markers must be better correlated with angiographic and transcranial Doppler data in a larger population. The preliminary data are promising.[26]

13.6 TREATMENT MODALITIES

Current innovations in actual treatment modalities are actually new investigations of old ideas. A resurgence of interest surrounds the use of hypothermia for intracranial pressure control and treatment of head injury and stroke and many metabolic support concepts are close to clinical trials. Most neuroprotectant agents have failed to win clinical approval, as outlined in Chapter 4.

13.6.1 Hypothermia in Intracranial Pressure Management

Induced hypothermia for control of ICP is currently under study as a neuroprotective tool after traumatic brain injury.[27–32] Multiple theories about its mechanism exist. Possible candidates include reduction of metabolic rate, reduction of increased ICP, decrease in cerebral edema formation, attenuation in the opening of the blood–brain barrier, inhibition of inflammatory response, and a decrease in the release of glutamate, nitric oxide, and free radicals associated with traumatic brain injury.[27–30]

Most studies revealed modest gains with little or no statistical improvement in outcome. Evidence for this modality is still lacking and indicates an increased risk of pneumonia for this treatment.[31] However, active studies continue because of the strong momentum. Many researchers feel that studies have been done inappropriately and that the treatment modality has clinical value. Most studies will take patients with a Glasgow coma scale (GCS) of eight or less within 24 hours of injury and cool to anywhere from 32 to 35°C. Patients are maintained at these temperatures for 2 to 7 days.[28–30] Rewarming procedures can also be harmful, so proper protocols must be elucidated. The most favored procedure is slow or passive rewarming instead of active rewarming.

13.6.2 Stroke Management

Stroke management in the ICU involves a combination of problems. Most strokes are ischemic. Treatment in the acute stage is based on prompt triage so that patients may be considered candidates for tissue plasminogen activator (tPA; see Chapter 12). However, the 3-hour window is very short and only a small percentage of eligible patients reach hospitals in the required time. Outside the 3-hour window, treatment involves hyperdynamic and ICP management for large strokes (discussed in other sections) and risk factor management.

Our university and others recently started using tPA for ischemic strokes presenting within a 3- to 6-hour window after the onset of symptoms.[33] The tPA is administered intra-arterially, specifically to the site of the clot. Direct administration allows higher doses, but still carries risk of intracranial hemorrhage. This use is not currently FDA-approved, and is discussed further in Chapter 12.

Another method of clearing cerebral arteries in the acute setting involves the use of a device called the concentric retriever. The retriever underwent a Phase I trial and has just begun a Phase II trial called "mechanical embolus removal in cerebral ischemia" (MERCI).[33] An interventional radiologist uses the device by deploying a self-expanding coil into an acute clot. The coil becomes entangled with the clot and is removed with the introduction catheter. The first trial involved seven centers across the United States. The device was used up to 8 hours after symptom onset. More than 50% of the patients had clot removed and half experienced good functional recoveries (see Chapter 12 for further discussion).

Hemorrhagic strokes constitute 15% of all strokes and usually have poor outcomes. Current treatment includes ICP management with intraventricular catheters and physical therapy. Few patients are candidates for clot removal. Ongoing trials are aimed at halting intracerebral hemorrhage as soon as it is diagnosed. Recombinant factor VIIa is administered in the emergency room for rapid hemostasis. This treatment is very early in the development phase. Efforts have also been made to remove clots medically due to their deleterious effects on the brain. Columbia University has begun a trial to place intraventricular catheters in addition to direct intraventricular thrombolytic therapy. The goal is to reduce the clot burden, thus decreasing the time that toxic blood elements are in contact with viable brain tissue.

13.6.3 Metabolic Enhancement

Several classes of agents can affect stroke or head injury. Since the glutamate hypothesis related to secondary damage following either of these events was developed, a large number of pharmacological agents have been tested for their neuroprotective capabilities. However, they have not shown efficacy, suggesting that perhaps neuroprotection is a somewhat wider area than glutamate alone (see Chapter 4). Another category of treatment in addition to tPA for vessel restoration, hypothermia, and neuroprotection, is metabolic enhancement. This concept involves getting more energy to the ischemic or hypoxic damaged brain through alternative sources other than glucose.

For example, intravenous pyruvate in high doses has been suggested for stroke treatment because pyruvate provides rapid uptake via monocarboxylate transporters and can be utilized immediately in mitochondria without conversion as long as oxygen is present.[34] Lactate supplementation has also been suggested, in addition to creatine, magnesium, nicotinamide and other natural substances. If these metabolic substrates and cofactors can reach an ischemic or hypoxic region, then perhaps neurons can be saved by the additional metabolic support.

13.7 CONCLUSIONS

Intensive care progress specific to neurosciences has developed as a comprehensive care scheme. Many of the ideas discussed are not necessarily new and were revisited recently in attempts to improve management of difficult neurological entities. ICP management is far from resolved. Perhaps treatment for these injuries will lie in replacement therapies after the damage has been done (see Chapters 2 and 3), but the goal of the intensive care specialist must remain to reverse or minimize neurological injury in the acute setting.

New research venues in these settings must be identified because the situation involves many invasive devices and catheters that may provide specimens from the human brain environment. However, considerable expense is involved, and further definition of patient candidates for neuroscience ICU treatment will be critical.

REFERENCES

1. Ullman, J., Cebrovascular pathophysiology and monitoring in the neurosurgical intensive care unit, in *Intensive Care in Neurosurgery,* Andrews, B.T. (Ed.), Thieme Medical, New York, pp. 29–43, 2003.
2. Kett-White, R. et al., Multi-modal monitoring of acute brain injury, *Adv. Tech. Standard Neurosurg.,* 27, 87–134, 2002.
3. Haitsama, I.K. and Maas, A.I., Advanced monitoring in the intensive care unit: brain tissue oxygen, *Curr. Opin. Crit. Care,* 8, 115–120, 2002.
4. Avezaat, C.J.J, van Eijndhoven, J.H.M., and Wyper, D.J., Cerebrospinal fluid pulse pressure and intracranial volume–pressure relationships, *J. Neurol. Neurosurg. Psychiatr.,* 42, 687–700, 1979.
5. Kiening, K.L., Schoening, W.N., Lanksch, W.R., and Unterberg, A.W., Intracranial compliance as a bed-side monitoring technique in severely head-injured patients, *Acta Neurochir.,* 81, 177–180, 2002.
6. Lin, E.S., Poon, W., Hutchinson, R.C., and Oh. T.E., Systems analysis applied to intracranial pressure waveforms and correlation with clinical status in head injured patients, *Br. J. Anaesth.,* 66, 476–482, 1991.
7. Portnoy, H.D. and Chopp, M., Cerebrospinal fluid pulse waveform analysis during hypercapnia and hypoxia, *Neurosurgery,* 9, 14–27, 1981.
8. Christensen, L. and Borgeson, S.E., Single pulse pressure wave analysis by fast Fourier transform, *Neurol. Res.,* 11, 197–200, 1986.
9. Piper, I.R. et al., Systems analysis of cerebrovascular study in head-injured patients, *J. Neurosurg.,* 73, 871–880, 1990.
10. Takizawa, H., Gabra-Sanders, T., and Miller, J.D., Changes in the cerebrospinal fluid pulse wave spectrum associated with raised intracranial pressure, *Neurosurgery,* 20, 355–361, 1987.
11. Paramore, C. and Turner, D.A., Relative hazard of ventriculostomy infection and replacement, *Acta Neurochir.,* 127, 79–84, 1994.
12. Smythe, P.R. and Samra, S.K., Monitors of cerebral oxygenation, *Anesthes. Clin. N. Am.,* 20, 293–313, 2002.
13. Hemphill, J.C. et al., Carbon dioxide reactivity and pressure autoregulation of brain tissue oxygen, *Neurosurgery,* 48, 377–383, 2001.

14. Dings, R. et al., Clinical experience with 118 brain tissue oxygen catheter probes, *Neurosurgery*, 43, 1082–1095, 1998.
15. Doppenberg, E.M., Zauner, A., Watson, J.C., and Bullock, R., Determination of the ischemic threshold for brain oxygen tension, *Acta Neurochir.,* 71, 166–169, 1998.
16. Scheufler, G. et al., Does tissue oxygen tension reliably reflect cerebral oxygen delivery and consumption? *Anesth. Analg*., 95, 1042–1048, 2002.
17. van den Brink, W.A. et al., Brain oxygen tension in severe head injury, *Neurosurgery,* 46, 868–876, 2000.
18. Sarrafzadeh, A.S. et al., Detection of secondary insults by brain tissue pO_2 and bedside microdialysis in severe head injury, *Acta Neurochir. Suppl.,* 81, 319–321, 2002.
19. Magnoni, S. et al., Lack of improvement in cerebral metabolism after hyperoxia in severe head injury, *J. Neurosurg.,* 98, 952–958, 2003.
20. Chih, C.P., Lipton, P., and Roberts, E.L., Do active cerebral neurons really use lactate rather than glucose? *Trends Neurosci.*, 24, 573–578, 2001.
21. Woertgen, C., Rothoerl, R.D., and Holzshuh, M., Comparison of serial S-100 and NSE serum measurements after severe head injury, *Acta Neurochir.,* 139, 1161–1165, 1997.
22. Romner, B., Ingebrigsten, T., and Kongstad, P., Traumatic brain damage: serum S-100 protein measurements related to neurological findings, *J. Neurotrauma,* 17, 641–646, 2000.
23. Rothoerl, R.D., Woertgen, C., and Holzschuh, M., S-100 serum levels after minor and major head injury, *J. Trauma,* 45, 765–767, 1998.
24. Woertgen, C., Rothoerl, R.D., and Metz, C., Comparison of clinical, radiologic, and serum marker as prognostic factors after severe head injury, *J. Trauma,* 47, 1126–1130, 1999.
25. Spinella, P.C. et al., S-100β protein-serum levels in healthy children and its association with outcome in pediatric traumatic brain injury, *Ped. Crit. Care Med.,* 31, 939–945, 1999.
26. McGirt, M. et al., Serum von Willebrand factor, matrix metalloproteinase-9, and vascular endothelial growth factor levels predict the onset of cerebral vasospasm after aneurysmal subarachnoid hemorrhage, *Neurosurgery,* 51, 1128–1134, 2002.
27. McIntyre, L.A. et al., Prolonged therapeutic hypothermia after traumatic brain injury in adults, *JAMA*, 289, 2992–2999, 2003.
28. Zhi, D., Zhang, S., and Lin, X., Study on therapeutic mechanism and clinical effect of mild hypothermia in patients with severe head injury, *Surg. Neurol.,* 59, 381–385, 2003.
29. Polderman, K.H. et al. Effects of therapeutic hypothermia on intracranial pressure and outcome in patients with severe head injury, *Intens. Care Med.,* 28, 1563–1573, 2002.
30. Hayashi, S. et al. Effect of early induction of hypothermia on severe head injury, *Acta Neurochir.,* S81, 83–84, 2002.
31. Gadkary, C.S., Alderson, P., and Signorini, D.F., Therapeutic hypothermia for head injury, *Cochrane Database of Systematic Reviews,* CD001048, 2002.
32. Tokutomi, T. et al., Optimal temperature for the management of severe traumatic brain injury: effect of hypothermia on intracranial pressure, systemic and intracranial hemodynamics and metabolism, *Neurosurgery,* 52, 102–112, 2003.
33. Gobin, Y.P. and Lavine, S., Expanding tPA's use in stroke treatment, *Presbyterian Neurosci.,* 7–10, August, 2003.

14 New Directions and Therapeutics in Surgical Spine Treatment

Dennis A. Turner and William J. Richardson

CONTENTS

0-8493-1482-8/05/$0.00+$1.50

14.1 INTRODUCTION

Spine therapeutics and surgery for spine diseases have undergone long development periods — spine surgeries, particularly laminectomies, have been performed for over a century. Hypotheses concerning spine diseases are now undergoing radical revamping based on new information about the pathogenesis of pain syndromes, basic biology of bone and disc tissue, cellular replacement therapy and stem cells, development of the potential for arthroplasty instead of joint fusion, and refining of knowledge about spine biomechanics and appropriate forms of stabilization.

Additionally, concepts about origins of mechanical and axial pain (as opposed to spinal cord or nerve or root-mediated pain) are continuing to evolve, as is the role of surgery in treatment of axial spine discomfort. Spine surgery developed primarily as an empiric set of treatments for both axial and neurogenic pain and deficit, and is now subjected to more hypothesis-based testing and rationalization of existing therapies for validity. Judgment questions about when to suggest surgery and how to adequately inform patients of possible treatment options continue to resist agreement among spine specialists, particularly because essentially all spine surgery is considered optional and appropriate or helpful only for symptomatic relief under limited clinical circumstances.

This chapter will review many of these new concepts in the treatment of spine diseases, particularly treatments involving surgical approaches. In addition to advances in understanding of spine biology and development of new procedures, judgment decisions about when to perform spine surgery will be reviewed, as will clinical trial needs and formats, particularly the differentiation of clinical trials for device approval versus trials for rationalization of therapy. Chapter 16 will discuss clinical trials further.

14.2 NEW CONCEPTS IN BIOLOGY OF NORMAL DISC TISSUE AND BONES OF THE SPINE

As magnetic resonance imaging (MRI) scans become more ubiquitous, new aspects of the natural history and evolution of spine appearance have evolved. The appearance of a "normal" spine involves well-hydrated disc joints with intact height that appear bright on T2-weighted scans,[1,2,3] cylindrical vertebral body appearance, a triangular-shaped spinal canal posterior to the vertebral body with sufficient room for spinal cord and peripheral nerve roots, and specific alignment in both the sagittal and coronal planes. The alignment includes a cervical lordosis, thoracic kyphosis and lumbar lordosis, and various indicators can be used to indicate appropriate overall alignment as well.

Significant medical and radiological knowledge has accumulated regarding development of the spine (for example, in abnormalities such as spinal schism, spina bifida, and myelomeningocele) and growth of the spine in childhood (growth plates, etc.). However, much of this knowledge continues to be challenged by newer hypotheses and clinical findings, particularly concerning the discs and facet joints and their relationships to clinical disorders of the spine. While MRI scans and radiographs show excellent detail of bones, joints, some ligaments, and nervous elements of the spine, in many cases little or no correlation exists between the findings and patient discomfort or neurological abnormalities.

14.2.1 Disc Cell Biology

Because most of the spine is developed from notochord, it is natural to consider what remnants of notochord persist through development.[2,3] Chordoma, the most prominent neoplasia associated with notochord, may develop at any time and usually at the midline of the vertebral and cranial axes as a remnant of notochord that has transformed into a fast-growing tumor. One current concept is that initially (until mid-childhood) remnant non-neoplastic notochordal cells remain within the nuclei of disc joints, and these cells contribute to disc hydration. However, after these notochordal cells are no longer present (presumably based on a developmentally regulated schedule), the discs may slowly begin to desiccate due to the lack of cells that enhance and promote hydration. This desiccation may lead to one of the prominently noted MRI features of the spine, namely that exuberant disc hydration (usually visualized on a T2-weighted image) decreases substantially with age. By the age of 40, many patients have highly desiccated discs that appear dark on MRI scans.

Interestingly, much is known about collagen, extracellular matrices, and other structural aspects of discs and how these elements are involved in a slow process of degeneration over a lifetime. Disc cells (including both notochordal cells and chondrocytes) prefer a three-dimensional matrix culture growth system instead of a flat two-dimensional culture system. They respond to both mechanical and osmotic shock in culture and *in vivo*.[2,3] Ongoing work suggests that replacement of notochordal cells (or some equivalent from stem cells) may in some ways be able to repair disc joints biologically.[4–11]

This repair may be limited by vascular changes in endplates with the development of arthritis and perhaps by more limited diffusion of metabolic substrates into disc nuclei as a function of age. Thus, degeneration involves changes in the cellular composition of discs, alterations in diffusion, blood supply to the endplate, collagen disruption, and dehydration, all of which eventually lead to the dark disc appearance on T2-weighted MRI scans and then to narrowing sufficient to give rise to typical osteoarthritic appearances of the endplates.

Any of these alterations could potentially be subject to treatment. Treatment of discs at most levels can be accomplished via percutaneous needle approaches that have been used for many years for various treatments such as chymopapain and percutaneous discectomy approaches.

14.2.2 Spine Bone Biology

For most fixation approaches to the spine (such as pedicle screws), the quality of bone may be the limiting feature affecting hardware attachment. For example, in osteoporosis, the pull-out strength of screws is markedly reduced, limiting the ability to provide adequate fixation of the spine in some instances. Many new medical treatment approaches are now used to enhance bone density in osteoporosis, as well as enhance bone healing in fusion (discussed next). One interesting quality of bone is the need for stress and pressure to augment healing. Presumably bone healing follows minute electrical gradients created by stress; small electrical currents can substantially enhance fusion rates.[12–14]

Many different types of bone fusion electrical stimulators (external to the body and internal near the bone graft) are commonly used because of this enhanced healing process. Likewise, if a hardware construct is too rigid, it is possible that a bone graft may not be subject to sufficient stress (because of shielding by the hardware) for adequate healing. New constructs that allow some slipping with time are touted as potentially increasing bone fusion rates due to persistent steady pressure placed on grafts, theoretically enhancing healing and graft incorporation.

Medical treatments to increase bone density and quality include daily or weekly dosing of bisphosphates. Future options are annual injectable dosing and parathyroid hormone (PTH). Both can stabilize bone density (but not necessarily increase it beyond the baseline) or restore it.

14.2.3 Biomechanics of Spine Joints

Abnormalities such as scoliosis, lack of maintenance of normal lordosis in the cervical and lumbar regions, and exaggerated kyphosis in the thoracic region have been known for years to change the biomechanical properties of the spine. If the upper torso is not centered over the spine and pelvis, the body may compensate with a secondary tilt at some level, to one side or in the anterior–posterior direction.

In recent years, sagittal balance has become recognized as important, particularly to maintain upright posture comfortably.[15] The previous use of fixation devices that led to distraction of the posterior elements of the spine rather than compression in the lumbar region produced the now-recognized flatback syndrome with straightening of the lumbar lordosis.[15] Guidelines now cover when to consider surgery for scoliosis and sagittal imbalance problems in terms of progression of abnormal curves over time and patient ability to compensate by using other joints. However, most adult cases of coronal or sagittal imbalance are primarily treated to relieve axial or extremity pain, rather than treating existing or threatened neurological impairments, leading to the elective (and optional) nature of most corrective deformity-related spine surgeries.

14.3 JOINT AND BONE DEGENERATION

The spine, like all other joints of the body, is subject to degenerative changes, usually termed osteoarthritis, degenerative joint disease (DJD), or spondylosis. These terms imply a slowly progressive set of changes of bones and joints, as the wear and tear of daily life (and also trauma) take their toll on joint surfaces. The literature describes a large number of changes in disc joints, particularly changes in the end plates and associated vasculature, alterations in cell population in the discs, decreased hydration of discs (leading to decreased disc space height and laxity in the annulus), and eventual near-complete loss of discs that causes end-plate surfaces to grate on each other.[16–18] Likewise, the facet joints (as synovial joints) undergo degenerative processes, possibly leading to loss of joint function that can facilitate the development of anterolisthesis or spondylolisthesis.[19]

Although these changes are well documented from both radiological and pathological examinations of spine joints, what remains unclear is how these changes may lead to various types of discomfort in certain individuals. Clearly, from a patient standpoint, a wide range of variation in expression of pain related to joint degeneration exists. In some cases, radiology studies demonstrate severe arthritic changes without accompanying pain or discomfort. Even mild arthritis in some individuals may be associated with seemingly severe axial pain. This discrepancy between subjective complaints and objective studies remains unexplained, and is not commonly accounted for in clinical studies of spine surgery performed in many cases primarily to relieve axial discomfort. This topic is further explored in the following sections on disc changes and potential etiologies of back pain.

14.3.1 Are Disc Changes Involution or Degeneration?

The concept of development inherently includes a series of genetically programmed changes in cells, leading to growth, proliferation, and maturity. As part of this programmed series, many features such as puberty can be delayed many years, as can maturity of several parts of the brain, particularly the frontal lobes. In this sense of genetic programming of organisms, development is usually considered a "normal" process. However, it is now clear that many aspects of later life, particularly limitation of cell division (for dividing cells) through telomere control, are also genetically programmed in much the same way.

As part of development, cell populations change, and programmed cell death is a critical aspect of maturing and eliminating unwanted and unneeded cells, particularly in the nervous system. Thus, development, maturation, and in many ways the entire life spans of most organisms involve genetically programmed sequences of intrinsic alterations in cell populations, their birth and death, and expression of patterns of proteins. Programmed senescence is commonly termed "involution." It represents a slow winding down of processes — the inverse of development.

Consider the hypothesis that the primary cells for disc maintenance may be remnant notochordal cells, particularly for hydration and extracellular matrices, and that the growth end plate is the crucial determinant of vascular supply and nutrient diffusion into the discs.[2] If these crucial cells are programmed to disappear at an early age (perhaps before age 10), the subsequent disc changes popularly termed "degeneration" are inevitable and preset, but may take years to unfold. Perhaps subsequent wear and tear or trauma may partially influence this process of programmed disc changes, but if the basic process is already genetically mapped, subsequent clinical influences may be minimal. The associated question is the degree to which these programmed changes really represent degeneration or are in fact normal processes akin to growing larger or developing typical frontal lobe properties, such as senses of timing, insight, long-term planning, and initiative that appear gradually over many years during childhood and early adulthood.

Regardless of whether disc changes are truly programmed or degenerative, many patients will want to reverse or treat the process, similar to the way they want to reverse many other aspects of human maturity and senescence. The drive to augment oneself and be normal appears to be inherently human and is universally accepted, as evidenced by the wide variety of medical and consumer products aimed at enhancing or maintaining function as we age.[20] However, mixed with this philosophy of enhancement, is the ambiguity as to whether disc changes produce discomfort or limit function. This ambiguity is highlighted by current studies that indicate disc arthroplasty does not necessarily lead to improvement in axial pain, similar to many clinical studies on disc fusion that argue for careful patient selection.[21–25] Thus, potential elements and structures within individuals and their spines that can lead to axial pain are still in dispute and often difficult to identify clinically, representing a considerable challenge for disc arthroplasty in general and its clinical role in particular.[8]

14.3.2 What Causes Back Pain?

A recent review by Lutz assessed hypotheses of the origin of axial spine pain over the past 100 years and the evolution of related concepts.[26,27] The main etiologies presupposed to lead to back pain included neural, muscular, osseous, disc-related, and psychogenic causes. The interesting finding was that at the turn of the 20th century, the most common causes presupposed in the medical literature were neural, muscular, and osseous. After the advent of discectomy in the 1930s, most attention has been directed to the disc joint as the main culprit. In this context, it is important to discriminate clear neurogenic pain due to nerve pressure or damage from axial pain. This distinction is not commonly made in the literature.

A practical rule to differentiate the two types of pain is to draw an imaginary parasagittal plane through the sacroiliac joint for the lumbar region, and similarly for the thoracic and cervical regions. Pain that is primarily medial to this plane may be considered as predominantly axial, and pain that is lateral to this plane (particularly in the sciatic notch and below) is usually predominantly neurogenic (related to the lumbosacral plexus or L5–S1 roots).

Can disc joints be responsible for axial or neurogenic pain? Clearly, many cases of herniated disc can lead to nerve pressure and hence cause radiating arm or leg pain that is adequately treated with discectomy. In many patients, axial pain continues after discectomy, so discectomy is not commonly considered a sufficient treatment for axial spine discomfort. However, some studies suggest that much axial pain can be related to other structures. For example, lumbar discectomy procedures have been done with local anesthesia so that the ability of individual structures to cause discomfort can be specifically addressed. In these procedures, significant back pain can be caused by pressure on or incision of the spinal fascia adjacent to the spinous processes, and some by muscle tension, whereas minimal discomfort is associated with removal of the lamina.

Interestingly, pressure on normal discs (those without previous injury or herniation) does not lead to discomfort; pressure on discs associated with damage or herniation can lead to leg pain (in the absence of nerve root pressure or touch). These now classic human experiments have led to the concept of radicular nerve ingrowth into the annulus after damage (from small collaterals as the ganglion passes the disc joint lateral to the facet), leading to a referred pain syndrome associated with disc pressure. This referred pain syndrome can simulate true nerve pressure, but without nerve pressure signs consisting of motor, reflex, and sensory alterations specific to that nerve root.

Additional studies have used hypertonic saline injections into the spinal fascia adjacent to the spinous processes, to initiate severe axial pain, similar to that associated with a typical episode of acute myofascial strain of the cervical or lumbar region. Likewise, local anesthetic blocks of trigger points within the fascia can at times relieve paraspinal discomfort, suggesting that ligamentous damage can certainly lead to severe pain. This is similar to the incisional pain near the midline associated with even small laminectomy incisions. Thus, from several types of studies, significant back pain appears to be associated with stretch, damage, irritation or contusion of spinal fascia and other soft-tissue mechanical structures, and less with posterior bony structures. Clearly, however, significant axial pain can be associated with vertebral body collapse (as in compression fractures) or with metastatic disease involving and expanding vertebral bodies. This discomfort can be relieved in some cases by vertebroplasty (injecting methacrylate cement into the vertebral body for stabilization)[28–30] or by radiation treatment of bones involved in metastatic disease.

Because axial pain can be clearly associated with neural, ligamentous, and osseous causes in some cases, what about discogenic discomfort? The standard diagnostic test for assessing discogenic pain is the discogram — placing a small needle laterally through the annulus into the nucleus and injecting a small amount of saline to expand the disc. Presumably this expansion then places tension on the annulus. Unfortunately, this test has proven to be highly nonspecific. Many patients (and volunteers) develop axial pain with the saline injection, making it difficult to differentiate degeneration-associated axial pain from spontaneous discomfort associated merely with the injection of a normal disc. Likewise, spine fusions (usually done anteriorly to avoid the incisional-related back pain of a posterior procedure) may or may not relieve axial neck or back pain, and outcomes vary widely and are highly subjective.[4,25] Presumably the advent of disc arthoplasty may clarify the role of disc degeneration in the treatment of axial pain but results of preliminary studies to date are mixed at best.[24,25]

Thus, the conclusion remains that all the major possible etiologies for axial pain can lead to significant axial pain under some circumstances including neural, ligamentous, soft tissue (including muscular), osseous, discogenic, and psychogenic causes. The clinical ability to differentiate these entities has been helped somewhat by the high sensitivity of MRI, but MRI also has a great lack of specificity. In many cases, asymptomatic abnormalities are treated as symptomatic because the two types of abnormalities are difficult to separate.

14.3.3 Differentiation of Axial and Extremity (Neurogenic) Pain Syndromes

If axial spine discomfort is hard to sort out and usually multifactorial, what about neurogenic causes of extremity pain? In most cases, radiculopathy due to nerve root pressure can be reliably determined clinically, and approximately 80 to 85% of extremity pain can be relieved by simple nerve root decompression. Common causes of nerve root compromise include herniated discs and lateral recess stenosis. However, many patients may report referred pain syndromes with extremity pain, but without clear localizing deficits; nerve root decompression is often not helpful.

Additionally, patients with long-standing nerve root pressure or those whose pressure has been spontaneously relieved (i.e., resolved herniated discs) may still experience radicular pain despite demonstrated nerve decompression. In these patients (at least 15% of the population), it is commonly presumed that internal or intrinsic nerve root changes or secondary alterations of the central nervous system (CNS) are responsible for the persistent pain. Subsequent treatments include medicines that function primarily at CNS level (e.g., amitryptiline or neurontin) and direct CNS stimulation.

In these cases, the classic neurosurgery hypothesis of mass effect (i.e., pressure on the affected nerve root) fails and an alternative hypothesis of internal damage or CNS conditioning and retention of the pain memory is required. The latter is clearly demonstrated by nerve root section proximal to the area of damage — an older technique that was very unreliable in relieving long-term radicular pain syndrome.

14.4 EXISTING SPINE OPERATIVE PROCEDURES

Operative procedures fall into two main groups: (1) those directed at treatment of extremity pain or involving central decompression of nervous elements, and (2) those directed at spine stabilization or treatment of axial pain. Examples of these will be discussed, primarily as the context to exploring new treatment options currently being assessed. Because of the nature of surgical procedures and the fact that U.S. Food and Drug Administration (FDA) approval applies only to drugs and devices, surgical approaches are not submitted for FDA approval; surgeons are generally free to experiment with surgical techniques.

14.4.1 Laminectomy and Anterolateral and Posterolateral Decompression

The spinal canal is roughly round or triangular on cross-section and bounded anteriorly by the vertebral body, laterally by the pedicles and facets, and posteriorly by the lamina and ligamentum flavum. Depending on the location of an offending lesion, these separate bony elements may be removed to gain access to the spinal canal. By trial and error, the least invasive of these approaches has proven to be laminectomy, which involves unilateral or bilateral (and often spinous process) removal from a posterior approach. While this procedure has the theoretical disadvantage of removing the posterior interspinous ligament (a tension band), the joints (anteriorly the disc joint, laterally the paired facet joints) appear to provide sufficient mechanical stability. The incidence of postlaminectomy instability varies from 5 to 15% in multiple series, depending on the reason for the laminectomy, position and angulation of the facets, and presence of degeneration, particularly of the facets. Laminectomy has been used for over 120 years to gain access to the spinal canal and remains one of the most common spine operations performed.

Other approaches require different access routes, such as lateral extracavitary, anterolateral (i.e., thoracotomy in the thoracic spine), or anterior. In these cases, more bone must usually be removed for access into the spinal canal. In some cases, the planned bone removal is clearly sufficient to warrant considering a stabilization procedure as part of the primary procedure, including placement of bone, hardware, or both to add stability to the spine. Such stabilization is usually done to maintain normal spine alignment rather than correct a spine deformity, and is usually labeled *in situ* fusion. In the future, arthroplasty of the involved joint may be performed, although facet joint replacements lag far behind anterior disc joint replacement prostheses.[8,19] All these procedures are commonly performed and well documented in contemporary spine texts, including details of the technique.

14.4.2 Midline versus Lateral Spinal Canal Syndromes

Spinal canal approaches include those primarily for access to nerve roots laterally or for access to the central aspect of the canal. Two primary syndromes of involvement are well recognized. The first syndrome is lateral impingement on a nerve root, which is likely to give rise to radicular pain and loss of function (motion and sensation). Extensive dermatome maps were compiled in the 1930s from root sections for relief of cancer-associated pain. These dermatome maps include common sensory patterns and motion loss associated with a root section, and can be extended to many forms of root involvement. Common mechanisms leading to lateral syndrome include herniated discs, lateral recess or foraminal stenosis, and nerve root tumors.

The second syndrome is central and consists of myelopathies in the cervical and thoracic regions (due to spinal cord involvement)[31–33] and multiple root pressure in the lumbar region (as is noted in neurogenic claudication).[34] Common reasons for spinal canal involvement include stenosis from degenerative changes, intramedullary (within the spinal cord) or extramedullary (outside the spinal cord) tumors, and internal conditions within the cord or nerve roots.

14.4.3 Mechanics and Principles of Spine Fusion

Spine fusion without instrumentation dates back to the 1940s, when healing fractures only with immobilization was insufficient. The period of immobilization was often 3 to 6 months, usually in an in-patient setting. Fusion was considered to augment healing. The advent of anterior cervical spine and anterior lumbar spine fusion procedures in the 1950s brought on an era of fusion of degenerated joints not previously subjected to trauma. That development led to fusion of degenerated joints both for neurological decompression and also to treat axial pain.[3,8,13,18]

For a time, posterior fusion alone was considered an alternative to posterior laminectomy and discectomy, based on the hypothesis that immobilization of the segment would eventually lead to improvement of the radiculopathy. As treatment of severe scoliosis and other spine deformities became common, it was realized that external bracing and noninstrumented fusion had severe limitations and relapses ensued. Therefore, Harrington and others developed instrumentation for spine fixation.

Between 1980 and 1990, a large number of new instrumentation devices to augment fusion arrived on the market. Interestingly, many were used without specific FDA approval for the spine indication — most of the hardware for posterior instrumentation, such as thoracic and lumbar pedicle screws and lateral mass cervical screws. These screws were adapted from long bone use, together with plates and rods with which to connect them. However, because of the policy of "grandfathering," many types of anterior instrumentation such as anterior cervical plates, and anterior thoracic and lumbar instrumentation became FDA-approved. More recently, lumbar pedicle screws have become approved for limited indications such as severe L5–S1 spondylolisthesis. However, for many years, the use of instrumentation for nonapproved uses was somewhat in limbo legally, even though it was commonly used clinically.

The general principle of instrumented fusion is to provide internal support for vertebrae, thus improving fusion rate and providing stability while healing occurs, or correcting a deformity so a fusion occurs in a better anatomic position. For example, noninstrumented lumbar lateral fusion (intertransverse process) appears to show a radiographic fusion rate of 65%; the same fusion approach with pedicle screws shows a 95% radiographic fusion rate.[13,35] Pedicle screws also demonstrate another principle, that of 3-column support, since the screws traverse the posterior elements into both posterior and anterior aspects of the body. This support, depending on bone quality and degree of fixation, offers considerable initial rigidity. However, another principle is that if the bony fusion fails to fully develop, the hardware will eventually loosen and/or break. Thus, assessing hardware stability can determine whether or not fusion has occurred. In addition to the hardware implanted in the spine, an external orthosis such as a cervical collar, lumbar corset, or brace may be used for additional support. Bone stimulators also may augment fusion, either as implantable or externally applied devices.[12,14] Since bone growth follows lines of stress and secondarily generated electrical force lines, the addition of an artificial electrical field, even of small magnitude, appears to augment bone growth. These general principles will be elaborated further in the sections on novel treatment.

Current spine instrumentation measures approved by the FDA include the anterior cervical plate, posterior Harrington and other devices for deformity correction, anterior thoracolumbar plates and devices (Z-plate and Kaneda instrumentation), and pedicle screws for spondylolisthesis at L5–S1. Additionally, a variety of horizontal and vertical cages are available for implantation in the disc space.[36] All these devices achieve immobilization to augment fusion instead of joint replacement devices to augment motion.

Because these devices by nature provide rigid constructs with immobilization, the inevitable force usually transferred through joints (now removed or immobilized) must be transferred to adjacent segments. Adjacent segment stenosis may affect all spine segments, in which potentially degenerative changes may progress faster next to immobilized segments because of the need to shoulder more stress of motion. Theoretically, spine arthoplasty should alleviate this accelerated degenerative change through augmentation of motion.[6,8]

14.4.4 Kyphoplasty and Vertebroplasty

Interventional procedures are available to treat collapsed and painful vertebral bodies. Common indications include osteoporotic compression fractures associated with severe axial pain and metastatic cancer involvement of the vertebral body.[28–30] Vertebroplasty involves bilateral injection of methacrylate into vertebral bodies via a transpedicular percutaneous approach. Kyphoplasty requires placement of a balloon into a vertebral body for restoration of height, then insertion of methacrylate to maintain height. Both procedures are usually done under fluoroscopic control.

Complications include transfer of emboli of methacrylate into the lungs and posterior extrusion of methacrylate into the spinal canal if a posterior vertebral body defect is present.[29] These procedures are commonly performed to treat subjective indications of severe axial pain, usually a few weeks after a fracture to allow the initial pain to resolve, but less than 6 months after the fracture. A possible (and fortunately very rare) complication is infection in the methacrylate that requires removal of the entire body, presumably through drilling away the dense plastic. While the approaches are well established, few controlled studies have focused on their clinical value and usefulness. Newer substances for injection such as hydroxyapatite cement and bone morphogenetic protein (BMP) are future directions for these techniques.[30]

14.4.5 Outmoded FDA-Approved Procedures

A large number of outmoded approaches and devices, often with FDA-approved indications, are no longer routinely performed. Examples include chymopapain, most of the percutaneous discectomy approaches (such as laser discectomy), and many types of instrumentation including various cages. Once a device is approved for use by the FDA, its manufacturer rarely stops production except in cases of significant safety concerns. With many of these approaches such as percutaneous discectomy, only safety studies were performed. Few or no efficacy studies were required (see Chapter 1).

The efficacy studies were all performed after introduction to the market. After 2 to 3 years of use and study, the approaches were by and large abandoned. Thus, FDA approval is only the initial step in gaining true market approval of a device or approach because FDA approval studies are performed usually by a manufacturer or clinical enthusiast and are usually poorly controlled. After FDA approval is obtained, the clinical skeptics can perform independent studies if a clinical need exists. In many instances, these secondary studies are marginal at best in terms of demonstrating efficacy.

Another reason devices become outmoded is replacement by better ones. Several generations of anterior cervical plates are now available. The newer versions offer easier use, locking mechanisms to prevent screw back-out, lower profiles, and assorted sizes — often at increased cost. The newer versions clearly eclipse many older versions, including the original Caspar plate that was difficult to apply in practice and led to many complications.

14.5 NEW SURGICAL PROCEDURES IN DEVELOPMENT

Evolution of surgical procedures is natural, and in general such transformations do not require FDA approval (only devices or instruments do). In addition, many new procedures have been modified or enhanced by the availability of novel devices and instrumentation. Recent introductions to the market include frameless, fluoroscopically guided computer systems to aid hardware placement. Examples include the Stealth Medtronic, Fluoro-Nav, General Electric and InstaTrak systems.

Since these devices are surgeons' aids and not therapeutic devices in their own right, they require a different category of FDA approval. The evolution of devices for stabilization has been constant. Devices on the market now include a full series of occipital-to-cervical fusion sets, cervical plates, adjustable screws, anterior plates, improved thoracic and lumbar posterior instrumentation, and a large variety of horizontal and vertical cages. Such devices will not be further detailed, but there is now an evolution from fixation and fusion hardware to arthroplasty for maintenance of motion.[1,8,17] Interestingly, the spine is one of the last bastions of fusion treatment of joint malfunction because decreased motion at one or a few segments usually does not detract significantly from overall spine function.

Several orthopedic concerns have transformed bone and joint surgery: advanced knowledge of joints including degeneration and replacement, bone and fusion biology and enhancement of both, and outcome measures to estimate whether surgical procedures are worthwhile. The surgical procedures and advances introduced earlier will be discussed in sequence.

14.5.1 Improvements and Rationalization of Laminectomy

Technical aspects of laminectomy have been debated for many years. One of the current discussions centers on what may be considered a minimally invasive laminectomy. One suggestion is a bilateral hemilaminectomy for lumbar stenosis, sparing the spinous process and interspinous ligament, but resecting part of each facet to achieve bilateral lumbar decompression. This is in contrast to a facet-sparing laminectomy, where the spinous process and medial aspect of the lamina are resected, but the facets are only undermined rather than trimmed.

This is one example of the small, technical changes that are constantly debated, although the outcome from overall surgery may still require further studies to fully define. As is common in many surgical disciplines, this differential approach lies in the category of surgeon preference, rather than relying on substantive biomechanical or clinical outcome studies to support one approach or another.

Most laminectomy procedures are performed to treat degenerative changes such as cervical or lumbar stenosis to achieve root or central (spinal cord or cauda equina) decompression. In general, laminectomies are elective procedures with usually subjective endpoints. An excellent example is decompression for cervical stenosis associated with cervical myelopathy. Treatment of this disorder has a long history. Over time, the recognition of the syndrome and MRI diagnosis have consistently improved, but no natural history studies have accompanied the improvements.[31–33]

Earlier studies in patients with severe myelopathies indicate progressive worsening over time, but the degree of worsening and the influence of surgery on the disorder remain poorly characterized. In addition, many surgeons now suggest decompression for asymptomatic cervical stenosis without clinical signs suggestive of myelopathy. As in the case of carotid stenosis and associated symptoms of either transient ischemic attacks or stroke, a clear delineation is required to assess how patients fare without surgery and determine the impacts of surgical decompression over the short and long runs, especially compared to the risks of surgery.

Preliminary randomized studies suggest that, at least in milder myelopathy patients, surgery produces no net benefit.[37] Although practitioners agree on the need for a full randomized study of all forms of cervical myelopathy,[38] the format and mechanisms of such a study remain highly contentious. This is one area where rationalization of the need and outcome for surgery is important to obtain. However, market forces and lack of collaboration among surgeons tend to forestall such a common multicenter action.

There is a perceived need among surgeons for increased information about outcomes from surgery and what types of surgery to perform. For example, when should a fusion be added to a laminectomy? Additional questions relate to improved patient education: which symptoms are really treatable with surgery and how can we optimally advise patients on surgical outcomes and risks?[39]

14.5.2 New Developments in Fusion Technology

The outcomes of surgical fusions have dramatically improved with advanced hardware over the past 20 years. Further improvements are expected with the use of bone-specific growth factors, particularly BMPs.[35,40,41] Most of the large number of growth factors are involved with promoting bone growth and development. Currently approved forms of BMPs involve rigid applications such as within cages to avoid dispersion to tissues where bone growth is not wanted.[42] New formulations for posterior lateral (intertransverse process) fusions to replace the onerous iliac crest autograft placed on the transverse processes to enhance fusion are in clinical trials.

Because one of the issues with instrumented fusions is rigidity, and hence the tendency to transfer stress to adjacent segments, one consideration is the return to noninstrumented fusions with enhanced rates of fusion augmented by BMPs. Other new techniques include gene therapy for transfection of BMPs and other growth factors, again with the primary goal of enhancing fusion.[41,43] Such approaches do not restore motion unless there is consideration of joint enhancement, such as with stem cell replacement of chondrocytic joint surfaces or replacement of notochordal stem cells for disc replacement.[1,7,8] Clearly, a large number of possibilities exist because mesenchymal stem cells are readily available and likely easier to induce to specific fates than neural stem cells due to the markedly fewer fates possible.[10]

14.5.3 Cervical Arthroplasty

The primary consideration for joint replacement has focused on disc joints because they are more accessible and larger than the facet joints, which are synovial. Synovial joint replacements are available for small joints, such as finger joints in rheumatoid arthritis. They have also been considered for additional replacement of a disc joint at a single segment and also the two facet joints. The Frenchay artificial cervical joint has now been in use in clinical trials for several years, but has not yet received widespread approval.[44]

As with many prototypes, certain issues must be resolved: anchoring the artificial joint within adjacent bodies and long-term functioning of the joint, but it is expected that further prototypes will be developed. Ideally, an artificial joint would be used with every anterior cervical procedure, instead of allografts or autografts, to preserve motion segments and prevent or forestall the long-term progression of degenerative changes at adjacent levels.

14.5.4 Lumbar Disc Arthroplasty

Lumbar disc joint replacement likewise is primarily under consideration for early disc degeneration — a condition treated presently with anterior lumbar fusion.[25] A disc replacement should be as easy to insert as an anterior cage. It should tightly hold into adjacent endplates and be able to withstand considerable wear and tear over time due to the large stresses imposed on lumbar disc joints. Current prototypes are not yet ready for widespread clinical use and they need revisions and better clinical trials to achieve full clinical integration.[24,25]

14.5.5 Stabilization without Fusion

To provide stability while avoiding the complications of fusion surgery, a pedicle screw system with an elastic synthetic compound is used to control motion. Early studies are promising, but long-term results are not yet available. One indication for this type of nonfusion fixation is segmental instability with stenosis.[45]

14.6 JUDGMENT DECISIONS AND PATIENT-INFORMED CONSENT

Because most spine surgery falls into the elective and optional category of medical treatment, considerable judgment about when to propose a procedure is required on the part of a surgeon. Judgment and sufficient understanding are also required of a patient so that he or she can truly give informed consent. Clearly, different surgeons have different indications as to when they propose various procedures. Wennberg initially demonstrated the wide variation that exists even on a small scale by performing a regional analysis of the rates of spine surgery.[46]

Subsequently studies of large Medicare regions throughout the U.S. showed less variation due to averaging across multiple practitioners within each region.

It is clear that some surgeons may consider a procedure worthwhile if 90% of the patients to whom it is offered may benefit; another surgeon may view a 60% likelihood of improvement as an adequate measure. Although most surgeons assess such decision making on a patient-by-patient basis, the preference can persist over many patients to create an average. One factor in such decision making relates to patients who do not improve with operative intervention. The larger the percentage in this group, the more a surgeon will be challenged by poor results and over time will likely adjust his or her judgment. However, the origins of judgment tend to be based partly on training and experience, so such adjustment may not necessarily occur.

14.6.1 What Spine Problems Are Susceptible to Operative Procedures?

Spine surgery is clearly very common in the U.S., particularly compared to the U.K., suggesting that different standards and indications are applied in the two regions. Thus, the question arises: is spine surgery in the U.S. suggested too often and for less than reasonable indications? Since most surgery is for relief of pain (clearly a subjective outcome), does the frequency also reflect the tendency in the U.S. for a desire among patients to be completely pain-free? These are thorny ethical issues that many spine surgeons in the U.S. find uncomfortable to contemplate, although they relate to other forms of optional medical care such as plastic surgery or enhancement of bodily functions.[20]

14.6.2 Appropriate Patient Information and Basis for Informed Consent

Patients commonly have misconceptions regarding spine surgery and these misconceptions are rarely discussed. One approach has been to develop a patient video that provides both information and describes surgery from a patient's perspective.[39] Two common misconceptions were identified. The first was that "the MRI scan will tell the doctor what is wrong with me." Patients do not appear to recognize the high number of asymptomatic abnormalities and may not understand that clinical interpretation of MRI scans requires integrating patient symptoms with MRI findings. The second misconception was that "surgery is 98% effective for back pain." Thus, patients commonly perceive that MRI scans are critical. In reality, for most evaluation purposes, an MRI scan represents one piece of clinical information that may or may not be concordant with the patient's symptoms.

These results highlight the nearly ubiquitous use of MRI scans for assessing spine abnormalities due to common availability and a low degree of invasiveness. However, from a surgeon's perspective, it is common now for patients to treat MRI findings as their primary concerns instead of subjective symptoms such as back and leg pain. In other words, considerable patient concern surrounds even asymptomatic findings on MRI scans, particularly osteoarthritis and age-appropriate degenerative changes.

The video presentation developed to educate patients involves an unbiased explanation as to which symptoms actually improve with surgery and the realistic likelihood of improvement.[39] Because spine surgery is optional and best considered helpful for specific problems, this likelihood can be estimated for each predominant symptom such as radicular pain or axial pain, depending on the specific surgery. The likelihood of worsening (risk) can also be explained. An excellent aspect of the video presents various patients' perspectives indicating whether their symptoms were sufficiently bothersome to make surgery worthwhile. Contrasting views are presented. In the example of lumbar stenosis with walking-related pain, the ability to walk only a limited distance without pain may be a severe disability for one patient while another may not be bothered by this limitation. Presenting opposite patient perspectives regarding interest in elective surgery for the same basic problem allows a patient to understand that (1) surgery is not necessary in an absolute sense but can be helpful for certain problems and (2) surgery involves a certain rate of helping and a certain rate of worsening from unexpected problems. Thus, a balanced approach to making a surgical decision includes placing a patient's symptoms (and underlying condition) into perspective. A patient should understand that the surgical treatment is not perfect but can be helpful, depending on the symptoms and the patient's underlying concerns regarding the procedure.

The critical knowledge needed for an individual to make a good decision regarding surgery varies considerably from patient to patient and a surgeon's recommendation may weigh heavily in the decision. The basic principles upon which the surgeon must touch to obtain a fully informed consent include describing patient-specific outcomes in specific terms the patient can fully understand, the salient risks of the surgery (in basic categories, using the concept that certain individuals may be affected if a problem arises), the lack of guarantee (except that the surgeon, of course, will try to avoid problems), and the need for recovery time.

Included in this discussion should be the patient's perception about the surgery (some patients are very worried about "going under the knife") and the underlying disorder. These are all complex factors that feed into the decision. One critical feature of the decision to elect or decline surgery that should be emphasized to patients is that surgery is irreversible, so both the patient and surgeon must be committed to caring for any postoperative problems. Likewise, the possibility that a positive benefit from surgery may or may not occur, depending on the outcome identified preoperatively, is a difficult principle to explain to patients who always expect to be fully relieved of their problems.

Figure 14.1 illustrates facets of the problem of a lumbar herniated disc. The patient's perspective involves two main concerns: the resulting leg pain and whether the underlying disc problem (regardless of pain) could lead to future back problems (i.e., is the problem serious?). From the patient's perspective of surgery, residual back pain (not considered to be helped by the procedure) may be the most troubling problem, and far outweigh the relief of leg pain at a later date. From the surgeon's perspective, a lumbar discectomy is considered an excellent procedure, but caveats are usually relayed to the patient in terms of relative relief of back versus leg pain. From society's perspective, the surgery may or may not be worth performing, depending on whether the patient promptly returns to work and whether his or her

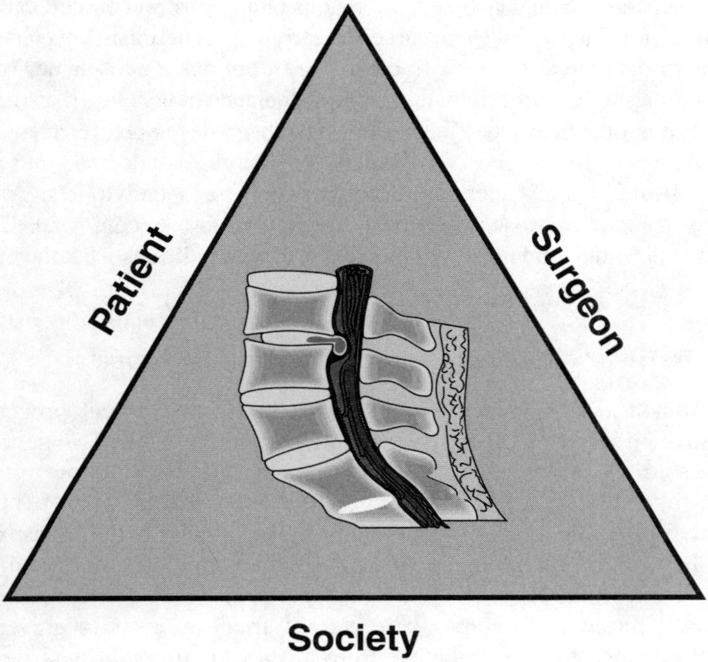

FIGURE 14.1 (See color insert following page 146.) Multiple perspectives on lumbar disc herniation. The center component is one of the most common spine problems: lumbar disc herniation. There are three different perspectives on the disease and surgical treatment, depending upon the vantage point. The patient, the surgeon, and society all have different views of definition of outcome and whether a surgical procedure is worth undertaking.

care will consume exorbitant resources. What is clear is that the same medical problem and the same surgical procedure are valued very differently by patients, surgeons, and society.

14.6.3 When Is a Surgical Procedure "Necessary?"

Patients often ask whether it is "necessary" to have surgery when it is suggested as an option. "Necessary" is defined medically as clinically indicated and implies serious consequences if the surgery is not performed. Examples of clearly "necessary" surgery include life-saving procedures to treat acute epidural hematoma with deteriorating mental status or performing a carotid endarterectomy with clear neurological symptoms and severe carotid stenosis. In both situations, the clinical condition is obvious or significant medical evidence has accumulated. Hence, the surgical procedure may be highly recommended and in some cases, life-saving.

Many conditions and diseases have obvious consequences if surgery is not done, for example, a severe infection that may worsen. Most spine surgery does not fall into the "necessary" category based on obvious clinical criteria or medical evidence. The context into which most spine surgery clearly falls is optional, but clearly helpful under appropriate circumstances. In other words, options are available. Whether or not to opt for a surgical procedure depends on a balance of benefits, risks, and patient opinion. From a medical view, indications have been developed for medical conditions in which surgery may be appropriately considered an option and thus medically indicated. In this context, cervical decompression for spondylotic myelopathy is cited as a standard approach that most surgeons usually recommend. Even then, many patients demur, and no strong medical evidence indicates what the best course of action may be.

14.6.4 Is There a Role for Prophylactic Spine Procedures?

A high standard of evidence is required to perform surgical procedures for preventative purposes on otherwise asymptomatic individuals. Two neurosurgical conditions are examples where the evidence strongly points to performing procedures, the purpose of which is only to treat potential future (but likely) occurrences rather than make the patient symptomatically better. These examples are clip ligation of symptomatic (previous subarachnoid hemorrhage) cerebral berry aneurysms to prevent rerupture (and possibly worse consequences such as death) and treatment of symptomatic carotid artery stenosis to prevent stroke. Studies performed to show whether to treat these two conditions (berry aneurysms and carotid stenosis) in the absence of premonitory symptoms have been problematic.[47,48]

The major factor determining whether the low rate of conversion of asymptomatic to symptomatic status is worth a procedure is the actual risk of the procedure. Thus, when the risk that a patient will become symptomatic is low and a procedure has tangible risks, considerable long-term data on the natural history of the disease and the impact of the disease on health status are required to make a convincing case for preventative surgery.

Asymptomatic cervical spinal stenosis and cervical spondylotic (compression) myelopathy with stenosis (symptomatic stenosis) provide an interesting contrast. Many patients have MRI scans that show degrees of spinal cord compression or deformity due to cervical degenerative (spondylotic) stenosis without clear myelopathic symptoms. In contrast, some patients who have abnormal MRI scans showing both stenosis and abnormal signals within the spinal cord usually demonstrate some forms of spinal cord-related symptoms. While we can argue for offering surgical decompression to patients with symptoms (to stabilize their symptoms), can the argument be extended to prevent an unknown event (development of cervical myelopathy) from occurring?

No natural history data are available on the progression of cervical stenosis, the likelihood of developing myelopathy, or the determination whether patients fare better than this natural history after undergoing cervical decompression.[37,38] As in the case of asymptomatic carotid artery stenosis, a large randomized, follow-up study with a clear natural history arm would be required to delineate whether any net benefit would accrue from an aggressive (presymptomatic) surgical approach to cervical stenosis. Because cervical stenosis and cervical myelopathy are common problems, such a study is feasible if interest among surgeons and a commitment for long-term follow-up to assess outcomes in patients are present.

14.7 WHEN SHOULD CLINICAL TRIALS BE PERFORMED AND BY WHOM?

Clinical trials in general and randomized clinical trials in particular remain very controversial in spine surgery, except for small studies evaluating new technology.[49,50] Because spine surgery is very common, it produces a large impact on society in terms of costs and standard issues arise as to how to effectively perform unblinded trials of surgical treatments. Although other trial formats are possible (open label or cohort trials), randomized trials still offer the most efficient route (in terms of patients and money) to answer clinical questions provided sufficient interest and funding are available. Chapter 15 gives further information on possible clinical trial formats for making decisions about surgical procedures and outcome measures.

Although objective outcome measures are known for many spine syndromes, in most cases spine surgery is performed for subjective relief of pain and quality-of-life measures are critical to determine whether a patient perceives the same benefit from the procedure as do surgeons (Figure 14.1). Thus, careful selection of objective, measurable outcome measures and issues related to quality of life (i.e., SF-36 [36 questions to assess general health status], etc.) are important for assessing the impact of a surgical procedure on a patient's life.

14.7.1 Clinical Trial Formats

Randomized, effectively powered pivotal studies are usually not required for FDA approval of devices. Most devices achieve approval in other ways such as grandfathering and relating to older devices (see Chapter 1). The clinical trials performed for device approval are usually performed by "enthusiasts," who have often been involved with development of the device and are interested in bringing it to market. Because enthusiasts believe strongly that their device is both efficacious and worth marketing (otherwise why would they do the work?), they are biased because of this outlook and also by substantial financial incentives. FDA approval studies ideally should be performed by clinicians who have no stakes in the outcome.

After a device is FDA-approved, secondary studies are commonly performed by "skeptics" who have no stakes in the results to assess more critically the worth of a device in practice. In many cases, devices become outmoded because of these secondary studies and overall lack of market approval. For example, percutaneous discectomy procedures have by and large been abandoned because of lack of efficacy. Thus, should efficacy and safety be determined by external, critical randomized trials before new hardware becomes FDA-approved?

14.7.2 FDA Approval Requirements

The FDA regulates marketing of devices and drugs, for considerations of safety and to assess manufacturing standards, to ensure that optimal practices are used in the initial testing and production. The FDA is particularly careful about surgical implants because of their permanent nature. All spinal instrumentation components are characterized as significant risk devices. The typical steps in device approval are that a company obtains an investigational device exemption (IDE) for initial clinical testing, which certifies that both appropriate animal studies and manufacturing requirements have been met (see Chapter 1).

After initial feasibility clinical studies, pivotal safety and efficacy studies are performed for final approval, unless a device is substantially equivalent to one already on the market. It is important to remember that the FDA does not regulate surgeons; it only regulates manufacturers and marketing. It can require labeling of any device or drug to indicate FDA-approved uses, however. It is important to explain to patients that nonapproved uses are not necessarily experimental and that it is up to the individual physician or surgeon to decide what is appropriate.

For many years pedicle screws (and other posterior instruments such as cervical screws) used for spine fixation (for fusion augmentation) were not FDA-approved for that purpose. The final consensus was to recommend them, indicating to patients that in the surgeon's experience such devices (such as pedicle screws) were optimal for the condition even though they were not specifically approved by the FDA for such use. "Off-label" uses of drugs or devices by individual physicians or surgeons are common and accepted, but a practitioner should have a good rationale for such use if questioned.

14.7.3 Enthusiasts and Skeptics

Improvement in treatment is a natural extension of patient care and the perception of less-than-optimal treatment schemes in current practice (or less-than-optimal outcomes in patients) is a strong force for developing better or newer approaches to treatment. Such alternative approaches could be new surgical methods, new devices, or new treatment protocols.

Generally, one individual or a group of clinical investigators has sufficient energy or enthusiasm for development of a new concept in spine surgery, and tends to be the force behind development, clinical testing, and often FDA approval of a new device. However, in the process of developing a new approach or device, the enthusiasm is high because the new approach is better than conventional treatment (or else it would not be worth developing). This investigator enthusiasm could also extend to a new device from which the investigator may receive royalties or other financial compensation for development and eventual use. Clearly, the enthusiastic developer does not necessarily maintain an objective viewpoint on the overall worth of a new approach or device. However, in spite of this bias toward presumed usefulness or efficacy, the enthusiast is usually responsible for the FDA approval process in terms of providing data sufficient to ensure approval.

Thus, by the time the FDA has approved a new device (as opposed to a new drug), only one clinical trial may have been performed by the developer or enthusiast who has considerable subjective bias in favor of the device and who focuses more on safety than efficacy. It is then commonly left to the market to sort out (after FDA approval) the value of the new device. Usually at the time of product introduction, a consortium of independent investigators (skeptics) has begun to organize, and often they do not share the blind acceptance of the worth of the new introduction to the market. In many cases, clinical trials to assess the true worth of the approach or device will then be conducted by the skeptics, often leading to abandonment of the product (despite the existence of FDA approval). This has happened many times. For example, percutaneous discectomy approaches (such as chymopapain) are still FDA-approved and available, but are rarely used because of a lack of demonstrated efficacy or unacceptable side effects.

As noted, FDA approval data are usually insufficient to satisfy a skeptic's thirst for both efficacy and safety information, although usually preliminary safety information may be known. One consideration may be to have skeptics (those not involved with the development of an approach or device) available to assess efficacy and safety prior to FDA approval. Another suggestion is for FDA approval to require scientific data beyond those provided by the company or enthusiasts, particularly a controlled trial performed by independent investigators.

14.8 CONCLUSIONS

There are many significant areas of conflict related to advances in spine therapy. For example, if arthroplasty were to become more widespread, interest in fusion enhancement might drop rapidly because of a decreased need for fusion procedures. Although many new devices are constantly in development, the actual value in practice of these new approaches and devices remains unknown, particularly in relation to existing therapies.[51]

In many cases, the enthusiasm for something new (and the equation of this enthusiasm to an improved outcome) can overbalance a rational approach to deciding value. Significant judgment questions remain about spine surgery in the United States, particularly procedures performed for treatment of axial pain syndromes. The answers in terms of how to decide the location and type of pain generator for subsequent surgical attack remain murky. Thus, spine surgery is at a very active point in its history. Because technology constantly provides new equipment and approaches, rational decision making based on the efficacy of new technologies is almost completely lacking.

Due to the rapid flux of devices, techniques, and surgical approaches, rationally designed clinical trials often are too cumbersome and cannot keep pace with technology development because the time requirements for initiating and concluding studies may range from 5 to 7 years. If a surgical technique or device under study is a "fad" and becomes rapidly outmoded, any trial is of historical interest only. As a result of this concern, few trials are performed and practitioners have scant data upon which to build clinical decisions for patients. Improved premarketing clinical trials may limit device development, but could bring more order into the highly complex clinical environment of spine surgery.

REFERENCES

1. An, H.S., Thonar, E.J., and Masuda, K., Biological repair of intervertebral disc, *Spine*, 28, 15, S86–S92, 2003.
2. Gruber, H.E. and Hanley, E.N., Recent advances in disc cell biology, *Spine*, 28, 186–193, 2003.
3. Martin, M.D., Boxell, C.M., and Malone, D.G., Pathophysiology of lumbar disc degeneration: a review of the literature, *Neurosurg. Focus*, 13, 1–6, 2002.
4. Alini, M. et al., A biological approach to treating disc degeneration: not for today, but maybe for tomorrow, *Eur. Spine J.*, 11, S215–S220, 2002.
5. Bao, Q.B. and Yuan, H.A., Nucleus replacement: new technologies, *Spine*, 27, 1245–1247, 2002.
6. de Kleuver, M., Oner, F.C., and Jacobs, W.C.H., Disc replacement for chronic low back pain: background and a review of the literature, *Eur. Spine J.*, 12, 108–116, 2003.
7. Ganey, T.M. and Meisel, H.J., A potential role for cell-based therapeutics in the treatment of intervertebral disc herniation, *Eur. Spine J.*, 11, S206–S214, 2002.
8. Gunzburg, R. et al., Arthoplasty of the spine: the long quest for mobility, *Eur. Spine J.*, S63–S64, 2002.
9. Phillips, F.M. et al., Biological treatment for intervertebral disc degeneration, *Spine*, 28, S99, 2003.
10. Roughley, P.J. et al., The potential and limitations of a cell-seeded collagen/hyaluronan scaffold to engineer an intervertebral disc-like matrix, *Spine*, 28, 446–454, 2003.
11. Wallach, C.J., Gilbertson, L.G., and Kang, J.D., Gene therapy applications for intervertebral disc degeneration, *Spine*, 28, S93–S98, 2003.
12. Kucharzyk, D.W., A controlled prospective outcome study of implantable electrical stimulation with spinal instrumentation in a high-risk spinal fusion population, *Spine*, 24, 465–468, 1999.
13. Pilitsis, J.G., Lucas, D.R., and Rengachary, S.R., Bone healing and spinal fusion, *Neurosurg. Focus*, 13, 1–6, 2002.
14. Yuan, H.A. et al., A double-blind study of capacitively coupled electrical stimulation as an adjunct to lumbar spinal fusions, *Spine*, 24, 1349–1356, 1999.
15. Danisa, O., Turner, D.A., and Richardson, W.J., Posterior reductive ("egg-shell") osteotomy and fusion for reduction of kyphotic posture of the lumbar spine, *J. Neurosurg. (Spine),* 1, 50–56, 2000.
16. An, H., Emerging techniques for treatment of degenerative lumbar disc disease, *Spine*, 28, S24–S25, 2003.
17. Guyer, R.D. and Ohnmeiss, D.D., Intervertebral disc prostheses, *Spine*, 28, S15–S23, 2003.
18. Szpalski, M., Gunzburg, R., and Mayer, M., Spine arthroplasty: a historical review, *Eur. Spine J.*, 11, S65–S84, 2002.
19. Berven, S. et al., The lumbar facet joints: a role in the pathogenesis of spinal pain syndromes and degenerative spondylolisthesis, *Sem. Neurol.*, 22, 187–196, 2002.
20. Caplan, A.L, Is better best? A noted ethicist argues in favor of brain enhancement, *Sci. Am.*, 289, 104–105, 2003.
21. Bao, Q.B., The artificial disc: theory, design and materials, *Biomaterials*, 17, 1157–1167, 1996.
22. Bertagnoli, R. and Kumar, S., Indications for full prosthetic disc arthroplasty, *Eur. Spine J.*, 11, S131–S136, 2002.
23. Itoh, S., Development of an artificial vertebral body using a novel biomaterial, *Biomaterials*, 23, 3919–3926, 2002.

24. Verhaegen, M.J. et al., Artificial disc replacement with the modular type SB Charite III: 2-year results in 50 prospectively studied patients, *Eur. Spine J.*, 8, 210–217, 1999.
25. Zigler, J.E. et al., Lumbar spine arthroplasty: early results using the ProDisc II: a prospective randomized trial of arthroplasty versus fusion, *J. Spinal Dis. Techniques,* 16, 352–361, 2003.
26. Lutz, G.K., Butzlaff, M., and Schultz-Venrath, U., Looking back on back pain: trial and error of diagnoses in the 20th century, *Spine*, 28, 1899–1905, 2003.
27. Deyo, R.A. and Weinstein, J.N., Primary care: back pain, *NEJM*, 344, 363–370, 2001.
28. Garfin, S.R. et al., New technologies in spine: kyphoplasty and vertebroplasty for the treatment of painful osteoporotic compression fractures, *Spine*, 26, 1511–1515, 2001.
29. Genant, H.K. et al., Treatment of painful osteoporotic vertebral fractures with percutaneous vertebroplasty or kyphoplasty, *Osteoporosis Int.*, 12, 429–437, 2001.
30. Jasper, L.E. et al., An *ex vivo* biomechanical evaluation of a hydroxyapatite cement for use with kyphoplasty, *Am. J. Neuroradiol.*, 22, 1212–1216, 2001.
31. Chiles, B.W., Leonard, M.A., and Choudhri, H.F., Cervical spondylotic myelopathy: patterns of neurological deficit and recovery after anterior cervical decompression, *Neurosurgery,* 44, 762–770, 1999.
32. Kumar, V.G., Rea, G.L., and Mervis, L.J., Cervical spondylotic myelopathy: functional and radiographic long-term outcome after laminectomy and posterior fusion, *Neurosurgery,* 44, 771–778, 1999.
33. Singh, A. et al., Clinical and radiological correlates of severity and surgery-related outcome in cervical spondylosis, *J. Neurosurg. (Spine),* 94, 189–198, 2001.
34. Speciale, A.C. et al., Observer variability in assessing lumbar spinal stenosis severity on MRI and its relation to cross-sectional spinal area, *Spine*, 27, 1082–1086, 2002.
35. Boden, S.D. et al., Use of recombinant human bone morphogenetic protein-2 to achieve posterolateral lumbar spine fusion in humans: a prospective, randomized clinical pilot trial, *Spine,* 27, 2662–2673, 2002.
36. Godde, S. et al., Influence of cage geometry on sagittal alignment in instrumented posterior lumbar interbody fusion, *Spine*, 28, 1693–1699, 2003.
37. Kadanka Z. et al., Approaches to spondylotic cervical myelopathy, *Spine,* 27, 2205–2211, 2002.
38. Rowland, L.P., Surgical treatment of cervical spondylotic myelopathy: time for a controlled trial, *Neurology,* 42, 5–13, 1992.
39. Phelan, E.A. et al., Helping patients decide about back surgery: a randomized trial of an interactive video program, *Spine*, 26, 206–212, 2001.
40. Sandhu, H.S. et al., BMPs and gene therapy for spinal fusion, *Spine*, 28, 585, 2003.
41. Wang, J.C. et al., Effect of regional gene therapy with bone morphogenetic protein-2-producing bone marrow cells on spinal fusion in rats, *J. Bone Joint Surg.,* 85-A, 905–911, 2003.
42. Burkus, J.K. et al., Is INFUSE bone graft superior to autograft bone? An integrated analysis of clinical trials using the LT-CAGE lumbar tapered fusion device, *J. Spinal Dis. Techniques,* 16, 113–122, 2003.
43. Cha, C.W. and Boden, S.D., Gene therapy applications for spine fusion, *Spine*, 28, S74–S84, 2003.
44. Wigfield, C.C. et al., The new Frenchay artificial cervical joint, *Spine*, 27, 2446–2452, 2002.
45. Schwarzenbach, O. et al., The dynamic neutralization system for the spine: a multi-center study of a novel non-fusion system, *Eur. Spine J.*, 11, S170–S178, 2002.
46. Fisher, E.S. and Wennberg, J.E., Health care quality, geographic variations and the challenge of supply-sensitive care, *Persp. Biol. Med.*, 46, 69–79, 2003.

47. Sundt, T.M. Jr. and Whisnant, J.P., Subarachnoid hemorrhage from intracranial aneurysms, *NEJM*, 299, 116–122, 1978.
48. NASCET Trial Collaborators, Beneficial effect of carotid endarterectomy in symptomatic carotid stenosis patients, *NEJM*, 325, 445–453, 1991.
49. Carey, T.S., Randomized controlled trials in surgery: an essential component of scientific progress, *Spine*, 24, 2553–2555, 1999.
50. Fairbank, J., Spine update: randomized controlled trials in the surgical management of spinal problems, *Spine*, 24, 2556–2558, 1999.
51. Weinstein, J.N., Boden, S.D., and An, H.S., Emerging technology: should we rethink the past or move forward in spite of the past? *Spine*, 28, S1, 2003.

15 Clinical Research in Surgery

Ricardo Pietrobon and Dennis A. Turner

CONTENTS

15.1 INTRODUCTION

Clinical trials are the key interfaces of basic science findings, product development, rationalization of existing therapies, and their translation into effective clinical therapy (see Chapter 1). Thus, all translational research depends upon clinical trial data. However, clinical trial design is a specialty field in its own right and few basic scientists, clinician investigators, and clinicians have formal training in this field, particularly in neurosurgery and orthopedics. The performance of clinical trials and

0-8493-1482-8/05/$0.00+$1.50

the outcomes of those trials form the most severe bottleneck of new therapy development, particularly in surgical fields and device development, where many alternative pathways for product introduction exist and no specific criteria are available for trial format.[1,2] Thus, particularly in product development (unlike drug development), the manufacturer may omit clinical trials of efficacy determination altogether to focus only on safety issues, rather than risk a failed clinical trial that could doom FDA approval because alternative pathways to approval exist.

Various motivations should be considered when initiating a clinical trial. As noted in many chapters of this book, promising new therapies emerging from basic science studies can appear ready to apply to initial human feasibility studies (in a true translational approach). An enthusiastic investigator may promote these initial clinical studies for a variety of reasons or a sponsor (such as a corporate entity banking on a marketable product) may initiate human product development.

The motivations in both cases are usually financial and the trial outcome is a key ingredient in financial success. Both parties are under considerable pressure to bias the trial outcome. Clearly clinical trials in such situations are best performed by independent investigators not connected to the enthusiastic investigator or corporate entity in an effort to decrease this bias. In many cases, FDA approval is initially based only on biased clinical data. Only later (after market introduction) are clinical trials initiated by skeptical groups of independent investigators (often with negative outcomes).

Other motivations include improvement in knowledge and data on existing therapy. This was the case with the multiple carotid endarterectomy trials,[3,4] in which the basis for an existing therapy was challenged by both neurologists (who felt the therapy was dangerous at best and lacked efficacy compared to the natural history) and vascular surgeons (who felt surgical therapy was an unassailable necessary path and unethical to withhold). The uncertainty within the medical community regarding the relative benefits of treatments was sufficient that the level of equipoise favored randomized controlled trials. Since these trials involved an existing (insurance-funded) therapy, the cost of the trials mainly involved obtaining and analyzing data.

Another situation is development of a new surgical therapy, not necessarily based on a product, but an extension of an existing procedure or a new concept. Examples from the 1980s are embryonic allograft and adrenal autograft implantations in the basal ganglia for Parkinson's disease (see Chapter 8). In such cases, no corporate sponsor exists and no insurance reimbursement may be possible because of the experimental nature (and unknown benefits and risks) of the procedure.

Eventually the NIH-funded clinical trials of embryonic allograft neural transplantation required a sham control (a partial burr hole) because of the enormous bias on the part of patients favoring improvement from such dramatic surgery, particularly if the patients funded most or all of the procedures. Even a sham control involves considerable ethical issues and the issue of the relative worth of partially blinding patients to treatments is still debated today.

Even after FDA approval or a large controlled trial, considerable issues remain in the extrapolation of the results to clinical treatments performed by a wide range of practitioners on a variety of patients, often far beyond the initial disease indications. These issues may often require further open label trials based on a community setting for acquiring data on uses and risks in the population at large, but often the quality of data obtained from such open label trials is poor and difficult to truly analyze. The community of surgeons has considerable interest in such trials because of political concerns related to the focus of academic centers on randomized clinical trials. However, participation by a wide group of academic and private practice surgeons is difficult, and new trial formats clearly are needed to assess how treatments are really used in practice.

These primary motivations for considering and initiating clinical trials highlight the needs for well-designed clinical trial formats, for innovations at the FDA approval level, and in translational research as a whole. This chapter will review the key concepts involved in the design of a surgical study and the multiple choices faced by a designer. It starts by reviewing the importance of the current concept of "surgical practice based on evidence" in modern surgery. This is followed by an evaluation of some of the multiple clinical outcomes that may be measured in a surgical investigation and the most common epidemiological designs: randomized controlled trials, outcome studies, and population-based studies.

Although they are important for surgical research, designs such as decision analysis, surveys, health economic, and qualitative studies will not be covered. A special attempt will be made to demystify some of the concepts inherited from classical "evidence-based medicine" without further judgment. In particular, the role of randomized controlled trials for evaluating treatment efficacy will be scrutinized — emphasizing differences between surgical and nonsurgical research while stressing the importance of making a surgical study both feasible and generalizable to a "real-world" patient population.

15.2 WHY EVIDENCE-BASED SURGERY?

Evidence-based surgery is a current movement based on the application of scientific method to the whole body of surgical practice, including long-established surgical traditions that may never have been subjected to systematic scrutiny. In scope, it is similar to evidence-based medicine because it pursues "the conscientious, explicit and judicious use of current best evidence in making decisions about the care of individual patients." The roots of evidence-based medicine and surgery can be found in the work of Professor Archie Cochrane, a British medical researcher whose book *Effectiveness and Efficiency: Random Reflections on Health Services* (1972)[5] and subsequent advocacy increased acceptance of the evidence-based concept. Using scientific techniques from other fields such as meta-reviews of the existing literature, risk–benefit analysis, and randomized controlled trials, evidence-based surgery aims for the ideal that all surgeons should make "conscientious, explicit, and judicious use of current best evidence" in making decisions about patient care.

Practicing evidence-based surgery implies both clinical skill and expertise in retrieving, interpreting, and applying the results of scientific studies. It also involves communicating the risks and benefits of different courses of action to patients. Critics of evidence-based surgery claim that surgeons already follow this procedure; that good evidence is often deficient in many areas; that lack of evidence of benefit and lack of benefit are not the same; and that the more data are pooled and aggregated, the more difficult it is to compare patients in studies with the patient in the office. Despite its problems, evidence-based surgery does not aim to exclude the individual clinical experiences of surgeons. The intent is simply to enhance their experience with information from scientifically sound studies.

The next sections discuss evidence-based surgery from the perspective of an active clinical researcher attempting to design and conduct a study — not from the perspective of a surgeon trying to understand the literature. After a discussion of trial formats and the clinical outcomes that can be measured, fallacies in this approach will be evaluated in detail.

15.3 CLINICAL OUTCOMES

This section describes the main clinical endpoints that can be measured in a surgical study, including disease (pathology, impairment, and disability) and patient satisfaction. The outcome of a surgical procedure varies considerably, depending on the viewpoint (see Figure 14.1). The same surgical procedure (for example, a lumbar discectomy) may be viewed from the perspective of the patient, the surgeon, or society.

The value of surgery and the outcome may be viewed completely differently by these three involved parties. For example, the surgeon may view the outcome as a complete success if the patient's leg pain is resolved. The patient may view the same surgery as a complete failure if his or her back pain is not resolved (despite leg pain improvement), even though total resolution was not an expectation of the procedure. Society as a whole may view the surgery as not worth funding if patients do not commonly return to work, even though this expectation may be far different from expectations of the medical community. These differing viewpoints naturally lead on occasion to opposing philosophies as to the worth or overall benefit of a particular medical or surgical approach.

15.3.1 Pathology

Pathology has been defined as the interruption of normal cellular processes and the simultaneous homeostatic efforts of an organism to regain a normal state.[6–8] Active pathology can result from infection, trauma, metabolic imbalance, degenerative disease process, neoplasia, vascular, or other etiologies reflecting the basic mechanisms of disease occurrence. Examples of such processes include cellular disturbances consistent with the onset of disease processes such as spinal osteoarthritis and cerebrovascular accidents. Although surgical research has focused on pathology since the 19th century, largely following the Virchow tradition,[9,10] pathology is not linearly associated with the final clinical outcomes noticed by patients and surgeons. For example, the same degree of lumbar osteoarthritis noted on radiographic studies can affect patients' symptoms in different proportions, from completely asymptomatic to highly symptomatic. It is therefore necessary to consider pathological findings with other outcome measures such as impairment.

Pathological measures are often primary methods (separate from the symptoms associated with the primary disease) for understanding whether a treatment is working. For example, measurement of the size of a brain tumor on MRI scans can form a primary data source with which to compare various chemotherapy treatment regimens. However, these treatments may or may not improve other types of outcomes or patient survival. In spine studies, radiographic fusion is often used as a surrogate marker for success of a fusion procedure, even though this marker does not appear to correlate with patient outcomes in most other respects. Clearly, the use of pathological measures may be an important basis to decide on treatment efficacy at a basic level, but these measures likely require supplementation with other types of outcomes to decide whether a treatment on the whole is worthwhile at patient level.

15.3.2 Impairment

Impairment is a medically evident loss of function or abnormality at the tissue, organ, or body system level. Active pathology may result in some type of impairment, but not all impairments are associated with active pathology (e.g., congenital loss or residual impairments resulting from trauma). Impairments can also occur at the primary site of the underlying pathology (e.g., muscle weakness around an osteoarthritic knee joint), although they may also occur in secondary regions (e.g., cardiopulmonary deconditioning secondary to inactivity). Impairments can usually be objectively specified by an observer such as a physician or surgeon, and are classified in a standard text, the *American Medical Association Guide to Impairment.*[11]

Although impairment is a measure that is closer to the outcome as observed by a surgeon, a patient's perception about his or her own outcome is not linearly associated with impairment. For example, a limitation in shoulder range of motion secondary to a cerebral vascular accident may greatly affect the life of an active patient and be of little importance to a sedentary elderly patient. It is therefore necessary to extend the concept of outcome to a classification that captures the real impact of a disease on a patient's life — disability.

15.3.3 Disability

Several schools of thought have defined disability and related concepts. We will focus our discussion on the disablement model developed by Saad Nagi, a sociologist,[12] the *International Classification of Impairments, Disabilities and Handicaps (ICIDH-1)*,[13] and its current revision, the *International Classification of Functioning, Disability and Health (ICF)*.[14] The three concepts have in common the view that overall disablement represents a series of related concepts that describe the consequences or impact of a health condition such as lumbar arthritis on a patient's body, his or her activities, and on the wider participation of the patient in society.[6–8,10]

In this social perspective, disability may or may not be linked to medical impairment and the degree of disability may vary widely for the same impairment. A common example is a finger amputation, an easily observed medical impairment. It may not constitute a disability for some occupations (manual laborer) but would produce complete disability for others (concert pianist, surgeon).

According to the conceptual framework of disability developed by Nagi,[6–8,10] disability is the expression of a physical or a mental limitation in a social context. Nagi specifically views the concept of disability as representing the gap between a person's capabilities and the demands created by the social and physical environment — a product of the interaction of the individual with the environment. This is a fundamental distinction of critical importance to scholarly discussion and research related to disability phenomena.

Independent of Nagi's work in the early 1970s, a group in Europe developed the first draft of what later became the World Health Organization's *International Classification of Impairments, Disabilities, and Handicaps*. Similar to Nagi's, this model differentiates a series of related concepts: health conditions, impairments, disabilities, and handicaps designated the ICIDH-1 concepts. We will not review the ICIDH-1 classification except to note that in principle this original system was designed as a model for coding and manipulating data on the consequences of health conditions. This classification system was revised, giving rise to the ICF (Table 15.1 and Table 15.2). The ICF has two sections, each with two complementary components. Part 1 covers functioning and disability including body functions, structures, activities, and participation. Part 2 covers contextual factors — environmental as

TABLE 15.1
Function and Disability Sections of ICF

Component	Body Functions and Structures	Activities and Participation
Constructs	Changes in physiological function Changes in anatomical function	Capacity: executing tasks in standard environment Performance: executing tasks in current environment
Positive aspects	Functional and structural integrity	Activity participation
Negative aspects	Impairment	Activity limitation and participation restriction

TABLE 15.2
Contextual Factors of ICF

Component	Environmental Factors	Personal Factors
Domains	External influences on functioning and disability	Internal influences on functioning and disability
Constructs	Facilitating or hindering impacts of features of physical, social, attitudinal worlds	Impacts of personal attributes
Positive aspects	Facilitators	Not applicable
Negative aspects	Barriers and hindrances	Not applicable

well as personal. Each component consists of various domains and, within each domain, categories that are the units of classification.

Disablement models such as Nagi's and the ICIDH-1 formulation present the disablement process as more or less a simple linear progression of response to illness and its consequences, thus leading to a static conceptualization of the whole process. This view fails to recognize that disablement is more often a dynamic process that can fluctuate in breadth and severity across the life course; it is anything but static or unidirectional.

More recent disablement formulations and elaborations of earlier models have explicitly acknowledged that the disablement process is far more dynamic. In these newer concepts, a given disablement process may lead to further downward spiraling consequences. Pope and Tarlov[15] use *secondary conditions* to describe any type of secondary consequence of a primary disabling condition. Commonly reported secondary conditions include pressure sores, contractures, depression, and urinary tract infections, but it should be understood that they can be pathologies, impairments, functional limitations, or additional disabilities. Longitudinal analytic techniques now exist to incorporate secondary conditions into research models and are beginning to be used in disablement epidemiologic investigations.

15.3.4 Patient Satisfaction Theories

Patient satisfaction is an important outcome measurement since it can influence the delivery of medical care at both the societal (total consumption of health care resources) and individual (patient participation) levels. Because patient satisfaction is a multidimensional concept, it is important to start by understanding its multiple definitions. Patient satisfaction is a complex concept that may incorporate sociodemographic, cognitive, and affective components. Although many theories for patient satisfaction have been proposed, few have been extensively tested and validated in different health care settings. Moreover, few studies have been conducted to explain associations between patient satisfaction and patient characteristics or subsequent patient behaviors. Although theories of patient satisfaction are difficult to categorize in an organized and easily comprehensible fashion, one may group these theories into intrapatient comparisons (disconfirmation theory) and differences between individual patients and health care providers (attribution theory) or other patients (equity theory).

Intrapatient comparison theories explain the satisfaction phenomenon by a match between patient expectations and perceptions of medical care. Differences between what is expected and what is perceived to occur will contribute to patient satisfaction or dissatisfaction. This theory is the dominant model of nonmedical customer satisfaction in which consumers compare their perceptions of a product or service against prior expectations. The resultant size and direction of the disconfirmation results in satisfaction or dissatisfaction.

Equity theories are based on the premise that patient satisfaction relates to whether patients believe they have been fairly treated. Equity occurs when patients compare their balances of inputs (time and money) and outputs (medical care and its results) with those of other patients. Patient satisfaction occurs when people perceive they are treated fairly; it may increase when patients perceive their outcomes as more favorable than those of other patients with the same conditions.

Attribution theories assume that any causes for failed expectations will be examined. Dissatisfaction may occur if a patient and provider assume different reasons for a failure. A related concept is gap analysis in which identification of differences between provider and patient perceptions of services occurs. Addressing potential gaps arising because providers focus primarily on delivery of medical care and patients focus on services used may increase patient satisfaction.

15.3.5 Measurement of Patient Satisfaction

Although patient satisfaction has become a common measurement in clinical settings, proper assessment of a patient's cognitive evaluation of and affective response to medical care provided is an extremely complex task. The difficulty in measuring patient satisfaction lies in the fact that satisfaction is a multidimensional concept with inputs or determinants that are not yet clearly defined. One of the major criticisms of patient satisfaction research relates to methodologic issues including lack of psychometric standards, reliability, and validity of surveys. Many patient satisfaction survey instruments have not undergone rigorous psychometric construction, which is essential in the evaluation of all complex psychological phenomena.

Many patient satisfaction surveys lack discriminatory values to assess specific aspects of medical care. As an example of these biased measurements, studies have shown that the use of a single global measurement to evaluate patient satisfaction generally results in high (95%) satisfaction ratings. Unfortunately, these single-item questions cannot distinguish satisfaction with overall medical care from satisfaction with specific aspects of care. These rudimentary instruments often cannot accurately measure the multifaceted nature of patient satisfaction and may actually reflect satisfaction with other issues of medical care.

Implementation of a survey instrument may also be associated with potential bias. Possible problems include mode of administration (e.g., telephone, personal interview, mail, and structured versus unstructured interviews), timing of survey, nonresponders, and use of proxies. Although no consensus about an acceptable response rate seems to exist, different methods of administration may produce different response rates. The response rate is important in that missing data from nonresponders may affect the validity of the results. The differences between responders and nonresponders are important because patients who are less satisfied with their medical care may be less likely to respond to satisfaction surveys.

The timing of the survey and use of proxies may also introduce bias. There is a greater likelihood of recall bias with increasing lengths of time between hospital discharge and administration of the survey, which may affect patient responses. Proxies (e.g., family members and friends) may not accurately reflect the views of patients. Patient satisfaction assessments are typically considered as nonparametric data and appropriate statistical analysis should be conducted.

15.4 SELECTING MOST APPROPRIATE CLINICAL TRIAL DESIGNS

Selecting the most appropriate trial design is one of the most important steps in the design of a surgical study. In some situations, a cohort study may not appropriately control for baseline differences between two or more treatment groups, whereas a randomized controlled trial may be unfeasible or may over-select a study population that would no longer be truly representative of the group of patients seen on a daily basis. Therefore, the choice of a study design implies knowledge that is extraneous to a purist and a theoretical view of the field. Other issues include pragmatic factors determined by the experiences of previous researchers attempting to conduct similar research studies, available time and funding, and the general expectations of peers in the field. The goal of the study is also critical, for example, feasibility studies for initial human trials of a device or surgical procedure versus pivotal trials for FDA approval.

15.4.1 Randomized Controlled Trials

Definition — A randomized controlled trial (RCT) is a design in which participants are allocated at random to receive one of several clinical interventions. Since treatments are randomly allocated, differences in baseline characteristics across groups tend to be balanced if the groups are large and, therefore, any differences in outcome can be attributed to the intervention.

Classification — RCTs can be broadly classified into the following categories according to exposure to the intervention:[16]

1. In parallel RCTs, each group of participants is exposed to only one of the study interventions. Most surgical RCTs fall into this category.
2. Cross-over RCTs use a design in which each participant is given all the study interventions in successive periods; the order of the interventions is determined randomly. It is interesting to note that conclusions from parallel studies are produced by comparisons across groups; in cross-sectional RCTs, the comparisons are made from participants. Two issues are crucial for the success of a cross-over trial. First, the condition treated must be chronic and stable. Second, the interventions must be of short duration to avoid contamination of the groups that will follow.
3. Factorial design is performed when two or more interventions are used separately and also in combinations and compared against a placebo. For example, when comparing treatments A and B for a certain condition, four groups would be formed: one to receive only treatment A, one to receive only treatment B, one to receive treatments A and B, and a group that would receive no treatment.

When to perform RCTs — Unlike nonsurgical trials, logistical problems largely determine the ideal situation for conducting an RCT. These factors include:

1. A sufficient number of eligible patients should be available. The evaluation should be based on data rather than a surgeon's opinion about the number of cases because surgeons tend to overestimate true numbers of patients.
2. The intervention should be stable, with the perspective that treatment will continue in a similar fashion until the trial is concluded. Since surgical RCTs tend to last longer than their nonsurgical counterparts, it is important to ensure that the therapy is not a simple fad, and that results will still be applicable when the study is completed. Otherwise, the results will be meaningless if the therapy is outmoded.
3. Participating surgeons should be truly involved in the study and fully believe in the presence of equipoise. Equipoise (state of genuine uncertainty) about the comparative efficacy of two different interventions is an ethical imperative for conducting a study.

Important points — If RCT is selected as the design of choice, some important concepts should always be kept in mind:

1. When a placebo is considered, it is important to ensure that no other intervention has been previously shown to be effective. If an efficacious therapy exists, it should be used in place of a placebo to avoid patient suffering.
2. It is important that the randomization be performed appropriately and include true randomization mechanisms such as random number generation. Mechanisms such as selection of treatments based on the first letter of a last name or day of the week are pseudorandom methods and should be avoided.
3. Surgical procedures and postoperative management should be identical across participating surgeons and institutions. Differences in surgical techniques will create clusters that can be impossible to control during comparative analysis across techniques.
4. Outcomes should be assessed by a researcher not involved with the previous stages of the study and, of great importance, they should never be assessed by the treating physician.
5. If an outcome can be interpreted by different individuals in different manners, such as the interpretation of radiographs, researchers should conduct previous inter- and intraobserver agreement studies to ensure that the measurement will not be biased by the opinion of a single evaluator.

15.4.2 Outcome Studies

Definition — *Outcome study* is a popular name commonly used in the surgical field to describe studies involving prospective cohorts of patients, particularly when standardized scales are used to compare functional outcomes across patients and treatments. Although the intervention is not randomized, outcome studies are valuable tools for determining associations between treatments and disease outcomes. Of particular importance is that, because of the absence of randomization and its issues of selection bias, outcome studies are fundamental in determining how a certain intervention applies to patients in more natural settings.

Outcome studies usually vary in lengths of follow-up, many constituting collections of all patients attending a certain clinic and thus enrolling all patients until their last clinical visits. As could be expected, durations of follow-up will vary and results can only be appropriately analyzed by time-to-event methods to obtain the chance of a particular outcome for an individual with a particular treatment. One weakness of outcome studies is that they are usually inefficient for measuring rare outcomes where a secondary data analysis of a population-based study would be more appropriate.

Classification — Outcome studies can be generally classified as:

1. Prospective outcome studies — An investigator chooses or defines a sample of subjects to be studied. This selection is usually based on a specific diagnosis, surgical procedure, or interest. The same set of outcome scales is then applied to all participants before and after the surgical intervention. Finally, the comparison across treatment groups is usually performed by comparing postoperative outcomes as measured by the scales adjusted for baseline scores and other potential confounders of the association between treatment and outcome.
2. Retrospective outcome studies — The design of a retrospective outcome study is essentially the same as that of a prospective study. Retrospective studies, however, are usually based on longitudinal cohorts of patients from a given clinic; outcome scales are widely spread across all diagnoses and procedures. Although the outcome measurements are usually somewhat less specific than measurements from prospective designs, retrospective outcome studies have the advantage of providing a much larger population immediately. Those conducting retrospective outcome studies should pay particular attention to important factors such as number of missing observations within measurements (missing values for specific variables) and missing observations within the clinic (patients who may have attended but failed to complete outcome questionnaires).
3. Multiple outcome studies and external controls — This design compares several separate outcome groups. It is typically used when two or more groups are treated with different surgical techniques. The most important factor is to ensure that the baseline variables (functional levels, disease classifications, and sociodemographic factors) of the outcome groups are as homogeneous as possible. Having groups who overlap reasonably in the areas mentioned above will ensure that these differences can be controlled during the comparison across the different surgical interventions.

15.4.3 Outcome Scales

Because outcome scales are at the cores of outcome studies, it is important to understand what they are and how they are validated. Validated outcome scales are characterized by their reliability, validity, and responsiveness to clinical change. These properties ensure that the data are collected and interpreted in a systematic and reproducible way, allowing comparisons across different patient populations.

Reliability — This is the property that determines whether the instrument measures the outcome of interest in a consistent and reproducible way. Reliability is assessed by measuring internal consistency and temporal stability. Internal consistency requires the items constituting a scale to be highly intercorrelated and measure the same concept or construct. Scales whose items are all highly intercorrelated are considered to be one-dimensional because they measure only a single construct. If a scale measures more than one construct, its items are expected to correlate in clusters, and the scale is multidimensional. Internal consistency is usually expressed by Cronbach's coefficient. Values above 0.7 are usually considered to express acceptable internal consistency. Score stability over time, on the other hand, refers to the consistency of scores obtained on different occasions by the same individuals.

One of the most common measures of temporal stability is test–retest reliability. For example, a scale demonstrates good test–retest reliability if patients with stable conditions tend to have similar scores over time. A common problem of test–retest reliability is that the assumption of a stable underlying condition often can be supported only if the time between the two evaluations is relatively short and if patients can be assumed to not be responding to items based on a recollection of their previous responses.

Validity — This is an indication that the scale primarily measures the construct it is intended to measure instead of another related construct. For example, a scale devised to measure neck pain or dysfunction should not capture dysfunction due to concomitant depression. The commonly reported types of validity are (1) face, (2) content, (3) criterion-related, and (4) construct validity. A scale is considered to have face validity if its content seems to measure what it is supposed to measure.

This evaluation is usually performed by the scale designers rather than the target population without any quantitative evaluation, and therefore it can be biased. A scale demonstrates content validity when the items reflect all the significant aspects of the construct to be measured. Again, taking neck dysfunction as an example, work-related disability is only one of the dysfunctions caused by the underlying disease and a scale presenting items exclusively about work dysfunction would capture the entire scenario. Thus, while such a scale may have adequate content validity as a measure of work dysfunction, it would lack content validity as a measure of dysfunction conceptualized more broadly.

Criterion-related validity implies that a scale is able to predict some criterion variable, such as the course of the underlying disease. Criterion validity can be applied to situations where the criterion follows (postdictive validity), precedes (predictive validity), or coincides with (concurrent validity) the measurement in question. Finally, construct validity refers to a scale's behavior in relation to other related assessment tools. For example, one can reasonably hypothesize that neck pain would be associated with impaired quality of life. A neck pain scale would therefore be considered to have construct validity if a correlation between the neck pain scale and a valid quality-of-life questionnaire could be documented.

Responsiveness — Responsiveness is the ability of an instrument to detect small but important clinical changes such as minimal clinically important differences. This index is the minimal score difference able to detect a "clinically important change," which is a subjective judgment made by a clinician or a patient independent of available treatment choices. Most experts would probably agree that it is important to define and assess the minimal clinically important differences for individual functional scales.

A crucial point in scale evaluation is that psychometric properties are not intrinsic to a specific instrument, but instead are highly susceptible to change as functions of the populations where they are used and how they are applied. As an example, athletes may perceive functional incapacities at levels of activity sedentary patients may never approach. In sum, it is important to consider how a scale may perform within a specific population of interest.

15.4.4 Secondary Data Analysis

Definition — Secondary data analysis is conducted by using data that has already been collected, usually with a broad purpose. Because it avoids the hurdles involved with primary data collection, the secondary data analysis approach provides a quick and efficient method of answering research questions.

In cases of research questions requiring very large populations, for example, national surveys, secondary data analysis is the only possibility because the cost of a prospective study would be prohibitive. Secondary data analyses can also serve as excellent resources for obtaining preliminary information on a research question that can later be further investigated through a prospective study with more specific clinical variables. Finally, large databases with variables that relate to latent variables such as disability or quality of life represent excellent resources for the formulation of outcome scales because their large numbers allow for the use of powerful statistical techniques such as item response theory.

Formulation of research questions — The formulation of research questions from secondary data can occur in one of two directions. First, researchers with preconceived research questions may look for large repositories of data in search of a database with the variables that may answer their questions. Second, researchers may navigate through a dictionary of available databases while trying to formulate research questions that are of interest to them, the clinical research community, and the public in general. In practice, research questions are usually formulated using a mix of the above-mentioned approaches.

Since the formulation of a research question depends on finding databases that will support the study question in both cases cited above, it is important to locate repositories containing information on multiple databases with easily retrievable information. Such a project was recently conducted by the Center for Excellence in

Surgical Outcomes of Duke University in Durham, NC. Known as QUESTFORM (QUESTion FORMulation), this web application aggregates detailed information on more than 50 different clinical databases. The first section of QUESTFORM contains extensive information about database characteristics, including primary purpose, validity of specific variables, details about data collection methods, year coverage, generalizability, total, and number of patient encounters.

QUESTFORM can be used in two modes. In the first mode, researchers can navigate through the data dictionaries of different databases while they are guided in the formulation of a well-formed epidemiological question. In other words, they are instructed on the selection of outcomes, primary predictor variables, confounders that can potentially distort the association between main effect and outcomes, and inclusion–exclusion criteria. A search tool for International Statistical Classification of Diseases and Related Health Problems (ICD) and Current Procedural Terminology (CPT) codes is provided to identify specific disease and procedure codes. Once the question is fully formulated, researchers can save the question in a graphical format known as a question diagram. Question diagrams can then be reviewed by other members of the research team for project feasibility (clinical epidemiologist), statistical approach (statistician), coding (statistical programmer), and literature review (participating students).

In the second mode, researchers can navigate through previously formulated question diagrams. These previous examples serve as templates that can be modified to generate new research projects. By navigating through previously formulated diagrams, researchers can learn from observing successful designs and also save time while creating new designs that bear structural similarities with previous question diagrams.

15.5 FALLACY BEHIND LEVELS OF EVIDENCE

Much has been said about RCTs for the evaluation of treatment efficacy because RCTs are considered the gold standard against which all other clinical research designs should be compared. In this section, we will argue that despite the several advantages of RCTs if one is to consider internal validity only, RCTs may lack external validity or feasibility in several surgical situations where other designs would be clearly more appropriate.

Because most RCTs are performed in academic medical centers, the number of surgeons involved in surgical trials is often limited and carefully monitored for quality. For example, the number of procedures performed prior to trial initiation and their outcomes and morbidities are often carefully audited to ensure that the surgical procedure is performed with the highest possible quality and consistency. Although this auditing ensures that the trial design covers a single surgical procedure (as far as can be specified), the same surgical procedure may actually be performed in many different ways and with far different outcomes once it is available to a wide range of nonacademic and academic centers after the trial conclusion. This can lead to severe problems with extrapolation or generalization of the trial results to the populations of both patients and surgeons.

15.5.1 Traditional Concept of Evidence-Based Medicine and Its Roots

Evidence-based medicine has traditionally claimed that RCTs provide better evidence when compared with other study designs. This is clearly demonstrated by Table 15.3. The clear preference for RCTs can also be distinguished by the designation of the Oxford Centre for Evidence-Based Medicine of nonrandomized studies as the "hurly-burly of real-world clinical care." But is the choice of a research design something to be made with disregard for the logistics surrounding the study?

15.5.2 Why Surgery Is Different

Despite the now classical statements about RCT design superiority, we believe that the choice of research design is multifactorial and that, despite their internal validity, RCTs may in some situations achieve results that are of quality inferior to the quality of their nonrandomized counterparts. Several problems can threaten the validity of an RCT:

1. Restrictive enrollment criteria implemented to enhance internal validity in clinical efficacy trials can have the unintended consequence of excluding cases that would make the study sample truly representative of a real-world population (e.g., those with comorbidities).
2. Evidence obtained solely from RCTs may be misapplied by policy makers, payors and/or practitioners who misunderstand the approach and misinterpret it as prescribing a narrowly formulaic ("cookbook") approach to healthcare.
3. Many patients are unwilling to be randomized to treatments, particularly when one assignment option involves, in their perspectives, inert or ineffective treatments.
4. When an RCT protocol involves a single sustained treatment, the design may fail to reflect usual practice in which shifts in treatment occur until a desired outcome is achieved and maintained.
5. Research documentation and reporting of critical phenomena, such as treatment delivery (fidelity), therapy process measures, and population reach are infrequent.
6. Relevant outcomes including functional status, quality of life, durability of change, potential negative or iatrogenic outcomes, cost of treatment, and client satisfaction may have been neglected in detriment of harder endpoints.
7. Overemphasis on treating or fixing presumably homogeneous disorders may detract from potentially more valuable efforts to understand what caused the problem originally, what contingencies now maintain it, how treatment influences biopsychosocial processes to produce desirable behavioral change, and what changes are needed to address more complex, comorbid problems.

TABLE 15.3
Oxford Centre for Evidence-Based Medicine Levels of Evidence (May 2001)

Level	Therapy/Prevention, Etiology/Harm	Prognosis	Diagnosis	Differential Diagnosis/Symptom Prevalence Study	Economic and Decision Analyses
1a	SR (with homogeneity) of RCTs	SR (with homogeneity) of inception cohort studies; CDR validated in different populations	SR (with homogeneity) of Level 1 diagnostic studies; CDR with 1b studies from different clinical centers	SR (with homogeneity) of prospective cohort studies	SR (with homogeneity) of Level 1 economic studies
1b	Individual RCT with narrow confidence interval	Individual inception cohort study with >80% follow-up; CDR validated in single population	Validating cohort study with good reference standards; CDR tested within one clinical center	Prospective cohort study with good follow-up	Analysis based on clinically sensible costs or alternatives; systematic reviews of evidence; includes multiway sensitivity analyses
1c	All or none	All or none case-series	Absolute SpPins and SnNouts	All-or-none case series	Absolute better-value or worse-value analyses
2a	SR (with homogeneity) of cohort studies	SR (with homogeneity) of retrospective cohort studies or untreated control groups in RCTs	SR (with homogeneity) of Level >2 diagnostic studies	SR (with homogeneity) of 2b and better studies	SR (with homogeneity) of Level >2 economic studies
2b	Individual cohort study (including low quality RCT; <80% follow-up)	Retrospective cohort study or follow-up of untreated control patients in RCT; derivation of CDR or validated on split sample only	Exploratory cohort study with good reference standards; CDR after derivation or validated only on split sample or databases	Retrospective cohort study or poor follow-up	Analysis based on clinically sensible costs or alternatives; limited reviews of evidence or single studies; includes multiway sensitivity analyses

2c	Outcomes research; ecological studies	Outcomes research		Ecological studies	Audit or outcomes research
3a	SR (with homogeneity) of case-control studies		SR (with homogeneity) of 3b and better studies	SR (with homogeneity) of 3b and better studies	SR (with homogeneity) of 3b and better studies
3b	Individual case control study		Nonconsecutive study or lacking consistently applied reference standards	Nonconsecutive cohort study or very limited population	Analysis based on limited alternatives or costs; poor quality estimates of data; includes sensitivity analyses incorporating clinically sensible variations.
4	Case series (and poor quality cohort and case control studies)	Case series (and poor quality prognostic cohort studies)	Case control study; poor or nonindependent reference standard	Case series or superseded reference standards	Analysis with no sensitivity analysis
5	Expert opinion without explicit critical appraisal or based on physiology, bench research, or first principles	Expert opinion without explicit critical appraisal or based on physiology, bench research, or first principles	Expert opinion without explicit critical appraisal or based on physiology, bench research, or first principles	Expert opinion without explicit critical appraisal or based on physiology, bench research, or first principles	Expert opinion without explicit critical appraisal or based on economic theory or first principles

RCTs = randomized clinical trials

SR = systemic reviews

CDR = clinical decision rule

SpPins = high specificity so a positive result rules in diagnosis

SnNouts = high sensitivity so a negative result rules out diagnosis

Source: www.cebm.net/levels_of_evidence.asp

8. Efforts to standardize treatments potentially support progression toward a restriction of treatment that will enable therapy to be delivered by paraprofessionals or by computers.

These points do not indicate RCTs have no place in surgical research, but mean that their advantages must be contrasted against their weaknesses. In other words, blindly applying the principles primarily developed for nonsurgical studies to surgical studies is, at least, a mistake.

15.5.3 Other Limitations of Surgical Trials

Particularly in small disciplines such as neurosurgery, the number of patients spread throughout the world may be insufficient for an adequate randomized trial of a treatment despite considerable interest in the outcome. Small patient populations can thus pose severe constraints particularly if they are widespread and hard to capture. The fragmented health care system in the United States likewise hampers patient access. One proposed solution is to limit surgical procedure reimbursement to patients enrolled in clinical trials. That would encourage both surgeons and patients to improve enrollment and enhance the number of trials. It is difficult to standardize surgical procedures to the point where the same technique is performed by many surgeons and studied at many locales, in contrast to studies of drug formulations and devices that can be standardized.

Because many device companies would rather avoid the expenses and the uncertain outcomes of randomized trials, alternate methods of FDA approval exist, but they severely limit the amount of data concerning device efficacy at the time of market introduction. Often fewer than 100 patients are studied adequately at one center. After market introduction, it is highly unlikely that a device company would sponsor an additional critical trial by skeptics, since negative trial results would lead to decreased revenues. As a result, companies have little incentive to perform trials of devices.

Since surgeons' incomes depend on procedures, they have little incentive to compare surgical treatments to nonsurgical treatments and such comparisons are nearly forbidden in surgical circles. Rather, most surgical trials compare one form of device or procedure to another, rather than to alternative treatments. Development of a community consensus among surgeons about equipoise and when the time is correct to initiate a study is also very difficult. In many cases, surgeons almost have to be forced into studying surgical procedures by outside influences and circumstances.

The funding of studies is also problematic. The National Institutes of Health (NIH) fund only a small number of surgical studies and even pilot studies are difficult to initiate without initial funding. NIH's determinations of what should be studied may also differ considerably from the topics surgeons would suggest, reflecting the discrepancy between society's needs and those perceived by medical practitioners. All these factors combine to create a difficult environment in which many small, retrospective studies are performed, many without any lasting merit or contribution. However, taking the next step toward a prospective trial is almost prohibitive in terms of the enthusiasm needed to obtain patient and surgeon enrollment, funding, and consensus within the surgical community.

15.6 CONCLUSIONS

The motivation for initiation of clinical trials varies, depending on the goal of the study and who will benefit from it. Surgical clinical trials are very different from those usually designed for medical trials, including FDA approval studies and rationalization of existing therapies. Small populations, particularly in neurosurgery, fractionated health care systems, and lack of understanding of clinical trial formats all contribute to difficulty in initiating clinical trials of substantial, lasting benefit. While all neurosurgery practitioners desire valid information about the treatments they suggest to patients, the path to that information is highly convoluted and limited. Nevertheless, all translational therapy depends on clinical trial format, which in some cases (such as stroke trials), can be the limiting feature of new therapy introduction.

REFERENCES

1. Choi, D.W., Exploratory clinical testing of neuroscience drugs, *Nat. Neurosci.,* 5 (Suppl.), 1023–1025, 2002.
2. Haines, S.J., Evidence-based neurosurgery, *Neurosurgery,* 52, 36–47, 2003.
3. North American Symptomatic Carotid Endarterectomy Trial Collaborators, Beneficial effect of carotid endarterectomy in symptomatic patients with high-grade carotid stenosis, *NEJM,* 325, 445–453, 1991.
4. Endarterectomy for moderate symptomatic carotid stenosis: interim results from the MRC European Carotid Surgery Trial, *Lancet,* 347, 1591–1593, 1996.
5. Cochrane, A., *Effectiveness and Efficiency: Random Reflections on Health Services,* Royal Society of Medicine Press, London, 1972.
6. Nagi, S.Z., A study in the evaluation of disability and rehabilitation potential: concepts, methods, and procedures, *Am. J. Public Health,* 54, 1568–1579, 1964.
7. Nagi, S.Z., Congruency in medical and self-assessment of disability, *IMS Ind. Med. Surg.,* 38, 27–36, 1969.
8. Nagi, S.Z., An epidemiology of disability among adults in the United States, *Milbank Mem. Fund Q. Health Soc.,* 54, 439–467, 1976.
9. Virchow, R. and Chance, F., *Cellular Pathology, as Based upon Physiological and Pathological Histology: Twenty Lectures Delivered in the Pathological Institute of Berlin during the Months of February, March and April 1858,* R.M. DeWitt, New York, 1860.
10. Virchow, R.L.K. and Smith, T.P., *Postmortem Examinations, with Especial Reference to Medico-Legal Practice,* P. Blakiston, Philadelphia, 1885.
11. Gocchiarella, L. and Anderson, G.B.J., Eds., *Guide to the Evaluation of Permanent Impairment,* 5th ed., AMA Press, Chicago, 2001.
12. Nagi, S.Z. and Marsh, J., Disability, health status, and utilization of health services, *Int. J. Health Serv.,* 10, 657–676, 1980.
13. World Health Organization, *International Classification of Impairments, Disabilities, and Handicaps: A Manual of Classification Relating to the Consequences of Disease,* WHO Publications Center, Albany, NY, 1980.
14. World Health Organization, *International Classification of Functioning, Disability, and Health: Short Version,* Geneva, 2001.
15. Pope, A.M. and Tarlov, A.R., Eds., *Disability in America? Towards a National Agenda for Prevention,* National Academy Press, Washington, D.C. 1991.

16. Matthews, J.N.S., *An Introduction to Randomized Controlled Clinical Trials*, Oxford University Press, London, 2000.

16 Neurosurgery Teaching Techniques and Neurosurgical Simulation

Jeffrey S. Henn and Dennis A. Turner

CONTENTS

16.1 INTRODUCTION

What is the process through which a medical student becomes a neurosurgeon? How do we teach the skills necessary for success? What role does a resident play in his or her own educational process? Before consideration of neurosurgical simulators, we should first reflect on these questions and understand the processes by which neurosurgeons are currently trained. Then we can consider ways in which the potential benefits of simulation and other new teaching techniques may contribute to the process.

0-8493-1482-8/05/$0.00+$1.50

The current process clearly has many limitations because of the long hours and number of years required for training and defining the need for additional manpower. Simple questions remain unanswered, such as whether neurosurgery training should be closed-ended (a fixed number of years as is currently the practice in the United States) or open-ended (until a job opens, possibly after many years of training, as is commonplace in other areas of the world).

16.2 CURRENT NEUROSURGICAL TRAINING

16.2.1 Apprentices to Practitioners

Traditionally, neurosurgical residency has been an apprenticeship process in which residents spend several years with experienced neurosurgeons and participate in all aspects of the profession. The environment for this apprenticeship is audited for sufficient operative cases and approximate measures of the adequacy of the learning experience through conferences and exam certifications. Residents learn operative techniques, surgical anatomy, clinical management, and the subtleties of interactions with patients and families. As with any apprenticeship, the resident gradually assumes more responsibility and autonomy.

Training is regularly supplemented with lectures, small-group learning sessions, and self-study. In addition, a significant portion of training occurs during interactions of residents. "Senior" residents help in the training of "junior" residents. Ultimately, neurosurgical residents are exposed to the information and clinical experience that will prepare them to leave training for the next phase of their practices. The goal is to produce high quality, independently functioning neurosurgeons, who are competent, adhere to high standards of professional conduct and patient care, and serve as assets to their communities.[1]

A critical part of the training process is evaluation of a resident's performance throughout the training period. Advancement is often dependent on completion of various milestones judged internally at the training institution or by national boards. Emphasis is now placed on evaluating residents from all specialties on six core competencies involving knowledge, judgment, behavior, and technical skill:

1. Patient care
2. Medical knowledge
3. Interpersonal and communication skills
4. Professionalism
5. Problem-based learning
6. Systems-based learning

The process of acquiring factual knowledge and clinical judgment occurs in several ways. The most direct method is a program of self-directed study via textbooks, journal articles, and Medline searches, but the method has a limited role for teaching clinical judgment. Another way of gaining factual knowledge and judgment is through apprentice relationships with attending physicians and senior residents during patient rounds and small group discussions, or from lectures and grand rounds. These types of learning programs typically include consideration of clinical examples, review of factual knowledge, and Socratian-type interactions between residents and mentors. The learning processes have recently been augmented through the advent of evidence-based medicine (see Chapter 15). The advantage of this type of learning is that significant information can be efficiently imparted by the mentor. However, the process relies to some extent on the mentor's experience and willingness to teach.

A potential role of neurosurgical simulation is its use in the form of a clinical database that can be used by students to improve their factual knowledge and, to an extent, their judgment and clinical decision making, by taking advantage of a wider array of cases than may be present at one institution or during one training period.[2,3]

Apprentices also tend to retain local practices and knowledge that may be sometimes parochial and limited by their mentors' views of the world. Such local preferences may then be treated as dogma, only to be perpetuated later as neurosurgical myth. Such biases can be retained for many years and may be difficult to recognize and outgrow.

16.2.2 Judgment in Neurosurgery

Is our current system of neurosurgery training only a Halstedian, apprenticeship process?[3] Neurosurgery clearly has two aspects: judgment — how to decide to whom to suggest a surgical procedure and on what rationale — and technical — how to perform a procedure with maximal efficacy and least risk. While considerable emphasis is placed on the technical aspects and learning, less stress focuses on the judgment side, which is the more critical in terms of subsequent liability and practice enjoyment.

Typically, judgment is taught Socratically, by questioning and answering, by building an internal database containing patient symptoms, common syndromes and their treatments, and knowledge about procedures, outcomes, risks, and recovery times. Judgment is usually considered as a case-by-case set of rules that can be internalized, but which are subject to basic hypotheses, principles, and background knowledge. Development of judgment requires time and experience. Honing and refinement take into account past successes and failures. For example, selecting patients appropriately for surgical approaches often involves taking into account past cases from personal experience along with outcomes cited in the evidence-based literature.

16.2.3 Technical Aspects of Neurosurgery Teaching

The process of learning neurosurgical technique includes mastery of complex three-dimensional anatomy, visuospatial perception, and motor skills. Intraoperative training is the most direct method for developing these skills. It constitutes the gold standard for learning neurosurgical technique and ultimately forms the foundation of a neurosurgeon's career. Typically performed through an apprentice relationship with senior surgeons, it reflects the model of surgical training pioneered by Halsted.

Anatomy and technique are learned through observation of a senior surgeon and directly in a supervised setting. The dual-headed microscope has greatly facilitated this type of learning. Intraoperative training, however, has several important limitations. The environment can be relatively limited by time and tension because of potential risks or bad outcomes and may not foster education as the primary goal. Additionally, anatomic exposure is necessarily limited to what is clinically warranted, and usually this is minimized to improve recovery time, hampering visualization in many situations. Ultimately, patient care must always take priority over education. In this regard, virtual reality simulators may help residents advance sooner by demonstrating more complete dissections and underlying anatomy.[3]

In contrast, technical aspects can usually be taught by direct supervision or by animal approaches such as implementing microvascular anastamoses in small rodents. Technical expertise is a combination of practice, supervision providing guidance, and the repetitive use of the hands as needed for motor learning. The technical side is currently handled by direct observation and apprenticeship, with progressive responsibility based on certain steps or levels. The levels can be indexed according to degree of difficulty (i.e., carpal tunnel, disc, and shunt procedures preceding craniotomies). Milestones that must be achieved before residents progress to the next level are documented.

One training tool is cadaveric dissection that provides residents the potential to improve anatomic understanding, visuospatial perception, and motor skills. Cadaveric dissections are interactive, three-dimensional, and relatively transferable to the operating room setting. However, dissection also has important limitations including significant costs (preparation, facilities, instructors, and equipment), limited availability of specimens, and the substantial differences between living and cadaveric tissues. Finally cadaveric dissection involves a substantial time commitment and is not amenable to repeated rehearsal of a specific procedure.

Animal dissection provides opportunities to improve visuospatial and motor skills, but the technique has both practical and ethical limitations and the anatomic differences are usually significant. Thus, current training involves a large amount of direct intraoperative assisting to allow direct visualization of human anatomy and exposure and small (but key) cadaveric and small animal dissection experience to augment the clinical experience gained over many years.

16.2.4 Training Objectives

The goal of training is to produce fully trained academic neurosurgeons who are clinically competent and have excellent technical skill, superb judgment, and thorough knowledge of related disciplines, including basic neuroscience, neurology, neuropathology. and neuroradiology. Technical competence is achieved through "gradual delegation of earned responsibility for investigative and operative care to penultimate levels."[1] The development of competence in clinical neurosurgery requires a trainee to:

1. Master the principles of surgery
2. Become familiar with the basic science and diseases of the nervous system

3. Develop the necessary technical skills to perform neurosurgical procedures
4. Learn to relate and work effectively with colleagues in medicine and surgery and other health care professionals and ensure the development of a keen sense of responsibility and compassion toward patients and their families
5. Understand the impacts of neurosurgery on society including medical ethics, health care economics, law, prevention of disease, and promotion of health
6. Develop an understanding of clinical and basic research techniques including biostatistics and epidemiology

Residents must assume graduated responsibility throughout the course of their residencies in terms of background knowledge, pre- and postoperative management, operative experience and independence in decision making. However, supervision is critical to a training program, and feedback from more experienced individuals is essential to education along with a constant and sincere effort to learn on the part of the resident.

Patient care is the core background to neurosurgical education and thus intimate knowledge of patients forms the basis for informed decision making and increased participation by residents in patient care decisions and management. The objectives for technical competence build upon progressive training experience in neurosurgical procedures, usually in an apprenticeship mode under direct supervision.

16.3 ARE THERE TRAINING DEFICIENCIES?

The current fixed length residency program is relatively short, compared to the large number of judgment skills, procedures, and care issues that must be adequately taught. It does not account for the varying learning rates of different physicians. For that reason, some flexibility in training length may be important to accommodate and overcome such differences. The current 80-hour work week limitation makes it difficult to take occasional night calls and allow sufficient patient follow-up to adequately assess a resident's judgment. These limitations are compounded by insufficient exposure to and understanding of less common cases because residents commonly handle more common situations.

The skills routinely taught in most residency programs are primarily aimed at a high quality clinical practice situations. They are not necessarily directed toward academic investigations in the fields of basic neuroscience, translational neuroscience, or clinical neuroscience. For many years, the minimum requirement for the adequate pursuit of quality basic neuroscience has been at least 3 to 5 years of experience beyond residency, at the graduate student (i.e., M.D.–Ph.D. combined degree), postdoctoral, or mentored faculty level. This additional time is critical for enabling clinician investigators to become sufficiently qualified to compete for external funding from federal (National Institutes of Health, Department of Veterans' Affairs, Department of Defense, etc.) or foundation sources.

Unless a clinician investigator is competitive in obtaining funding, it is unlikely that research of sufficiently high quality will continue to make valuable contributions to neurosurgery and the wider field of neuroscience. Developing and maintaining adequate clinical credentials and sufficient research experience to be truly competitive are difficult challenges at both the resident trainee and faculty levels. These abilities are under-emphasized. The difficulties are compounded by the additional requirements for teaching and mentoring, as well as family commitments and obligations (Figure 16.1). Ethically, family commitments cannot be handled by a substitute person and have a much higher priority than any of the other demands. Anyone can be replaced professionally unless, of course, his or her ego cannot tolerate replacement.

To meet translational and clinical neuroscience objectives, a master's in public health and epidemiology (M.P.H.) may provide a suitable path. This initial degree provides some training in clinical investigation and trial design and can provide a base upon which to build further training. Additional career pathways for clinical neuroscience investigation should include training in epidemiology and statistics as well as in clinical trial design and principles of translating neuroscience concepts into clinical utility. These training issues have been discussed in terms of capacity for translational research within academic medical centers, and the projected need for manpower to perform critical studies in the future (see Chapter 1).

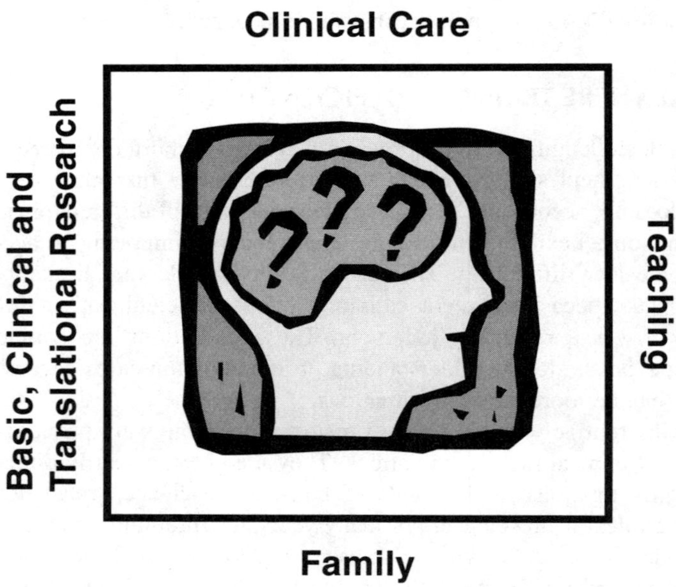

FIGURE 16.1 (See color insert following page 146.) Dilemmas of academic neurosurgery. The center figure represents a typical neurosurgeon trying to decide on a career in academia, or after embarking on such a career, deciding how to pay appropriate and adequate attention to multiple critical aspects of life simultaneously. He or she has no easy choices.

Since the general public is somewhat suspicious of the involvement of residents in procedures and "ghost surgeries (where a resident performs a procedure without supervision)," it is critical that adequate supervision be present at all times and that optimal use of technical training outside the operative suite be encouraged before residents are allowed to perform procedures on humans. Clearly, current training could be expanded to become less of an apprenticeship and more of a truly educational experience by optimizing knowledge-based and judgment-based approaches in every situation possible instead of viewing training as a rote technical exercise. Surgical simulation and practice judgment may go a long way toward supplementing traditional, wholly patient-based approaches.

16.4 HOW DO WE IMPROVE OUR TRAINING CAPABILITIES?

16.4.1 Judgment Training Developments

Web approaches for developing data to support judgments made in the course of considering procedures are currently in development. These approaches enable practitioners to learn and test the processes of making judgments including making rational decisions, basing decisions on evidence in the literature, and methods of presenting the decisions to patients. As discussed in Chapter 15, fallacies are present at all levels of evidence, particularly in small specialties such as neurosurgery, in which randomized controlled trials are rarely performed due to small patient populations. In cases like neurosurgery where it is not always possible to obtain high level evidence because of small patient populations, the ability to infer information from the existing literature is critical for optimal decision making. Development of improved patient encounter simulations may also help in understanding how differently patients may value a surgical procedure; for some patients, the negative aspects of surgery may outweigh any possible benefits.

16.4.2 Technical Training Advances

Technical training is currently limited by the availability of suitable patient encounters and the teaching skills of mentors in teaching settings. While cadaver surgical approaches are common methods of practicing skills and can be very useful, the tissue characteristics (for example, brain deformation properties) of a cadaver differ from those of an intact patient *in vivo*. There is considerable interest in surgical simulation, particularly in virtual reality immersion settings[4–7] similar to simulation devices used to train aviators. This type of simulated environment, usually involving virtual reality goggles and realistic touch and tactile feedback and various types of instrumentation, allows simulation of manipulation of tissues *in vivo*.

These techniques rely on sophisticated three-dimensional renderings and models of tissue deformations to mimic realistically the properties of the brain, skull, spinal cord, and nerves. While such approaches are very demanding computationally, limited views have been incorporated into endoscope viewers and are used during intraventricular endoscopy procedures[6,7] in which video simulations can be combined with appropriate tactile feedback.

16.4.3 Simulation Techniques

Simulation may be broadly defined as the use of technology to recreate key elements of an interactive experience, usually accomplished through a computer interface and routinely including visual representations and some degree of interactive control. Simulation can include any of several components of an experience including visual, auditory, tactile, or even conceptual. Depending on the skills to be trained, a simulator may include one or more of these elements. In addition, the user's experience will be influenced by background knowledge. Theoretically, an ideal simulator is one in which the user is unable to distinguish between the simulation and the actual experience.[2,4,8]

Current real-world limitations preclude the complexity this would necessitate, but our ability to suspend disbelief allows us to develop effective simulators despite their not being mistaken for reality. While an ideal simulator would potentially recreate every aspect of an experience and be identical to reality, this is neither necessary nor especially important. For example a simulator of neurosurgical anatomy can still have significant value even if it lacks a tactile component. Certainly the potential costs and implementation efforts required to produce a "complete" simulator would make the device prohibitive and necessitate compromise.

It is not necessary or even important for a simulation system to "fool" the user. By accepting this principle, we can greatly reduce the potential hurdles to developing an effective, accessible simulator. It is essential to recognize that prohibitive costs and limited access can effectively make the world's best simulators valueless. Several high-end simulators have followed this path. The opportunity to provide inexpensive, readily available simulation is clearly the promise of the personal computer. In fact, the best simulators may be inexpensive and work on standard personal computers. Such devices will actually be used and can accomplish some or all of the potential goals of simulation.

The potential benefits of simulation are most significant if the experience simulated is uncommon, poses high risks, or demands extensive rehearsal. Flight simulators have been tremendously successful because they allow pilots to gain valuable experience without risk to themselves, their passengers, innocent bystanders, and expensive aircraft. In addition to frequent rehearsal of fundamental techniques, flight simulators allow pilots to experience unusual and dangerous situations in completely controlled environments. Similarly, surgical simulators offer several potential advantages including the ability to frequently rehearse the steps of a surgical procedure, familiarity with normal and unusual anatomies, and the potential for improved reactions to adverse events. These advantages are useful for educating residents, maintaining certifications, and disseminating novel or very complex techniques. From the perspectives of training and public perception, the potential advantages of simulation are improved surgical skill, minimization of surgical errors, and alternatives to learning on patients.[2,3,4]

16.4.4 Surgical Neuroanatomy Representation

The potential for simulators to augment visuospatial and motor skills is significant. The most common simulators are based on graphical representations of surgical neuroanatomy. They are typically animation-based or rely on photorealistic images to recreate the visual experiences of surgery. The more advanced simulators allow tissues to be deformed in a physically realistic manner. This involves substantially increased computational requirements and general assumptions about tissue characteristics.

A popular technique for accomplishing this type of deformation is the use of mass-and-spring lattices to model surfaces of structures.[5] A surface is assumed to be composed of a series of masses connected to each other by springs. Mathematical calculations can then be used to determine the degree of deformation associated with a particular force. This technique is less useful when attempting to model cut surfaces. For this reason, the use of finite element analysis has become popular. While providing more physically realistic tissue behavior, the computational requirements become substantially higher.

The use of haptics in neurosurgical simulators lends an additional degree of realism and training potential. Haptics is defined as the use of tactile feedback in an interactive experience. Examples of haptics include force feedback joysticks, exoskeletons, and semiconstrained robotic arms. In each case, the visual feedback of an interactive experience is mechanically linked to a haptic device to provide an additional degree of realism and more accurately replicate the surgical experience simulated.

16.5 MECHANISMS OF TRANSLATIONAL NEUROSURGERY

Throughout this book, a number of new concepts and approaches to clinical neurosurgery have been described, many based on laboratory concepts that are in the process of translation to initial clinical experimentation. Many of these concepts and approaches will become clinical therapeutics at some level and many will eventually be rejected after they are tested. Great laboratory ideas often flounder in the setting of clinical applicability when unanticipated consequences arise. The formats of clinical trials most likely to be implemented with the advent of new therapeutics were outlined in Chapter 15.

Most of the different clinical trial formats require team approaches. One team member should be knowledgeable about the disease state to be treated and whether sufficient clinical interests or controversy exists to make a trial worthwhile. A team should also have a statistician or epidemiologist familiar with trial design who can help set up a suitable trial intended to answer the questions posed by the clinician. Such a trial may often involve multiple cooperating sites that are willing to rigidly follow a specific protocol. Various additional personnel are also needed to acquire and maintain data independently. A neurosurgeon should remain at the heart of any team, both to provide the clinical rationale and experience in clinical testing schemes and identify questions most suitable for study.

For devices there also needs to be a sponsor, such as an interested "enthusiastic" investigator, or more commonly, a corporate entity with commercial interests at heart. The sponsor is the liaison to the FDA for eventual approval and market release of the device (see Chapter 1). This group forms the minimum team needed to begin a clinical trial solution to a new product or device. However, one of the key components of such a team is the interested clinician, who can pose the critical problem, which the clinical study or device will resolve, and who is sufficiently enthusiastic to maintain the trial format through the large number of regulatory and funding hurdles. The clinician needs to convince the community of other clinicians that resolution of the problem requires a formal clinical trial, establishing equipoise and uncertainty about the relative worth of differing treatment options.

Who will be these clinicians? As in Chapter 1, there is a large question as to whether or not this breed of clinician-investigators may be waning, particularly in surgical specialties such as neurosurgery. Unless there is a short payoff time from new device products to clinical applicability, neurosurgeons tend to lose interest in the problem, so maintaining a focus in spite of a more distant time to application is critical. The critical question at this time focuses on whether we are training such a blend of clinician investigators who will be sufficiently innovative and enthusiastic to skeptically approach current and future therapeutics, yet possess the critical training in both the basic science questions and clinical trials, to be able to address directly these critical concerns. The answer is not clear, nor the path to achieve this goal.

Clearly, one of the goals of neurosurgery training is to improve the specialty in the future by seeing that trainees are more involved with new approaches instead of the history of neurosurgery and be prepared to address future treatments and directions more appropriately and scientifically.

16.6 CONCLUSIONS

Neurosurgical training is very traditional, and not necessarily aimed at developing clinician investigators, particularly those who have bents toward translational neurosurgery or clinical investigation. Neurosurgery traditionally is a technical- and procedure-based specialty instead of having a research base. Critical and skeptical approaches to investigation are not necessarily highly valued within the specialty, particularly when the outcomes of investigations may result in limitation of practice or curtailing of procedures if results are negative.

Interest within neurosurgery is generally much greater in the history and development of neurosurgery than the development of translational approaches, particularly if the translational timeline to clinical application is greater than 2 or 3 years. Neurosurgery as a specialty could respond to such issues by altering the traditional approach to training, encouraging skeptical and investigational approaches to both judgment and technical aspects of neurosurgery, and aiding innovation even if it means curtailing procedures that have shown limited efficacy. A large number of innovational approaches to training now allow practitioners to hone their judgment and expand their knowledge of neurosurgical applications. Advances in simulation techniques and other technical advances will result in improvements in operative procedures. Whether these new approaches will be embraced and integrated into our training programs will be a critical question for the future.

REFERENCES

1. Drake, C.G., Neurosurgery: considerations for strength and quality, *J. Neurosurg.*, 49, 448, 1978.
2. Larsen, O.V. et al., The virtual brain project: development of a neurosurgical simulator, *Studies Health Technol. Inform.*, 81, 256–262, 2001.
3. Spicer, M.A. and Apuzzo, M.L.J., Virtual reality surgery: neurosurgery and the contemporary landscape, *Neurosurgery*, 52, 489–497, 2003.
4. Henn, J.S. et al., Interactive stereoscopic virtual reality: a new tool for neurosurgical education, *J. Neurosurg.*, 96, 144–149, 2002.
5. Platenik, L.A. et al., *In vivo* quantification of retraction deformation modeling for updated image guidance during neurosurgery, *IEEE Trans. Biomed. Eng.*, 49, 823–835, 2002.
6. Radetzky, A. et al., ROBO-SIM: a simulator for minimally invasive neurosurgery using an active manipulator, *Studies Health Technol. Inform.*, 77, 1165–1169, 2000.
7. Riegel, T. et al., Relationships of virtual reality neuroendoscopic simulations to actual imaging, *Minimally Invas. Neurosurg.*, 43, 176–180, 2000.
8. Tarr, M.J. and Warren, W.H., Virtual reality in neuroscience and beyond, *Nature Neurosci.*, 5S, 1089, 2002.

Index

B

C

D

E

F

G

J

K

L

T